Introduction to Laser Spectroscopy

Introduction to Laser Spectroscopy

Halina Abramczyk

Chemistry Department
Technical University
Łódź, Poland

2005

ELSEVIER

Amsterdam – Boston – Heidelberg – London – New York – Oxford
Paris – San Diego – San Francisco – Singapore – Sydney – Tokyo

ELSEVIER B.V.
Radarweg 29
P.O. Box 211, 1000 AE Amsterdam
The Netherlands

ELSEVIER Inc.
525B Stree, Suite 1900
San Diego, CA 92101-4495
USA

ELSEVIER Ltd
The Boulevard, Langford Lane
Kidlington, Oxford OX5 1 GB
UK

ELSEVIER Ltd
84 Theobalds Road
London WC1X 8RR
UK

First edition 2005

Library of Congress Cataloging in Publication Data
A catalog record is available from the Library of Congress.

British Library Cataloguing in Publication Data
A catalogue record is available from the British Library.

ISBN: 0 444 51662 X

∞ The paper used in this publication meets the requirements of ANSI/NISO Z39.48-1992 (Permanence of Paper). Printed in The Netherlands

To my parents Salomea and Edward,
to my husband Andrew
and son Victor

Acknowledgements

I would like to thank Dr Gabriela Waliszewska who checked the work for accuracy. She assisted me in taking a typed manuscript and putting it in final form including plots, graphics, and other illustrations. Ph.D. student Iwona Szymczyk assisted in translation of large part of the material into English. Their cooperation is really appreciated, I can truthfully say this work would not have been completed without their assistance.

Professor Zbigniew Kęcki first inspired me with love to molecular spectroscopy while I was a Ph.D. student at Technical University of Łódź. Professor Jerzy Kroh later sharpened my teaching and research skills by creating a warm, friendly and scientific atmosphere while he was my group leader.

Finally, I would like to thank my husband, Andrew, my son Victor and my close, life long friends, who gave me encouragement to undertake this effort.

Preface

This book is intended to be used by students of chemistry, chemical engineering, biophysics, biology, materials science, electrical, mechanical, and other engineering fields, and physics. It assumes that the reader has some familiarity with the basic concepts of molecular spectroscopy and quantum theory, e.g., the concept of the uncertainty principle, quantized energy levels, but starts with the most basic concepts of laser physics and develops the advanced topics of modern laser spectroscopy including femtochemistry.

The major distinction between this book and the many fine books available on laser physics and time resolved spectroscopy is its emphasis on a general approach that does not focus mainly on an extensive consideration of time resolved spectroscopy. Books at the correct level of presentation for beginners tend to be focused either totally or mainly on the basic fundamentals of lasers and include only a minimal amount of material on modern ultrashort laser spectroscopy and its chemical, physical and biological applications. On the other hand, books that contain the desired material to a significant degree, are too advanced, requiring too much prior knowledge of nonlinear optics, quantum theory, generation of ultrafast pulses, detection methods, and vibrational and electronic dynamics. This book is intended to fill the gap. More advanced problems of modern ultrafast spectroscopy are developed in the later chapters using concepts and methods from earlier chapters.

The book begins with a qualitative discussion of key concepts of fundamentals of laser physics. Spontaneous and stimulated transitions, Einstein coefficients, properties of stimulated radiation, population inversion and amplification and saturation are discussed. Chapter 2 introduces concepts of longitudinal and transverse modes, the quality factor of a resonator, the relationship between line width of stimulated emission and resonator quality factor. Chapter 3 explains how ultrashort pulses are produced. This discussion is used to show the differences between modelocking, Q-switching and cavity dumping. Chapter 4 presents a brief description of lasers that are used as a source of radiation in every laser experimental set up. Chapter 5 provides all of the necessary material to understand modern concepts of nonlinear spectroscopy. It starts with basic concepts of phase matching methods, second and third harmonic generation, parametric oscillator and ends with a brief description of advanced topics such as stimulated Raman scattering, coherent anti-Stokes Raman spectroscopy (CARS), nonlinear dispersion phenomena affecting picosecond and femtosecond pulse duration, including group velocity dispersion (GVD) and self phase modulation (SPM). Chapter 6 develops the theoretical background of pulses amplification and presents the main design features of amplifiers concerning on regenerative amplifier and chirped pulse amplification (CPA). Chapter 7 shows how to measure ultrafast pulses and draws a distinction between autocorrelation techniques and modern frequency domain techniques including the FROG technique. Chapter 8 illustrates basic features of the experimental methods of modern time resolved spectroscopy including fluorescence decay, pump-probe transient absorption, and techniques based on stimulated Raman scattering, including CARS and photon echo. The close relationship between pulsed nuclear magnetic resonance and coherent optical experiments is emphasised. Chapter 9 covers topics of current research interest in the context of physical problems that occur in chemistry, biology, materials science, and physics. The examples given in this chapter illustrate some of the main streams in modern time dependent spectroscopy. The examples are not intended as a treatment of the most valuable results of modern spectroscopy. Rather, a conscious choice was made to use examples that are tied closely to the previous chapters, and they are used to amplify and expand the topics that are covered. With its coverage of ultrafast chemical and physical processes, chapter 9 has a distinctive flavor of modern laser spectroscopy. This flavor can be illustrated by briefly considering the content of several topics of this chapter. It starts with the theoretical background of femtochemistry and wave packet dynamics. Then some examples on spectroscopic application of wave packet dynamics are given, including excited-state vibrational coherence, vibrational coherence in "reacting" excited-state molecules such as bacteriorhodopsin and H-bond dynamics. It also has some classical examples of time resolved spectroscopy such as photoisomerization of *cis*- and *trans*-stilbene, intramolecular charge transfer, molecular reorientations, intermediates investigations and excited state proton transfer. Towards the end of the chapter, the ideas developed in chapter 8 are extended to discuss some examples of ultrafast coherent spectroscopy including vibrational dynamics, energy relaxation T_1 and phase relaxation T_2. The chapter ends with a brief presentation of dynamics of an excess electron and dynamics of solvated electron. Chapters 10 and 11 are related to the interaction between laser radiation and human tissue and the potential hazard they create as well as some basic applications of lasers in the real world, with medicine as the most important field. Chapter 12 will help those readers who have problems with understanding detection techniques as the fundamentals of different detectors, including CCD cameras, are explained on an elementary level.

This book will provide a solid grounding in the fundamentals of many aspects of laser physics, nonlinear optics, and molecular spectroscopy. It explicates a variety of problems that are key components to understanding broad areas of physical, chemical and biological science. It will give readers sufficient background information so that they can focus their future efforts on more specialized topics in laser molecular spectroscopy.

Contents

1. Basic Physics of Lasers 1
1.1. Spontaneous and Stimulated Transitions. Einstein Coefficients. Properties of Stimulated Radiation 1
1.2. Laser Operation Basics 7
1.3. Population Inversion 11
1.4. Amplification and Saturation 15

2. Distribution of the Electromagnetic Field in the Optical Resonator 19
2.1. Longitudinal Modes 19
2.2. Quality Factor of Resonator Q. Relationship between Linewidth of Stimulated Emission and Resonator Quality Factor 21
2.3. Transverse Modes 25

3. Generation of Ultrashort Laser Pulses 31
3.1. Modelocking. Relationship between Linewidth of Stimulated Emission and Pulse Duration 32
3.2. Methods of Modelocking. Active and Passive Modelocking 40
3.3. Q-Switching 49
3.4. Cavity dumping 52

4. Lasers 59
4.1. Ruby Laser 60
4.2. Molecular Gas Lasers from the Infrared Region 61
4.2.1. Lasers Operating on Rotational Transitions 62
4.2.2. Lasers Operating on Vibrational-Rotational Transitions: CO_2 and CO 63
4.3. Chemical Lasers 68
4.4. Solid-State Lasers 69
4.4.1. Neodymium Laser and other Rare-Earth Lasers 70
4.4.2. Solid-State Tunable Lasers (Vibronic Lasers) 74
4.4.3. Fiber Lasers 78
4.5. Gas Lasers for the Visible Range 82
4.5.1. Helium–Neon Laser 82
4.5.2. Ion–Gas Lasers. Argon and Krypton Lasers 83
4.6. Liquid Dye Lasers 84
4.7. Gas Lasers for the Ultraviolet Range 87
4.7.1. Excimer Lasers 87
4.7.2. Nitrogen Laser 90
4.8. Diode Lasers 90
4.8.1. Intrinsic Semiconductors. Doped Semiconductors. *n-p* Junction 90
4.8.2. Diode Lasers 94

5. Nonlinear Optics 107
5.1. Second Order Nonlinear Phenomena 110
5.2. Phase Matching Methods 113

5.3. Practical Aspects of the Second Harmonic Generation ... 118
5.3.1. SHG for Pico- and Femtosecond Pulses ... 120
5.4. Parametric Oscillator ... 125
5.5. The Third Order Nonlinear Processes ... 131
5.5.1. Stimulated Raman Scattering ... 132
5.5.2. Coherent Anti-Stokes Raman Scattering (CARS) ... 135
5.5.3. The Other Techniques of Nonlinear Stimulated Raman Scattering ... 137
5.6. Nonlinear Dispersion Phenomena Affecting Picosecond and Femtosecond Pulse Duration – Group Velocity Dispersion (GVD) and Self Phase Modulation (SPM) ... 139

6. Pulse Amplification ... 147
6.1. Introduction ... 147
6.2. Theoretical Background ... 147
6.3. Design Features of Amplifiers ... 150
6.4. Regenerative Amplifier ... 152
6.4.1. The Pockels Cell ... 154
6.5. Chirped Pulse Amplification (CPA) ... 156

7. The Measurement of Ultrashort Laser Pulses ... 161
7.1. Autocorrelation Techniques ... 162
7.2. FROG Techniques ... 170

8. Selected Methods of Time-Resolved Laser Spectroscopy ... 175
8.1. Fluorescence Decay ... 176
8.2. The Pump-Probe Method ... 183
8.3. CARS as a Time-Resolved Method ... 189
8.4. Photon Echo ... 191
8.4.1. Spin Echo in NMR ... 191
8.4.2. Optical Resonance ... 198
8.4.3. Quantum-Classical Description of the Photon Echo ... 201
Weak Field Approximation ... 202
Rotating-Wave Approximation ... 203
Strong Field Approximation. Rabbi Frequency ... 204
8.4.4. Practical Advantages of Photon Echo Applications ... 209
8.5. Quantum Beats ... 212
8.5.1. Quantum Description ... 212
8.5.2. Examples of Quantum Beats Applications ... 214

9. Ultrafast Chemical and Physical Processes ... 219
9.1. Femtochemistry. Wave packet dynamics. Theory ... 221
9.2. Femtochemistry. Spectroscopic Application of Wave Packet Dynamics ... 229
9.2.1. Excited-State Vibrational Coherence ... 231
9.2.2. Vibrational Coherence in "Reacting" Excited-State Molecules. Bacteriorhodopsin ... 232
9.2.3. H-Bond Dynamics ... 235
9.3. Photoisomerization ... 240
9.3.1. Photoisomerization of *cis*- and *trans*-Stilbene ... 240
9.4. Intramolecular Charge Transfer ... 242
9.5. Molecular Reorientations ... 243
9.6. Investigation of Intermediates ... 245
9.6.1. Photoreduction ... 245
9.6.2. Carbenes ... 246
9.6.3. Excited-State Proton Transfer ... 247

9.7. Ultrafast Coherent Spectroscopy. Vibrational Dynamics ... 249
9.7.1. Energy Relaxation T_1, and Phase Relaxation T_2 ... 251
9.8. Dynamics of an Excess Electron. Solvated Electron ... 257
9.9. Excess Electron Spectroscopy ... 260

10. Lasers in Medicine ... 271
10.1. Introduction ... 271
10.2. Photochemical Interactions ... 274
10.2.1. Photodynamic Therapy ... 274
10.2.2. Sensitizers ... 275
10.2.3. Photochemistry of Sensitizers ... 277
10.3. Thermal Interaction ... 278
10.4. Photoablation ... 279
10.5. Plasma-Induced Ablation ... 280
10.6. Application of Lasers in Medicine ... 280

11. Potential Hazards Associated with Inappropriate Use of Lasers ... 285
11.1. Radiation Hazards ... 286
11.1.1. Eye Hazards ... 287
11.1.2. Skin Hazard ... 290
11.2. Other Hazards ... 291

12. Detectors ... 293
12.1. Detectors Types and Detectors Characterizing Parameters ... 294
12.2. Photoemissive Detectors ... 297
12.3. Semiconductor Detectors ... 300
12.4. Multichannel Detectors PDA and CCD ... 301

Subject Index ... 311

– 1 –

Basic Physics of Lasers

1.1. SPONTANEOUS AND STIMULATED TRANSITIONS. EINSTEIN COEFFICIENTS. PROPERTIES OF STIMULATED RADIATION

To understand the principle of lasers and their applications, it is essential to understand the interaction of radiation with matter. Quantum properties dominate the field of molecular physics and molecular spectroscopy. Both radiation and matter are quantized. Radiation corresponds to the *electromagnetic spectrum* presented in Fig. 1.1 that covers the range from long wavelengths of meters to short wavelengths of a fraction of angstroms. *Visible light* is only a narrow part of the electromagnetic spectrum. For a given frequency of radiation ω, the photons of that radiation have quantized energy E with only one value given by the famous *Planck relationship*

$$E = \hbar\omega \tag{1.1}$$

The energy of atoms or molecules is also quantized, which means the energy levels can have only certain quantized values. In the molecular system we can distinguish electronic, vibrational, rotational or additional levels attributed to the interaction with an external magnetic field. Electrons or molecules can jump between the quantized energy levels. These transitions are stimulated by the photons of radiation. During the transitions the atoms or the molecules absorb or emit radiation. These processes can be classified as *stimulated absorption, stimulated emission* and *spontaneous emission* (Fig. 1.2).

Stimulated absorption denotes a process in which an atomic or a molecular system subjected to an electromagnetic field of frequency ω absorbs an energy of $\hbar\omega$ from the photon. As a result of the absorption the atom or the molecule is raised from the state n to the upper state m of higher energy. Stimulated absorption occurs only when the energy of the photon precisely matches the energy separation of the participating pair of quantum energy states

$$E_m - E_n = \hbar\omega, \tag{1.2}$$

where E_n and E_m are the energies of the initial state n and the final state m, ω is the circular frequency of the incident radiation, and $\hbar$ is Planck's constant ($\hbar = h/2\pi$).

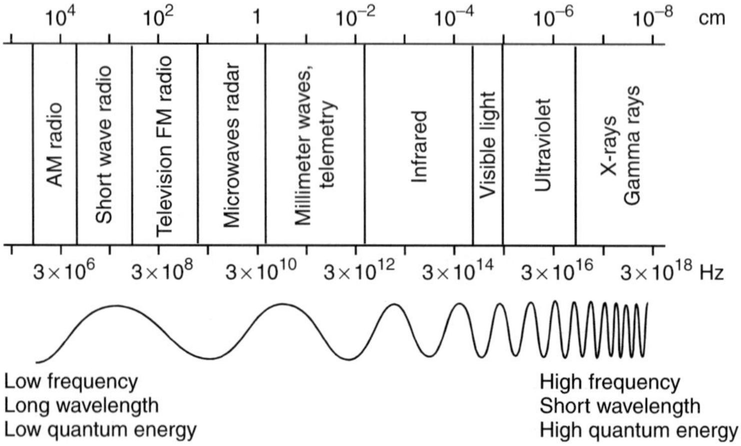

Fig. 1.1 Electromagnetic spectrum.

If the condition (1.2) is fulfilled, we say that the radiation is in *resonance* with the molecular transition.

Almost immediately (usually after nano- or picoseconds) most excited molecules return from the upper state, *m*, to the lower state, *n*, through the emission of a photon. This process is known as spontaneous *emission*. Spontaneous emission is obviously a quantum effect, because in classical physics the system can stay in a definite energy state infinitely long, when no external field is applied. Spontaneous emission in the visible spectral range is called *luminescence* or *fluorescence*, when the transition occurs between states of the same *multiplicity*, or *phosphorescence*, when the initial and final state have different multiplicity. Spontaneous emission is

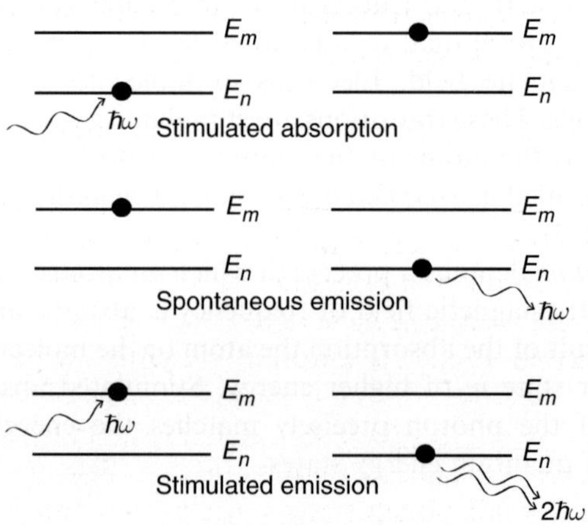

Fig. 1.2 Scheme of a two-level system illustrating stimulated absorption, spontaneous emission and stimulated emission phenomena.

not connected with the incident radiation in any way. The incident radiation does not affect the lifetime of a molecule in the excited state m from which the spontaneous emission is emitted. Moreover, spontaneous emission is not *coherent*, i.e., the emitted photons have no definite phase relation to each other. The emission comes from independent atoms or molecules, which are not related to one another in phase.

When the external electromagnetic field is strong, emission can take place not only spontaneously but also under stimulation by the field. This kind of emission is called *stimulated emission*, which like stimulated absorption, is induced by the external radiation. In stimulated emission a photon of the incident radiation interacts with a molecule that is in the higher energy state m. The interaction results in giving the energy quantum back to the radiation field by a molecule, followed by a simultaneous emission and dropping back from the upper level m to the lower level n. In another words, if a molecule is already in an excited state, then an incoming photon, for which the quantum energy is equal to the energy difference between its present level and the lower level, can "stimulate" a transition to that lower state, producing a second photon of the same energy. In contrast to spontaneous emission, stimulated emission exhibits phase coherence with the external radiation field. It indicates that the phases φ of the incident electric field $E = E_0 \cos(\omega t + \boldsymbol{k} \cdot \boldsymbol{r} + \varphi)$ and the emitted stimulated radiation are the same. Since the emitted photons have definite time and phase relations with the external field, they also have a definite time and phase relation to each other. Therefore the emitted light has a high degree of *coherence*. The coherence is a quantum effect and requires a quantum mechanical treatment of the interaction between the radiation and matter, which is beyond the scope of this book. However, the concept of coherence can also be understood by using a classical description of radiation. In the classical description the external electromagnetic field induces in a medium a polarization by forcing the dipole moments to oscillate in a phase consistent with the phase of the incident radiation. The oscillating dipole moments emit in turn radiation coherent with their own oscillations. A detailed description of the classical interactions between the field and the induced oscillators can be found in [1].

Stimulated transitions have several important properties:

a) the probability of a stimulated transition between the states m and n is different from zero only for the external radiation field that is in resonance with the transition, for which the photon energy $\hbar\omega$ of the incident radiation is equal to the energy difference between these states, $E_m - E_n = \hbar\omega$,
b) the incident electromagnetic radiation and the radiation generated by stimulated transitions have the same frequencies, phases, plane of polarization and direction of propagation. Thus, stimulated emission is, in fact, completely indistinguishable from the stimulating external radiation field,
c) the probability of a stimulated transition per time unit is proportional to the energy density of the external field ρ_ω, that is the energy per unit of the circular frequency from the range between ω and $\omega + d\omega$ in the volume unit.

The properties of stimulated absorption, stimulated emission and spontaneous emission discussed so far can be described using the following relations:

$$W_{nm}^{A} = B_{nm}\rho_\omega \qquad \text{for stimulated absorption,} \tag{1.3}$$

$$W_{mn}^{SE} = B_{mn}\rho_\omega \qquad \text{for stimulated emission,} \tag{1.4}$$

$$W_{mn}^{SPE} = A_{mn} \qquad \text{for spontaneous emission,} \tag{1.5}$$

where W_{nm}^{A}, W_{mn}^{SE} and W_{mn}^{SPE} denote the probabilities of the transitions for stimulated absorption, stimulated emission and spontaneous emission per time unit. The constants of proportionality B_{nm}, B_{mn} and A_{mn} are called the *Einstein coefficients*.

Let us establish the relations between B_{nm}, B_{mn} and A_{mn}. For this purpose let us consider an ensemble of quantum molecules in equilibrium with the field of their own radiation (absorbed and emitted photons), which is called *thermal radiation*. This type of equilibrium is often considered in physics, and a spectacular example of its application is Planck's formula for *blackbody radiation*.

Let us assume that the quantum molecules are represented by two-level quantum states that are non-degenerated. Since the system is in equilibrium with the radiation field, the number of the transitions per unit time from the upper state $m \to n$ has to be equal to the number of the transitions from the lower state $n \to m$

$$N_{m\to n} = N_{n\to m}. \tag{1.6}$$

The number of the transitions, N, depends on the transition probabilities per time unit W, and on the number of molecules in the initial state. Eq. (1.6) can be written as

$$W_{m\to n}n_m = W_{n\to m}n_n. \tag{1.7}$$

For the system in equilibrium the quantum levels population is described by the Boltzmann distribution

$$n_m = n_n \exp\left(-\frac{E_m - E_n}{kT}\right), \tag{1.8}$$

where k is Boltzmann's constant, and T is temperature in Kelvins. Moreover, the total probability of the emission $m \to n$ is the sum of spontaneous emission W_{mn}^{SPE} and stimulated emission W_{mn}^{SE}

$$W_{m\to n} = W_{mn}^{SPE} + W_{mn}^{SE}, \tag{1.9}$$

therefore using eqs. (1.4–1.5) one gets,

$$W_{m\to n} = A_{mn} + B_{mn}\rho_\omega. \tag{1.10}$$

The total probability of the absorption $W_{n\to m}$ is equal to the probability of stimulated absorption W_{nm}^{A} given by eq. (1.3)

$$W_{n\to m} = W_{nm}^{A} = B_{nm}\rho_\omega. \tag{1.11}$$

Inserting (1.10) and (1.11) into equation (1.7) one can write

$$(A_{mn} + B_{mn}\rho_\omega)n_m = B_{nm}\rho_\omega n_n. \tag{1.12}$$

Inserting eq. (1.8) into eq. (1.12) one obtains the radiation density ρ_ω:

$$\rho_\omega = \frac{A_{mn}}{B_{mn}} \frac{1}{\frac{B_{nm}}{B_{mn}} \exp\left(\frac{E_m - E_n}{kT}\right) - 1}. \tag{1.13}$$

Because the density of the thermal radiation energy ρ_ω, is described by Planck's equation for blackbody radiation

$$\rho_\omega = \frac{\omega^2}{\pi c^3} \frac{\hbar\omega}{\exp\left(\frac{\hbar\omega}{kT}\right) - 1}, \tag{1.14}$$

one can compare (1.13) and (1.14). From the comparison we obtain the following relations for the Einstein coefficients

$$B_{nm} = B_{mn} \tag{1.15}$$

and

$$A_{mn} = \frac{\omega^3\hbar}{\pi c^3} B_{mn}. \tag{1.16}$$

Readers who are interested in the derivation of Planck's equation are referred to textbooks describing this issue, e.g., [2].

When the levels m and n are degenerated, and their degeneration degrees are g_m and g_n, the relation (1.15) is modified to the form

$$g_n B_{nm} = g_m B_{mn}. \tag{1.17}$$

Thus, we have obtained the relations between the Einstein coefficients. First, equation (1.15) indicates that the probability of stimulated emission is equal to the probability of stimulated absorption. A practical hint from this relation is that materials characterized by strong absorption are expected to exhibit also a large stimulated emission. The relation (1.16) indicates that a material in which spontaneous emission does not take place, does not exhibit stimulated emission either. These simple relations determine the principal conditions which should be taken into account when looking for materials to be employed as the active medium in lasers (see section 1.2).

To gain a better insight into the interplay between these three processes let us calculate the coefficient $\omega^3\hbar/\pi c^3$ in equation (1.16) determining the connection between spontaneous emission A_{mn} and stimulated emission B_{mn}. For a wavelength $\lambda = 514\,\text{nm}$ characteristic for an ion argon laser Ar^+ corresponding to $\omega = 36.64 \cdot 10^{14}\,\text{rad} \cdot \text{s}^{-1}$ the coefficient $\omega^3\hbar/\pi c^3$ is $6 \cdot 10^{-14}$. This indicates that the coefficient of stimulated emission, B_{mn}, is much larger than the coefficient of spontaneous emission, A_{mn}. However, it does not indicate that the intensity of spontaneous emission is small in comparison with the intensity of stimulated emission. In fact, just the opposite is the case–at room temperatures ($T = 300\,\mathrm{K}$) stimulated emission is negligibly small. Indeed, one has to remember (eqs. 1.3–1.5) that stimulated emission depends on B_{mn} as well as on the thermal radiation density ρ_ω, which is negligibly small at room temperatures. Figure 1.3 shows the spectral characteristics of the thermal radiation density distribution ρ_ω for several temperatures. At temperature $T = 300\,\mathrm{K}$ the thermal radiation density, ρ_ω, in the visible range of $\omega = 36.64 \cdot 10^{14}\,\mathrm{rad \cdot s^{-1}}$ (514 nm) is very small, leading to the following ratio between spontaneous emission intensity and stimulated emission calculated from (1.14) and (1.16)

$$\frac{A_{mn}}{B_{mn}\rho_\omega} = \exp\left(\frac{\hbar\omega}{kT}\right) - 1 = 1.96 \cdot 10^{40}. \tag{1.18}$$

One can calculate from eq. (1.18) that only at temperatures as high as 40 000 K does the intensity of stimulated emission become comparable with the intensity of spontaneous emission.

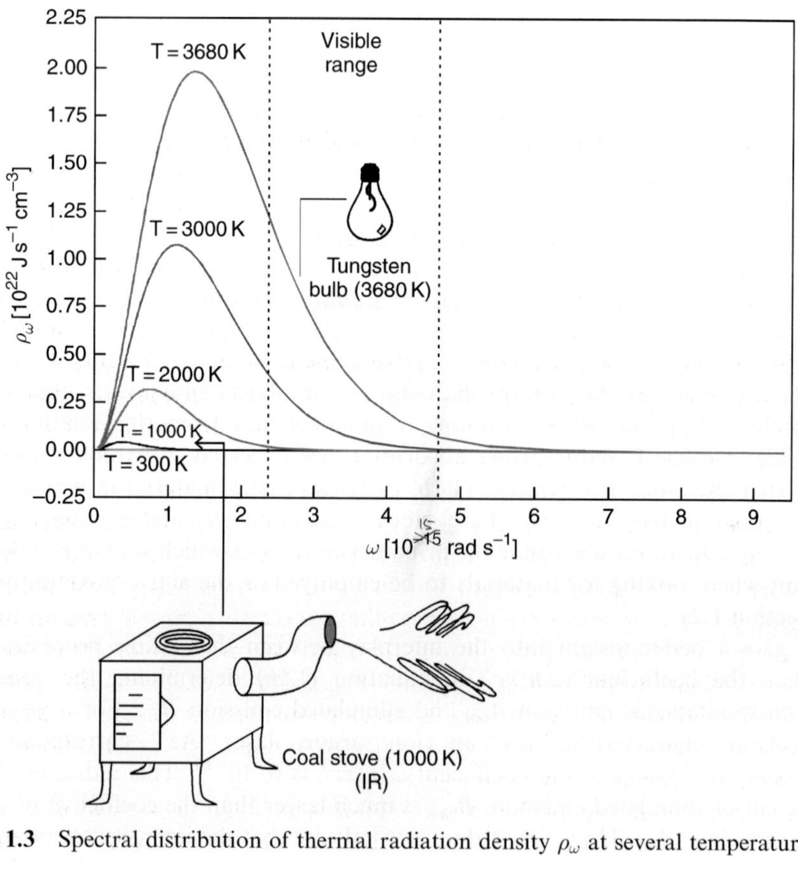

Fig. 1.3 Spectral distribution of thermal radiation density ρ_ω at several temperatures.

1.2. LASER OPERATION BASICS

It has been more than four decades since Maiman constructed the first ruby laser in 1960. Soon after the first laser was invented, a new type of laser technology began to emerge and its robust development still continues. The last decade brought dramatic advances in the control of coherent light, pulse energy increase, pulse length reduction, repetition rate increase, the generation of femtosecond X-ray tubes with laser produced plasma, developments in the field of detection, microscopy and imaging, and expansion of the optical spectrum into the IR and UV-XUV ranges. Lasers have developed into one of the most important tools in fundamental investigations and practical applications.

People who have been dealing with the laser technologies created their own specialized language to describe phenomena occurring in lasers, which has not necessarily made it easier for understanding by chemists, biologists or medical doctors who use lasers as tools for research and applications. Because lasers are ideal sources of light in photochemistry and molecular spectroscopy as well as information technology, communication, health, sciences, biotechnology, metrology, micro- and macro-production, an understanding of laser principles is an essential prerequisite for understanding how laser beams interact with physical and chemical matter as well as biological tissues. First, one would like to know how they act as well as to understand what the brand-specific technological parameters provided by the laser companies for the commercially available systems mean in order to have the ability to compare and evaluate the offers of different companies. The description of the laser principles proposed in this and the following chapters should facilitate this. We will replace a precise lecture with a popular way of presentation in the hope that it will help the reader to understand lasers better.

To understand the idea of laser operation one should consider two basic phenomena: stimulated emission and the *optical resonance*. The phenomenon of stimulated emission was described above. To understand optical resonance we will concentrate on laser construction. The main part of the laser is an *active medium* in which the laser action occurs. The active medium is a collection of atoms or molecules that can absorb and emit light. Stimulated absorption and stimulated emission always occur side by side. The laser is a device that emits light, so the number of absorption transitions in the active medium must be smaller than the number of emission transitions. However, according to the Boltzmann distribution (eq. 1.8) there are always fewer molecules at the higher energy level E_m than at the lower level E_n (Fig. 1.4a), which means that the total number of elementary absorption events is larger than that of emission events. The net result is that the incident radiation is absorbed. In order to force the active medium to emit we must create an inverted population – a temporary situation such that there are more molecules in the upper energy level than in the lower energy level (Fig. 1.4b). The creation of a *population inversion* is a prerequisite condition for laser operation. To produce the population inversion an external source of energy is required to populate a specific upper energy level. We call this energy the *pump energy*.

Once a population inversion is established in a medium it can be used to amplify light. Indeed, this process can be compared to a chain reaction – in an inverted

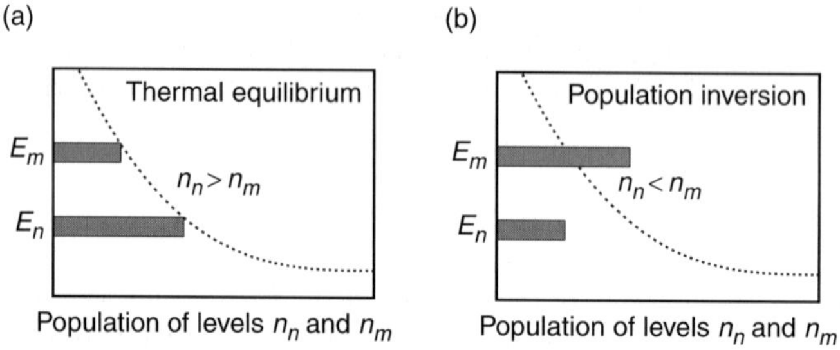

Fig. 1.4 Scheme illustrating the thermal equilibrium described by the Boltzmann distribution (a) and inversion population (b).

medium each incoming photon stimulates the emission of an additional photon (Fig. 1.2) that can be used for further stimulation. Thus, the applied signal gains energy as it interacts with molecules and hence is amplified. Unfortunately, the emitted photons are reabsorbed, some of the excited states are deactivated via spontaneous decay and they are lost for the effectiveness of stimulated emission channel. Spontaneous emission always tends to return the energy level populations to their thermal equilibrium with a depletion of the inverted level. Moreover, it triggers emission in many possible directions that are randomly distributed when the emission takes place from an isotropic medium. So, it is not so easy to maintain the amplification in an active medium, keeping the net gain of stimulated emission and overcoming all the effects that lead to the losses. Fortunately, a simple device can help much to solve this problem.

The effectiveness of the light amplification increases significantly when we put an active medium into a cavity between two mirrors, Z_1 and Z_2, characterized by a high degree of reflectivity (Fig. 1.5). This cavity is called the *optical resonator*. Confining light in the one dimensional box surrounding an active medium, with suitably chosen size and shape, creates efficient *positive feedback* allowing the device to work as an oscillator.

Let us consider now the idea of the optical resonance presented in Fig. 1.5. If we place a source of radiation emitting at a wavelength λ between the mirrors Z_1 and Z_2, separated by a distance L, the standing wave can be created as a result of reflection from the mirrors surfaces. The standing wave can be generated only when the following condition is fulfilled

$$L = n\frac{\lambda}{2}. \tag{1.19}$$

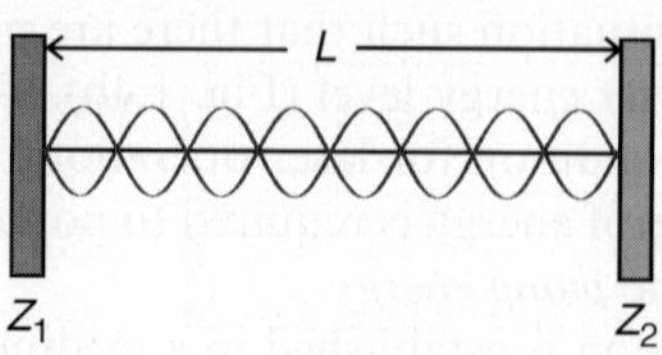

Fig. 1.5 Scheme illustrating the optical resonance.

The condition (1.19) indicates that the standing wave can be generated only when the integer multiple of the wavelength halves between mirrors Z_1 and Z_2 is comprised. For wavelengths that do not fulfill the condition (1.19), destructive interference will occur, causing the intensity of the standing wave to decrease to zero. Optical resonance was known much earlier than the discovery of the laser and it was used in *Fabry–Perot interferometers*.

Thus, to construct the simplest laser one should place an active medium between the Z_1 and Z_2 mirrors. An active medium is a substance in which the processes of stimulated absorption, stimulated emission and spontaneous emission occur under the external pumping energy (Fig. 1.6). As the active medium, also called the *gain medium*, a gas, liquid or solid may be employed. The area between the mirrors is named the *optical cavity* or the *laser cavity*, and the energy delivery process is called *pumping*. The mirror Z_1 is almost entirely non-transparent for the radiation inside the cavity (the reflectivity $R \sim 100\%$) and is called the *high reflector*. The mirror Z_2 has a larger transparency (e.g., $R = 90–99\%$), which allows the generated radiation to be emitted out of the laser cavity. This mirror is called the *output coupler*. The entire device consisting of the mirrors and the active medium is called the *optical resonator*.

Let us consider how the optical resonator operates. First, we must deliver the pumping energy to the active medium. Assume that the delivered energy does not affect the system equilibrium considerably, that is the energy level's population does not differ very much from the Boltzmann distribution. Pumping of the active medium causes a certain number of molecules to jump to a higher energy level and produces spontaneous and stimulated emission. The radiation of the emission could lead to light amplification and start a laser action. Unfortunately, the radiation is absorbed again by the medium, since the number of molecules n_n in the lower energy state n is still larger than the number of the molecules n_m in the higher energy state m. This indicates (eq. (1.12)) that the number of stimulated absorption transitions $B_{nm}\rho_\omega n_n$ is larger than the number of stimulated emission transitions $B_{mn}\rho_\omega n_m$ per time unit. Therefore, in thermal equilibrium it is impossible to invert this situation that simply results from eq. (1.8). Even at extreme conditions of temperature T going to infinity one can achieve only

$$n_n = n_m. \tag{1.20}$$

The optical resonator can start emitting only when we find a way to introduce the *population inversion* $n_n < n_m$ (Fig. 1.4b), since only then will the emission be larger

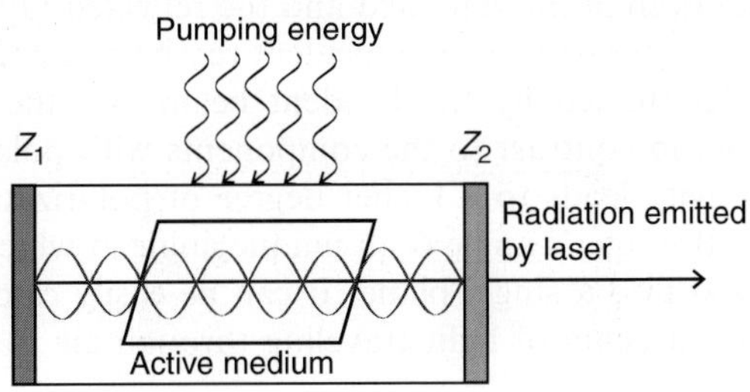

Fig. 1.6 Optical resonator.

than the absorption. This will be possible when the pumping delivers energy per time unit sufficient to cause deviation from the equilibrium Boltzmann population, leading to the population inversion. But pumping alone is not sufficient. The population inversion cannot be achieved with just two levels because the probability of absorption and stimulated emission is exactly the same, as shown in the previous paragraph. The molecules drop back by photon emission as fast as they are pumped to the upper level. We will show later that the molecular system must have at least three molecular levels. In order for the optical resonator to emit radiation, the gain due to stimulated emission has to be greater than the losses. The magnitude of the gain which overcomes the loss is called the *laser threshold.* When the population inversion is achieved and the gain is larger than the losses in the optical resonator, emitted radiation begins to travel between the mirrors. Only the part of the radiation which goes along the resonator axis will undergo amplification due to eq. (1.19), while the remaining part will be extinguished. Every next beam reflection from the mirrors and the passage through the active medium amplifies the beam. This effect is called *regenerative feedback*. The regenerative feedback is extremely useful because the optical resonator suffers from many losses. The absorption is only one of the reasons of the energy losses in the optical resonator. A significant part of radiation is lost as a result of scattering, refraction, diffraction, warming of the medium and mechanical instability of the resonator.

Due to the properties of stimulated emission that were discussed above, the generated radiation shows a high degree of spatial and temporal coherence, and is highly monochromatic and directional. The last feature indicates that the output beam is emitted into a small steric angle around the resonator axis. Laser light usually shows the high degree of polarization, which denotes that the electric field vectors are held in the same plane. One of the ways to achieve the high degree of polarization in the optical resonator is polarization by reflection. In such a spatial configuration, the beam traveling along the resonator axis arrives at the surface of an active medium at a certain angle, the *Brewster's angle*, as presented in Fig. 1.6.

Why does the radiation entering the surface of an active medium at the Brewster angle achieve so high a degree of beam polarization? Let us recall the principle of the reflective polarizer illustrated in Fig. 1.7. Light that is reflected from the flat surface of a dielectric (or insulating) material is partially polarized, with the electric vectors of the reflected light oscillating in a plane parallel to the surface of the material. If non-polarized light arrives perpendicularly to the surface, its polarization does not change, and the reflected and the transmitted beams remain non-polarized. However, if light arrives on the surface at angle α, the degree of polarization increases both in the refracted and the reflected rays. This results from the fact that only the components of the incident ray having polarization perpendicular to the plane determined by the incident beam and the normal to surface undergo the reflection, in contrast to the components with polarization parallel to the plane. This selectivity leads to a higher degree of polarization of the reflected and refracted beams. Brewster's angle is an unique angle at which the reflected light waves are all polarized into a single plane. It can be easily calculated utilizing the following equation for a beam of light traveling through air

$$n = \sin(\alpha)/\sin(\beta) = \sin(\alpha)/\sin(90^\circ - \alpha) = \tan(\alpha) \qquad (1.21)$$

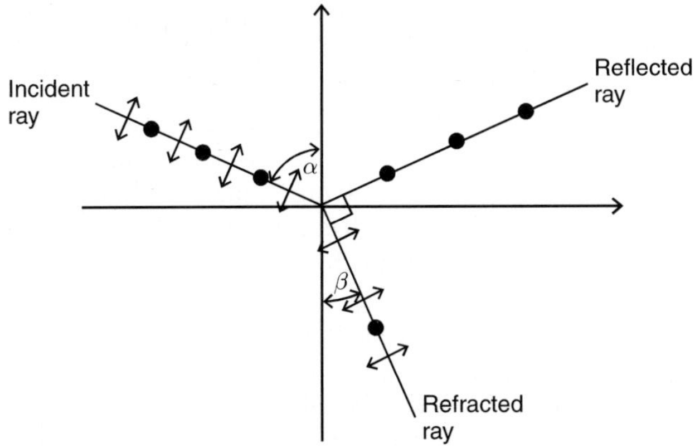

Fig. 1.7 Scheme illustrating the principle of the reflective polarizer, • ↔ denote polarization perpendicular and parallel to the plane of the drawing.

where n is the index of refraction of the medium from which the light is reflected, α is the angle of incidence, and β is the angle of refraction. Note that at the Brewster angle the component of the incident beam that is polarized in the plane of the drawing cannot undergo the reflection because the direction of oscillations at the point where the light enters the surface would be parallel to the direction of propagation of the reflected ray. If such a polarization of the reflected ray had existed, it would have been in conflict with the principle of electromagnetic theory that the light is a transverse wave, and not longitudinal. Thus, in the reflected ray direction no energy can flow for radiation with the polarization in the plane of the drawing. In contrast, the beam component with polarization perpendicular to the plane of drawing will undergo both refraction and reflection. The effect of polarization ordering is the strongest at Brewster's angle but it also occurs at other angles, although to a smaller degree. For details of the reflective polarizer the reader is referred to any textbook on optics [3].

Modern lasers commonly take advantage of Brewster's angle to produce linearly polarized light from reflections at the mirrored surfaces positioned near the ends of the laser cavity or an active medium cut at an angle equal to the Brewster angle. As a consequence of multiple light reflection in an optical resonator and multiple passages through active medium, the initially non-polarized light begins to become polarized gradually in one plane eliminating the second component.

1.3. POPULATION INVERSION

As discussed above, laser action may start when an active medium reaches a population inversion, which leads to predomination of emission processes over absorption. Now we will discuss how to achieve a population inversion. As mentioned above, in the two-level system E_n and E_m it is not possible to achieve the population inversion, and $n_n = n_m$ is an upper limit. Fortunately, two-level systems are a physical idealization rarely found in nature. In practice, molecules of an active medium in gases, liquids and solids have always more than two energy levels. In real lasers, an active medium typically involves a large number of energy levels with complex excitation and energy dissipation, including emissive and relaxation processes. However, the main features of the population inversion and pumping mechanisms can be understood by studying some simplified but fairly realistic models – three-level and four-level idealizations. Fig. 1.8 shows a typical energy diagram of a *three-level laser*, such as a ruby laser.

The three-level system is characterized by energies of E_0, E_1, E_2, *with* $E_1 > E_2$. Initially, all atoms or molecules are at the lowest level E_0. The excitation by delivering the pumping energy $\Delta E = E_1 - E_0$ transfers a certain number of the molecules from the level E_0 to the higher level E_1. Usually, the level E_1 represents a broad band so that the pumping can be delivered over a broad energy range. The excited molecules at E_1 return to the ground level E_0 choosing one of two possible paths: direct radiative or radiationless return to E_0 or indirect return via radiationless transition $E_1 \rightarrow E_2$ followed by a radiative decay to the ground state E_0. In realistic lasers most of the excited molecules choose the latter way of energy dissipation with fast radiationless transitions into the intermediate level E_2, which is a *metastable state*. The metastable state represents an excited state in which an excited molecule becomes "trapped" due to "*forbidden transitions*". The forbidden transitions mean breaking the selection rules of molecular transitions within the dipole approximation. For example, singlet-triplet electronic transition is forbidden within the dipole approximation although it occurs in real systems. The excited molecule can stay in the metastable state for a very long time – microseconds, and even milliseconds. For comparison the lifetime of majority electron excited states is only 10^{-9} s or less. The existence of a metastable state facilitates the achievement of population inversion and in fact is of paramount importance for laser action to occur. With regard to the fact that the lifetime in the state E_2 is much longer than that in the state E_1, it is possible to createę population inversion between the states E_2 and E_0 ($E_2 > E_0$) by pumping the system from E_0 to E_1. When the condition of population inversion $n_2 > n_0$ is achieved, the emission intensity in the optical resonator becomes larger than the absorption and stimulated radiation produces the laser output beam.

For small radiation intensities, when the medium saturation (that is the balance between the gain and the losses) has not been reached yet, stimulated emission can be written in a form analogous to the Lambert–Beer formula for absorption

$$I_l = I_0 \exp[-\alpha(\omega) l], \tag{1.22}$$

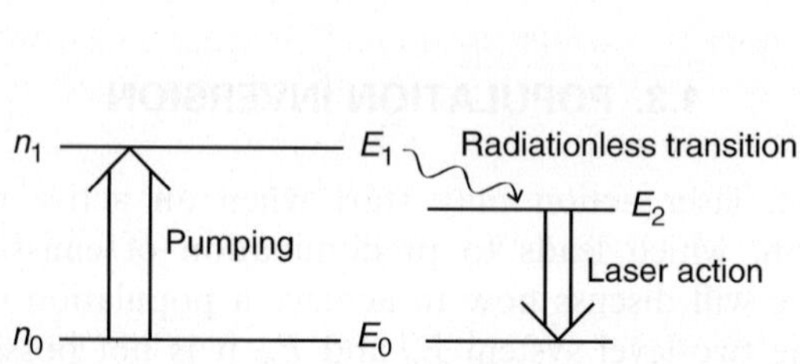

Fig. 1.8 Energy diagram of a three-level laser.

where I_0 is an intensity of incident light, I_l is an intensity after passing through the optical path l. For absorption processes, the coefficient $\alpha(\omega)$ is positive because $I_0 > I_l$ and is called the absorption coefficient. When the population inversion is created, $n_2 > n_0$, the system begins to emit radiation, so $I_l > I_0$ and the α coefficient in expression (1.22) has to be negative. In such a situation the Lambert–Beer expression can be written as

$$I_l = I_0 \exp[-B(\omega)l]. \tag{1.23}$$

where $B = -\alpha$ and is called a *small signal gain coefficient*. The expression (1.23) says that the light is gained $\exp(-B(\omega)l)$ times passing through an optical path length, l.

To make the situation more realistic, one should consider the losses in the optical cavity caused by scattering, refraction, diffraction, warming of the medium dispersion and absorption of active medium components that are inactive in laser transitions. Denoting losses as α_s ($\alpha_s > 0$), we can modify eq. (1.23) to the form

$$I_l = I_0 \exp[-(B(\omega) + \alpha_s)l]. \tag{1.24}$$

Initially, radiation in an optical resonator does not have properties typical for a laser radiation beam. An active medium, in which laser action has just started, emits a mixture of spontaneous and stimulated emission in each direction with light that is neither monochromatic nor polarized. However, with regard to the fact that the active medium is situated between two mirrors in an optical resonator, the process of radiation ordering begins.

First, because of the standing wave condition (1.19) the number of modes in the optical cavity is greatly reduced. Light becomes more monochromatic traveling along the optical resonator axis. Second, light becomes polarized and coherent. The *coherence* is the feature distinguishing the laser light from light originating from different radiation sources – both the high degree of front wave phase correlation and correlation in time. This feature is called *spatial* and *temporary coherence*. Third, multiple reflections on mirrors create positive feedback because the light in the resonator travels many times through an active medium. After $2n$ passages through the optical resonator of length l one obtains

$$G^{(2n)} = \frac{I_l^{(2n)}}{I_0} = (R_1 R_2)^n \exp[-2n(B(\omega) - \alpha_s)l], \tag{1.25}$$

where: $G^{(2n)}$ is the ratio of the intensity $I_l^{(2n)}$ after $2n$ passages through the optical resonator of length l and the initial intensity I_0, R_1 and R_2 are the reflection coefficients on the mirrors Z_1 and Z_2, respectively.

Lasers are divided into *continuous wave lasers* (*cw*) and *pulsed lasers*. In *cw* lasers the emitted light intensity is constant in time. This regime of work is achieved when the gain is equal to losses after the round-trip time through the resonator. The value of B in eq. (1.25), for which $G^{(2n)} = 1$, is called the *threshold gain*, B_t. When $G^{(2)} > 1$ (the gain achieved after the round-trip through the resonator), all following passages cause the amplification of light emitted by the laser. This condition is required for lasers working in a pulse regime.

The most typical example of a three-level laser with a metastable state as discussed above is a ruby laser. The active medium of this laser is a ruby crystal (Al_2O_3) in which

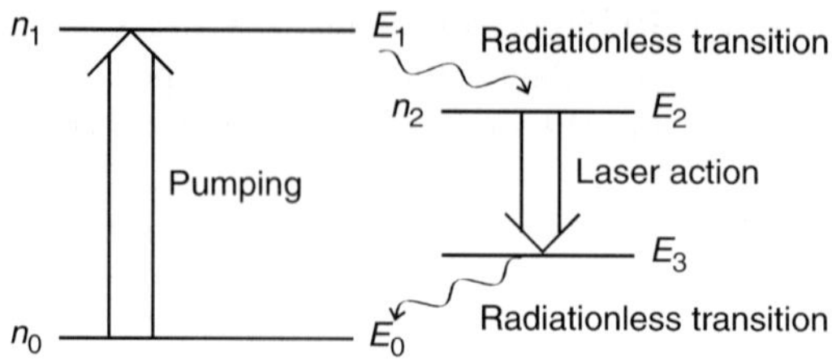

Fig. 1.9 Energy diagram of a four-level laser.

a small fraction of the Al^{+3} ions are replaced with chromium Cr^{+3} ions. The crystal is pumped by a flash lamp exciting the Cr^{+3} ions from the ground electronic level E_0 to the excited level E_1, which is actually made up of sublevel series resulting from interaction between an electron and vibrations of the crystal lattice. The fast radiationless transition from the E_1 state, characterized by a lifetime of about 50 ns causes the population of the metastable level E_2 of the Cr^{+3} ion, which has a lifetime of about 5 ms. If the flash lamp emits sufficiently intense pumping light, it is possible to produce population inversion that creates a laser action between the E_2 and E_0 levels (Fig. 1.8).

The population inversion can be achieved either by increasing the metastable upper state population or by decreasing the lower state population. The latter cannot be used in the three-level lasers because the laser transition takes place between the metastable excited state E_2 and the lowest energy level – the ground state E_0. It indicates that there is a competition between depletion and population of the E_0 state, which leads to much lower efficiency of the population inversion than can be achieved in the four-level lasers illustrated in Fig. 1.9.

In this case the level E_3 populated as a result of the laser transition $E_2 \rightarrow E_3$ is depleted quickly in a radiationless transition $E_3 \rightarrow E_0$. The four-level system permits a considerable increase in the efficiency of population inversion without delivering additional pumping energy.

The most typical example of the four-level laser is a neodymium laser Nd:YAG (the solid active medium of the *yttrium aluminum garnet* (YAG) crystal doped with neodymium Nd^{+3} ions) and a dye laser (liquid active medium). The ion argon Ar^+ laser often used in molecular spectroscopy laboratories is not the four-level laser. However, it may be treated as a special case of the four-level laser, with a preliminary pumping causing the ionization of argon atoms (Fig. 1.10).

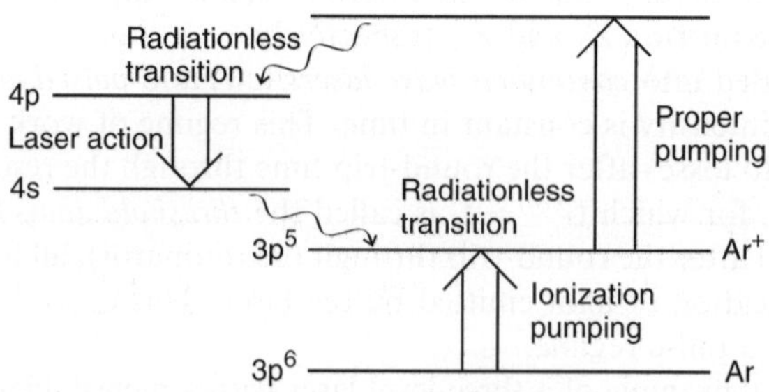

Fig. 1.10 Energy diagram for one of the laser transitions of an argon laser Ar^+.

1.4. AMPLIFICATION AND SATURATION

In thermodynamic equilibrium the population of energy states is described by the Boltzmann distribution. This indicates that at any temperature the majority of molecules usually stays in lower energy states. This statement is not always true for rotational levels exhibiting a high degree of degeneracy. Since the probabilities of stimulated transitions from the lower energy state to the higher energy state (absorption) and from the higher energy state to the lower energy state (stimulated emission) are equal, the total number of upward transitions is larger than the number of downward transitions, and the system absorbs the radiation energy

$$\frac{\partial \rho_\omega}{\partial t} < 0, \tag{1.26}$$

where ρ_ω is the energy density of the external field, that is the energy per unit circular frequency from the range between ω and $\omega + d\omega$ in volume units.

In order to amplify the incident radiation $\left(\frac{\partial \rho_\omega}{\partial t} > 0\right)$, the population inversion has to be achieved

$$\frac{n_m}{g_m} > \frac{n_n}{g_n}, \tag{1.27}$$

where n_m and n_n are the number of molecules at m and n energy levels, g_m and g_n are the degeneracy numbers of the energy levels.

When the population inversion is achieved, the traditional Lambert-Beer equation $I = I_0 e^{-\alpha l}$ loses its sense. It was derived for the assumption that the absorption coefficient α is constant, which is valid only in linear optics, when the incident radiation intensity is small and deviation from thermodynamic equilibrium is negligible. When larger intensity radiation is employed more and more molecules become transferred to a higher energy level and the state of saturation is achieved

$$\frac{n_m}{g_m} = \frac{n_n}{g_n}. \tag{1.28}$$

This indicates that the system is bleached, neither absorption nor emission of energy dominates. The Lambert–Beer equation $I = I_0 e^{-\alpha l}$ can be maintained formally if we assume that $\alpha = 0$. With further growth of the incident radiation intensity the system achieves the state called the population inversion, that is $\frac{n_m}{g_m} > \frac{n_n}{g_n}$ and the system begins to emit more energy than it absorbs (formally the absorption coefficient becomes negative $\alpha < 0$). Fig. 1.11 illustrates these three situations.

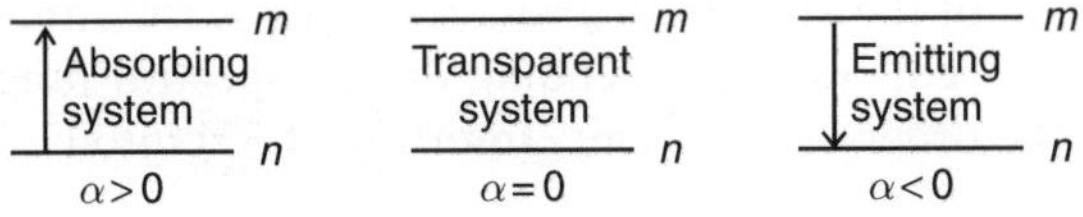

Fig. 1.11 Dependence of absorption coefficient on incident radiation intensity in nonlinear optics range.

Let us estimate what magnitude of intensities of the pumping beam have to be employed to reach the state of saturation. For simplicity let us consider the two-level system $n \rightarrow m$, although, as it was said earlier, only saturation $\left(\frac{n_m}{g_m} = \frac{n_n}{g_n}\right)$ can be reached, not the population inversion. Let $N_0 = n_n + n_m$ be the total number of molecules per volume unit, n_n and n_m are the numbers of molecules at the n level and m level. The population number at the m level is governed by

$$\frac{dn_m}{dt} = W(n_n - n_m) - \frac{n_m}{\tau}, \tag{1.29}$$

where

$$W = B_{mn}\rho_\omega. \tag{1.30}$$

In equation (1.29) the first term characterizes stimulated absorption and emission processes while the second term describes spontaneous emission or radiationless decay of the excited state characterized by a lifetime τ. Expressing n_m with $\Delta n = n_n - n_m$,

$$N_0 = n_n + n_m = (n_n - n_m) + 2n_m = \Delta n + 2n_m, \tag{1.31}$$

and introducing it into (1.29) one obtains

$$\frac{1}{2}\frac{d\Delta n}{dt} = -W\Delta n + \frac{N_0}{2\tau} - \frac{\Delta n}{2\tau} = -\Delta n\left(W + \frac{1}{2}\right) + \frac{N_0}{2\tau}. \tag{1.32}$$

Assuming that the stationary state is reached, $\frac{d\Delta n}{dt} = 0$, we get from eq. (1.32)

$$-\Delta n\left(W + \frac{1}{2\tau}\right) + \frac{N_0}{2\tau} = 0, \tag{1.33}$$

therefore

$$\Delta n = \frac{N_0}{2W\tau + 1}. \tag{1.34}$$

When the light intensity increases, W increases leading to Δn decreasing. The probability W is related to the density of the incident radiation according to (1.30). One can show that W is related to the radiation intensity I (that is the energy of photons per surface of 1 cm^2 during the time of 1 s) as follows

$$W = \frac{\sigma I}{h\nu}, \tag{1.35}$$

where σ is the absorption cross-section. The absorption cross-section σ is related to the absorption coefficient α by $\alpha = \sigma N$, where N is the number of molecules per 1 cm^3. The dimension of the absorption cross-section σ is cm^2, and the typical values are in the 10^{-12}–10^{-24} cm^2 range, depending on the spectral range. For the visible range they are on the order of 10^{-16} cm^2. Inserting (1.35) into (1.34) one gets

$$\Delta n = \frac{N_0}{2\frac{\sigma I}{h\nu}\tau + 1}. \tag{1.36}$$

Introducing

$$I_s = \frac{h\nu}{2\sigma\tau}, \tag{1.37}$$

one gets

$$\Delta n = \frac{N_0}{\frac{I}{I_s} + 1}. \tag{1.38}$$

The parameter I_s is called the *saturation parameter* and is an important measure of population inversion. When $I = I_s$, one gets $\Delta n = \frac{1}{2} N_0$, and $n_1 = \frac{3}{4} N_0$, $n_2 = \frac{1}{4} N_0$ from equation (1.38). This indicates that although the population inversion has not taken place yet, a significant fraction of molecules is already in the excited state E_2. The full saturation will be achieved by further increasing the intensity, when $I \gg I_s$, $\Delta n \to 0$ and $n_2 = n_1$. The parameter of saturation I_s has a simple physical interpretation, since the formula (1.37) can be recast to get

$$\frac{I_s \sigma}{h\nu} = \frac{1}{2\tau}. \tag{1.39}$$

The term $\frac{I_s \sigma}{h\nu}$ denotes the average rate of absorption per radiation energy unit, whereas $\frac{1}{2\tau}$ describes the average spontaneous relaxation rate. Therefore, the saturation parameter I_s denotes such an incident light intensity at which the rate of system pumping (upward transitions) is equal to the rate of spontaneous relaxation (downward transitions). When this happens, the saturation begins to develop. Assuming that $\sigma = 10^{-16}\,\mathrm{cm}^2$ and the relaxation rate is on the order of $\tau = 10^{-6}$ s, we get the saturation intensity I_s at about 1–2 kW/cm^2. The typical argon laser of 1 W used in spontaneous Raman scattering experiments emits a beam intensity on the order of $I = 1\,\mathrm{W}/(0.1\,\mathrm{mm})^2 = 0.1\,\mathrm{kW/cm}^2$. Thus, the intensity is not high enough to induce the population inversion and to pump e.g., a dye or sapphire laser. Higher powers of 5–8 W are needed to achieve this goal.

The condition (1.39) determines the saturation parameter for lasers working in the continuous work regime. The saturation condition for pulsed lasers is

$$I_s \tau_p = \frac{h\nu}{\gamma\sigma}, \tag{1.40}$$

where τ_p is the pulse duration, $\gamma = 1 + \frac{g_2}{g_1}$, when $g_2 = g_1$, $\gamma = 2$. This modification results from the fact that τ_p is much shorter than the lifetime τ characterizing the system relaxation, which is no longer essential.

REFERENCES

1. H. Haken, *Laser Theory*, Springer, Berlin/Heidelberg (1984)
2. K. Shimoda, *Introduction to Laser Physics*, 2nd edn., Springer Ser. Opt. Sci., vol. 44, Springer, Berlin/Heidelberg (1986)
3. W. Koechner, *Solid-State Laser Engineering*, Fourth Extensively Revised and Updated Edition, Springer-Verlag, Berlin Heidelberg New York (1996)

4. W. Koechner, *Solid-State Laser Engineering*, 5th Edition, vol. 1, Springer-Verlag, Heidelberg (1999)
5. F. Kaczmarek, *Introduction to Laser Physics*, PWN, Warszawa (1986), in Polish
6. A. Yariv, *Quantum Electronics*, 3rd edn., Wiley, New York (1988)
7. O. Svelto, *Principles of Lasers*, 3rd edn., Plenum, New York (1989)
8. J.R. Meyer-Arendt, *Introduction to Classical and Modern Optics*, 4th Edition, Pearson Education POD (1994)
9. M.G.J. Minnaert, *Light and Color in the Outdoors*, Fifth Ed., Springer-Verlag (1993)
10. K.D. Moller, *Optics*, Mill Valley, California: University Science Books (1988)
11. C. Mueller, M. Rudolph, *Light and Vision*, Time-Life, New York (1966)
12. H. Ohanian, *Physics, 2nd Edn. Expanded*, W. W. Norton (1989)
13. F.L. Pedrotti, Leno S. Pedrotti, *Introduction to Optics, 2nd Edition*, Prentice Hall (1993)
14. G. Waldman, *Introduction to Light: The Physics of Light, Vision, and Color*, Prentice-Hall (1983)

– 2 –

Distribution of the Electromagnetic Field in the Optical Resonator

2.1. LONGITUDINAL MODES

When an active medium, which is not placed in an optical resonator, is illuminated with monochromatic radiation, the emitted light is always non-monochromatic and spreads in all directions (Fig. 2.1a). The emission band (Fig. 2.1b) is wide and corresponds to spontaneous emission. However, when we place the same active medium in an optical resonator (Fig. 2.1c) and illuminate it with non-monochromatic radiation (e.g., from a flash lamp), the generated light has the properties characteristic for stimulated emission and is strongly monochromatic. The stimulated emission line is clearly narrower (Fig. 2.1d) than the spectrum corresponding to the non-monochromatic excitation.

This results, as we have already seen, from the fact that in the resonator only selected wavelengths λ, which fulfill the condition of the standing wave, undergo amplification

$$n\frac{\lambda}{2} = L, \tag{2.1}$$

where n is an integer. The condition (2.1) denotes that in the optical resonator of length L the standing wave at λ can be formed only when it contains the total multiple of the wavelength halves.

The light emitted from a laser is not entirely monochromatic as the condition (2.1) can be fulfilled for different combinations of the wavelength λ and the integer n. The standing wave arising along the resonator axis, characterized by the wavelength λ and the integer number, n, is called the *longitudinal mode n*. The next two longitudinal modes fulfill the following relations

$$n\lambda_n = 2L,$$

$$(n+1)\lambda_{n+1} = 2L, \tag{2.2}$$

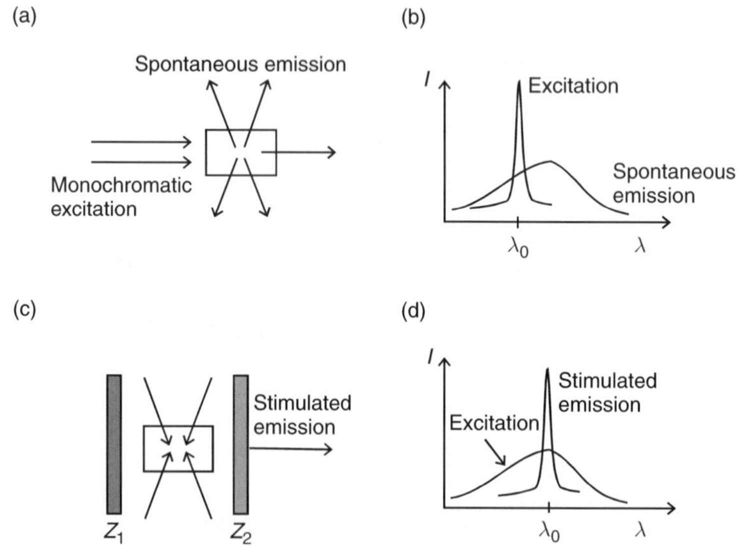

Fig. 2.1 Scheme illustrating the differences between properties of spontaneous emission and stimulated emission in an optical resonator.

and their frequencies $\nu\,(\nu = c/\lambda)$ differ by $\Delta\nu$:

$$\Delta\nu = \nu_{n+1} - \nu_n = \frac{(n+1)c}{2L} - \frac{nc}{2L} = \frac{c}{2L}, \tag{2.3}$$

which is equal to the reciprocal of the light round trip in the optical resonator. The total number of the longitudinal modes depends on spontaneous emission line width, the laser action threshold as well as on the spectral range. Indeed, assume that the laser action is related to a quantum transition $a \to b$ corresponding to the spontaneous emission of width $\delta\lambda$ (half width at half height) (Fig. 2.2). Let the threshold value at which the laser action begins correspond to half of the spontaneous emission intensity at maximum λ_0.

This indicates that the longitudinal modes determined by the largest $n_{\max}$ and the smallest $n_{\min}$ integers fulfill the condition

$$\begin{aligned} n_{\max}(\lambda_0 - \delta\lambda) &= 2L, \\ n_{\min}(\lambda_0 + \delta\lambda) &= 2L. \end{aligned} \tag{2.4}$$

Thus the total number of modes N is given by

$$N = n_{\max} - n_{\min} = 2L\left(\frac{1}{\lambda_0 - \delta\lambda} - \frac{1}{\lambda_0 + \delta\lambda}\right) \tag{2.5}$$

Thus,

$$N = \frac{2L}{\lambda_0^2 - \delta\lambda^2}[(\lambda_0 + \delta\lambda) - (\lambda_0 - \delta\lambda)] \cong \frac{4L\delta\lambda}{\lambda_0^2}. \tag{2.6}$$

where the term $\delta\lambda^2$ in the denominator was omitted since $\lambda_0^2 \gg \delta\lambda^2$. Equation (2.6) shows that the total number of the longitudinal modes N depends on the stimulated

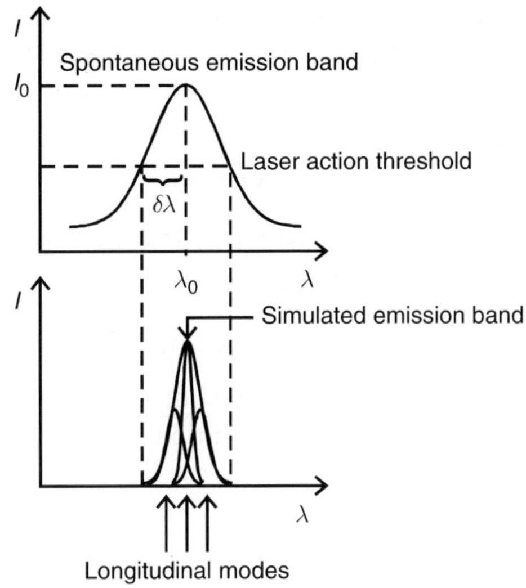

Fig. 2.2 Scheme illustrating the number of longitudinal modes.

emission linewidth, $\delta\lambda$, length of the resonator, L, and the spectral range (λ_0). The wider the emission line and the longer the resonator, the more longitudinal modes are generated in the resonator. This indicates that the light emitted from a laser becomes less monochromatic. This effect seems to be unfavorable in this context. However, we will show later that the large number of the longitudinal modes N is very useful in many cases, particularly for generation of very short laser pulses in the modelocking regime.

2.2. QUALITY FACTOR OF RESONATOR *Q*. RELATIONSHIP BETWEEN LINEWIDTH OF STIMULATED EMISSION AND RESONATOR QUALITY FACTOR

So far we have focused on the spectral width of spontaneous emission. Let us concentrate now on factors that lead to band broadening of the stimulated emission line. It is obvious from Fig. 2.2, that the total width of the stimulated emission line depends on the number of the longitudinal modes N for which the laser action occurs. However, the individual line of a single longitudinal mode n has also a certain width which is larger than the width predicted from the uncertainty principle. What is the reason of band broadening of the stimulated emission Δ_{las} for an individual mode? Generally, it depends on three main factors: a) resonator quality factor Q, b) degree of inversion population, c) pumping power P. The energy losses due to imperfection of an active medium as well as diffraction, scattering, heating of an active medium or mechanical instabilities of a resonator are the most important factors affecting band broadening of stimulated emission for individual modes. This imperfection can be characterized by a *resonator quality factor Q*. The quality factor of a system is commonly used in electronics and physics to describe dissipation of energy in oscillating systems. The quality factor Q is a measure of energy losses in a system. The smaller the losses in a system, the larger the quality factor.

Let us consider an example from classical physics: a harmonic oscillator and a damped oscillator (Fig. 2.3) described by equation

$$m\ddot{x} = -kx - \gamma\dot{x}, \tag{2.7}$$

where: $\dot{x}$ and $\ddot{x}$ are the first and the second derivatives with respect to time of the oscillator coordinate x, m is the oscillator mass, k is a force constant, γ the damping factor.

The quality factor of the system Q is defined by

$$Q = 2\pi \frac{\text{energy gained in system}}{\text{energy lost during 1 cycle}}, \tag{2.8}$$

so

$$Q = 2\pi \frac{A_0^2}{A_0^2 - A_0^2 e^{-2\beta T_0}}, \tag{2.9}$$

where A_0 is the amplitude at $t = 0$, and $\beta = \gamma/m$.

The model of the damped oscillator can be applied to describe phenomena occurring in the optical resonator. As a result of diffraction, reflections and other system imperfections, the optical resonator loses the accumulated energy, and the standing wave does not hold a constant amplitude. What is the relationship between the resonator quality factor Q and the individual line width of the stimulated emission Δ_{las}? We feel intuitively that the higher the resonator quality factor, the narrower the width. To be more precise, let us derive the mathematical relationship. Using the definition of the resonator quality factor Q (2.9) we get

$$Q = 2\pi \frac{A_0^2}{A_0^2 - A_0^2 e^{-2\beta T_0}} = \frac{2\pi}{1 - e^{-2T_0/\tau}}. \tag{2.10}$$

where the relationship $\beta\tau = 1$ between the damping factor β and the time τ has been employed. The time τ is a time after which the amplitude A_0 of the damped oscillators decays by $e = 2.718$ with respect to the initial time (Fig. 2.3). Assuming that $T_0 \ll \tau$, the term $e^{-2T_0/\tau}$ can be expanded in series

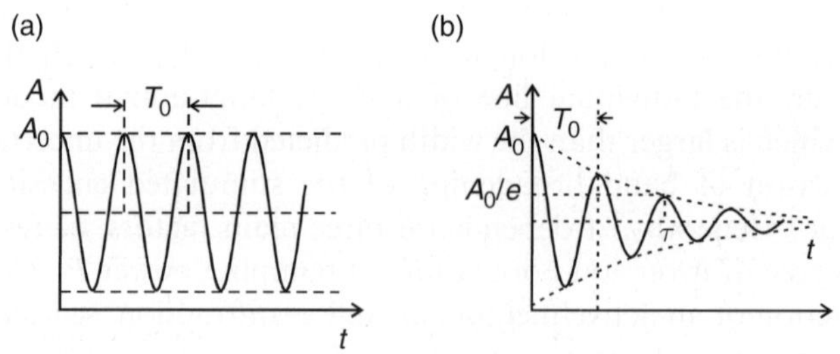

Fig. 2.3 Schematic illustration of (a) harmonic oscillator $A = A_0 \cos(\omega t + \phi)$, and (b) damped oscillator $A = A_0 e^{-\beta t} \cos(\omega t + \phi)$; $\beta = \gamma/m$; $\beta\tau = 1$, ω is a circular frequency.

$$e^{-2T_0/\tau} = 1 - 2T_0/\tau + \dots \tag{2.11}$$

Inserting (2.11) into (2.10) one gets

$$Q = 2\pi \frac{\tau}{2T_0} = \pi \nu_0 \tau = \frac{\omega_0 \tau}{2} = \frac{\omega_0}{2\beta}. \tag{2.12}$$

Equation (2.12) shows that the larger the damping rate β, the smaller is the quality factor, Q, of the resonator.

Now, we would like to know what the relationship is between the quality factor Q and the spectral width Δ_{las} of an individual mode. We will show that Δ_{las} is directly proportional to the damping rate β of the resonator, which is inversely proportional to the resonator quality factor Q

$$\Delta_{las} = 2\pi\beta = 2\pi \frac{\omega_0}{Q}. \tag{2.13}$$

The spectral distribution, Δ_{las}, in the frequency domain is related to the damped oscillator amplitude changes in the time domain by the Fourier transform (FT). The *Fourier transform* is defined by the integral

$$f(t) = \int_{-\infty}^{+\infty} f(\omega) e^{i\omega t} dt = \int_{-\infty}^{+\infty} f(\omega)[\cos(\omega t) + i \sin(\omega t)] dt \tag{2.14}$$

or

$$f(t) = \sum_{-\infty}^{+\infty} f(\omega)[\cos(\omega t) + i \sin(\omega t)], \tag{2.15}$$

where the real part of the FT is taken as the time domain output. The relationship between the time domain and the frequency domain is illustrated in Fig. 2.4 for the case when the function *f*(*t*) is the cosine function (such as for an harmonic oscillator). Figure 2.4 shows that the cosine function corresponds to the frequency spectrum represented by a very narrow line called the *Dirac delta* function.

The Dirac delta function $\delta(\omega - \omega_0)$ is defined as the following

$$\delta(\omega) = \begin{cases} 0 & \omega \neq \omega_0 \\ \infty & \omega = \omega_0 \end{cases}, \tag{2.16}$$

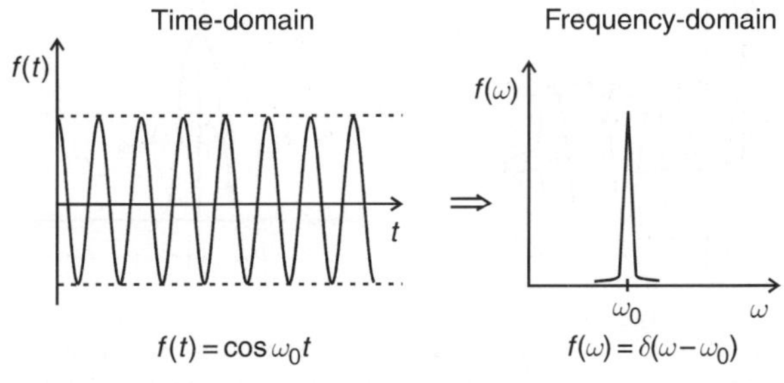

Fig. 2.4 Illustration of the time-domain and frequency-domain relationship.

$$\int_{-\infty}^{+\infty} \delta(\omega - \omega_0) f(\omega) d\omega = f(\omega_0). \tag{2.17}$$

It can be easily checked that the cosine function and Dirac delta function in Fig. 2.4 are connected through the Fourier transform. Indeed, inserting $f(\omega) = \delta(\omega - \omega_0)$ into (2.14) and using (2.17), we get

$$\mathrm{Re} \int \delta(\omega - \omega_0) e^{i\omega t} dt = \mathrm{Re}[e^{i\omega_0 t}] = \cos \omega_0 t. \tag{2.18}$$

For a damped oscillator, $f(t)$ is given by $f(t) = \cos \omega_0 t \cdot e^{-t/\tau}$. It can be shown that its Fourier transform is expressed by

$$f(\omega) = \frac{\sin[(\omega - \omega_0)\tau/2]}{[(\omega - \omega_0)\tau/2]}. \tag{2.19}$$

The functions $f(t)$ and $f(\omega)$ are illustrated in Fig. 2.5.
The function $f(\omega)$ described by (2.19) has a maximum at $x = (\omega - \omega_0)\tau/2 = 0$ corresponding to $\omega = \omega_0$, since $\lim\limits_{x \to 0} \frac{\sin x}{x} = \lim\limits_{x \to 0} \frac{(\sin x)'}{x'} = \lim\limits_{x \to 0} \cos x = 1$.

The line width of an individual mode Δ_{las} (Fig. 2.5) can be estimated, assuming that it is equal to the distance between the first side minima occurring for ω_1 and ω_2 corresponding to $x = \pm \frac{\pi}{2}$, or

$$\frac{\pi}{2} = \frac{\omega_1 - \omega_0}{2} \cdot \tau; -\frac{\pi}{2} = \frac{\omega_2 - \omega_0}{2} \cdot \tau. \tag{2.20}$$

From equation (2.20) we get

$$\Delta_{las} = \omega_1 - \omega_2 = \frac{2\pi}{\tau} = 2\pi\beta, \tag{2.21}$$

because the difference $(\omega_1 - \omega_2)$ is approximately equal to the bandwidth of an individual mode Δ_{las}. Inserting eq. (2.12) we get

$$\Delta_{las} = 2\pi\beta = \pi \frac{\omega_0}{Q}. \tag{2.22}$$

It has been shown that Δ_{las} is directly proportional to the damping rate β of the resonator and inversely proportional to the resonator quality factor Q.

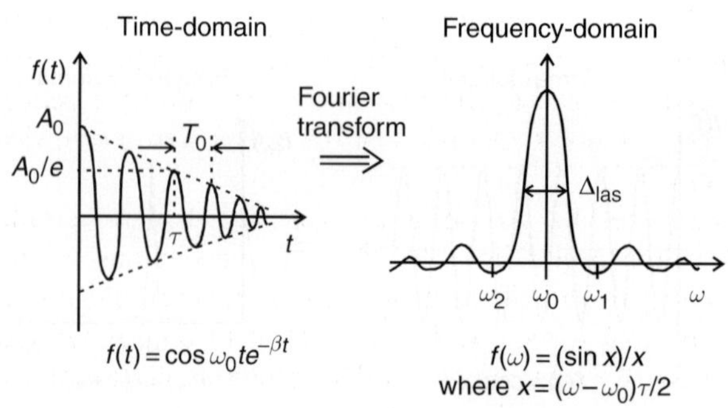

Fig. 2.5 Time domain and frequency domain for a dumped oscillator.

2.3. TRANSVERSE MODES

It is useful to think of the light inside a laser as formed of standing waves with distinct longitudinal modes along the laser cavity axis. However, this is oversimplification because lasers oscillate in different *transverse modes* as well. We observe an intensity distribution not only along the resonator axis, but also in the plane perpendicular to the direction of the laser beam propagation. The longitudinal modes are responsible for the spectral characteristics of a laser such as bandwidth and coherence length whereas the beam divergence, beam diameter, and energy distribution in the plane perpendicular to the beam propagation are governed by the transverse modes.

For many applications the *irradiance* [W/cm^2] of the laser beam is the most important parameter. It depends both on the laser output power and the size to which a laser beam can be focused. Thus, a beam profile, beam divergence, and beam diameter are important factors determining the irradiance. The depth of the hole and the hole diameter drilled by an industrial or medical laser depend on the spot size. Most medical surgeries require a laser with a well-characterized beam profile that must remain constant during the operation. Telecommunication is interested in laser beams of very low divergence that can be propagated over a long distance. These examples illustrate that the transverse intensity distribution is as important as the longitudinal distribution.

So far we have assumed that the optical resonator is a plane-parallel resonator (Fig. 2.6a). A plane wave generated in a plane-parallel resonator creates the standing wave with the wave reflected from the mirror. The intensity distribution along the resonator axis is described by the standing wave equation

$$q\frac{\lambda}{2} = L, \tag{2.23}$$

where q is an integer. To generate a standing wave of length λ the multiple of wavelength halves has to be contained in the optical resonator of length L. This indicates that for different integers q waves of different frequencies are produced. They are called longitudinal or axial modes and are characterized by q (the symbol n was used earlier, eq. (2.1)). However, one should be aware that in real systems the distribution of radiation intensity inside the resonator is much more complicated. First of all, the laser action can begins anywhere in the resonator (in the center or near the mirror) and the radiation arrives to the mirrors as a plane or a spherical wave. The plane of the mirror becomes the source of radiation and as a result a very complicated beam profile is produced. The resonator stability depends on its ability to maintain the light ray inside the resonator after multiple reflections from the mirrors. The configuration and geometry of the optical resonators decide about light reflections in the cavity.

The most common resonator configurations are: a) plane-parallel, b) confocal, c) hemispherical, d) unstable (Fig. 2.6). In plane-parallel and confocal resonators, action develops mainly in an axial area. In the confocal resonator the spherical wave front is well fitted to a mirror shape. In an unstable resonator the action develops in the whole active area of a laser. The unstable resonator does not have the ability to maintain light radiation inside for a longer time, but it permits to emit a pulse of gigantic power to be emited because the action develops in the whole active area.

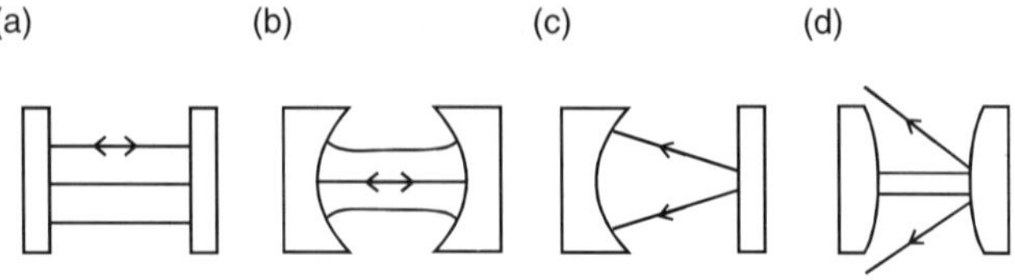

Fig. 2.6 Resonators types: (a) plane-parallel, (b) confocal, (c) hemispherical, (d) unstable.

So far we have been interested in the intensity distribution along the z axis. However, although the laser beam is highly directional, it has some intensity distribution along the x and y axis in the plane perpendicular to the propagation direction z (Fig. 2.7).

The intensity distribution in the xy plane perpendicular to the direction of the beam propagation is different for the different types of resonators presented in Fig. 2.6. The theory of the transverse modes has been treated in many papers and books [1–5]. For the confocal resonator (Fig. 2.6b) the intensity distribution is defined by the following equation [6]

$$\frac{4L}{\lambda} = 2q + (m + n + 1), \tag{2.24}$$

instead of that for the longitudinal modes

$$\frac{4L}{\lambda} = 2q. \tag{2.25}$$

The m and n are integers, q represents the number of wavelength halves held in the resonator of length L. The index q is very large in comparison with the indices of the transverse modes m and n, which are very small, usually 0, 1, 2, 4.

The distributions characterized by the integers m and n are called the *transverse mode* and are designated as TEM_{mn} or TEM_{pl} (TEM stands for the *transverse electromagnetic mode*). The modes defined with the same m (or p) and n (or l) values may differ by q describing the *longitudinal modes*. The transverse modes with different values of m (or p) and n (or l) subscripts differ in the intensity distribution in the plane perpendicular to the light propagation. The TEM_{00} mode is mathematically described by a *Gaussian profile*

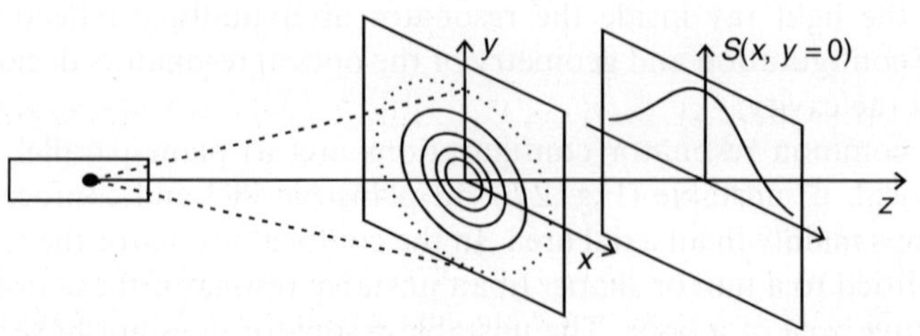

Fig. 2.7 Intensity distribution in the plane perpendicular to the propagation direction z.

$$E(r) = E_0 \exp\left(\frac{-r^2}{w^2}\right) \tag{2.26}$$

with the distribution of intensity

$$I(r) = I_0 \exp\left(\frac{-2r^2}{w^2}\right) \tag{2.27}$$

where w is the radial distance at which the field amplitude $E(r)$ has fallen to $1/e$ (and the intensity $I(r)$ to $1/e^2$ (13.5%) of its value in the center. This parameter w is often called the *beam radius* or "*spot size*", but one has to remember that the Gaussian function extends to infinity in the radial direction, leaving open the question of the beam diameter. In practice, however, this definition works well for Gaussian beams. For other profiles this definition is not useful and it is replaced in many cases by the "*second moment analysis*" [7]. This implies that for the TEM_{00} mode we can see a round spot with the maximum intensity on the beam axis decreasing to infinity in the radial direction in the plane perpendicular to the beam propagation.

The higher-order modes are obtained by multiplying the Gaussian profile by a polynomial. The transverse radiation distribution can be calculated using rectangular (m, n) or cylindrical coordinates (p, l), which are more convenient in some configurations of the resonator (Fig. 2.6). With cylindrical coordinates the *Laguerre* (*L*) *polynomial* is used

$$I_{pl}(r, \phi, z) = I_0 \rho^l [L_p^l(\rho)]^2 (\cos^2 l\phi) \exp(-\rho), \tag{2.28}$$

where $\rho = 2r^2(z)/w^2(z)$, r and ϕ are the polar coordinates in a plane transverse to the beam propagation. The lowest Laguerre polynomials are

$$L_0^l(\rho) = 1, L_1^0(\rho) = 1 - \rho, L_2^0(\rho) = 1 - 2\rho + \frac{1}{2}\rho^2. \tag{2.29}$$

The *Hermite* (*H*) *polynomial* is employed to describe the transverse radiation distribution with rectangular coordinates

$$I_{mn}(x, y, z) = I_0 \left[H_m\left(\frac{x(2)^{1/2}}{w(z)}\right) \exp\left(-\frac{x^2}{w^2(z)}\right)\right]^2 \times \left[H_n\left(\frac{y(2)^{1/2}}{w(z)}\right) \exp\left(-\frac{y^2}{w^2(z)}\right)\right]^2 \tag{2.30}$$

where z is the propagation direction of the beam. The lowest Hermite polynomials are

$$H_0(s) = 1, H_1(s) = 2s, H_2(s) = 4s^2 - 2. \tag{2.31}$$

Inserting (2.31) into (2.30) one can calculate a plot of transverse intensity distributions for the higher-order transverse modes, which are schematically illustrated in Fig. 2.8. For the TEM_{01} and TEM_{10} modes we observe two spots on the screen, for the

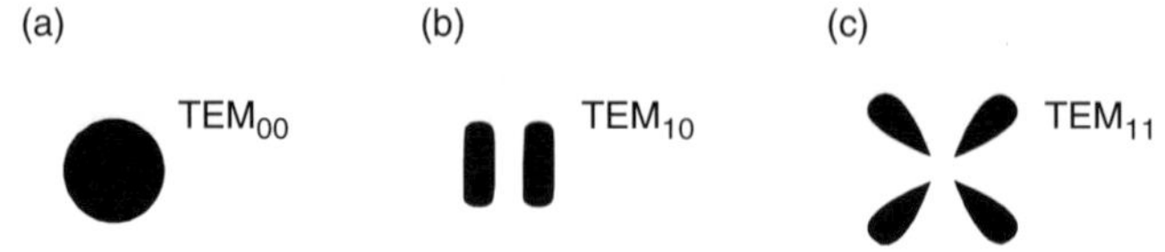

Fig. 2.8 Transverse mode patterns in rectangular coordinates: (a) TEM_{00}, (b) TEM_{10}, (c) TEM_{11}.

TEM_{11} modes – four spots with intensity maximum off-axis. The specific transverse intensity distribution depends on the resonator configuration presented in Fig. 2.6 because the beam size $w(z)$ depends on the resonator parameters such as curvature of the mirrors and the resonator length.

Figure 2.9 shows the distribution of the electric field intensity $I(x, y)$ in the xy plane for the confocal resonator. The fundamental mode TEM_{00} is symmetrical with respect to the resonator axis and its intensity has a Gaussian profile.

In most cases higher-order modes are undesirable. Most applications require the TEM_{00} mode because the intensity across the center of the TEM_{00} mode is Gaussian and permits us to focus higher intensities into smaller volumes. Because the modes higher than TEM_{00} have zero intensity in the center of intersection of the xy plane and the axis z, and are spatially much wider, the simplest way of TEM_{00} selection is inserting an aperture within a cavity. Any small loss in overall laser power resulting from restrictions to the TEM_{00} mode is compensated for by the greater efficiencies of processes such as dye laser pumping, cavity dumping, modelocking or frequency doubling.

Early methods of determining the transverse intensity distribution were essentially visual. Examining the transverse pattern of a *cw* laser beam on the lab walls or burn marks made by a pulsed beam are still used. They are partially replaced by profilometers – the instruments based on CCD (Charged Coupled Device) arrays or scanning a slit (or knife-edge) through the beam [7]. The limitation of a CCD is its

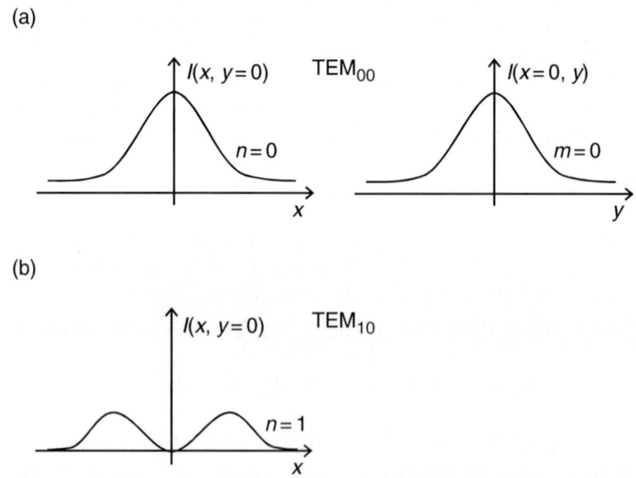

Fig. 2.9 The electric field intensity distribution for different transverse modes.

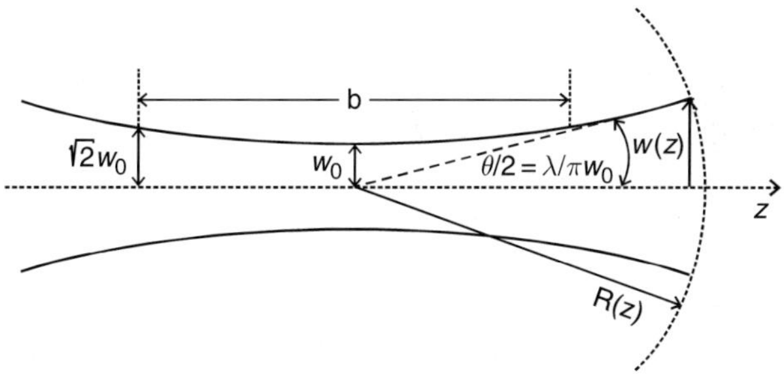

Fig. 2.10 Contour of a Gaussian Beam. Relation between the beam waist w_0, its divergence θ and the Gaussian beam spot size $w(z)$ at a distance z from the beam waist.

resolution (determined by the pixel size, 10–20 μm) and saturation of the array by powerful pulses. The scanning profilometers are based on the fact that the propagation of a Gaussian beam is fully specified by its beam waist and its divergence [8]

$$w_0\theta/2 = \lambda/\pi. \tag{2.32}$$

This implies that a Gaussian beam can be characterized by measuring its beam waist and its diameter $w(z)$ at one other location (Fig. 2.10). This method can also be used to examine deviations from the ideal Gaussian shape.

REFERENCES

1. A.G. Fox, T. Li, *Resonant modes in a maser interferometer*, Bell. Syst. Tech. J. 40(3) (1961) 453–488
2. G.D. Boyd, J.P. Gordon, Bell. Syst. Tech. J. 40 (1961) 489
3. G.D. Boyd, H. Kogelnik, Bell. Syst. Tech. J. 41 (1962) 1347
4. K. Kogelnik, T. Li, Appl. Opt. 5 (1966) 1550, see also Proc. IEEE 54 (1966) 1312, also H. Kogelnik, in *Lasers I*, ed. by A.K. Levine, Dekker, New York (1966), pp. 295–347
5. H.K.V. Lotsch, Optic 28 (1968/1969) 65, 328, 555, Optic 29 (1969) 130, 662, Optic 30 (1969) 1, 181, 217, 563
6. F. Kaczmarek, *Introduction to Laser Physics*, PWN, Warszawa (1986), in Polish
7. S. Matthews, *Good Fundamentals*, Laser Focus World 38(10) (October 2002) 82–87
8. W. Koechner, *Solid-State Laser Engineering*, 5th Edition, vol. 1, Springer-Verlag, Berlin Heidelberg (1999), p. 200
9. K. Shimoda, *Introduction to Laser Physics*, 2nd edn., Springer Ser. Opt. Sci., vol. 44, Springer, Berlin Heidelberg (1986)
10. A. Yariv, *Quantum Electronics*, 3rd edn., Wiley, New York (1988)
11. O. Svelto, *Principles of Lasers*, 3rd edn., Plenum, New York (1989)
12. H. Haken, *Laser Theory*, Springer, Berlin Heidelberg (1984)
13. J.S. Kruger, *Electro-Opt. Syst. Designs* 12, September 1972
14. R.J. Freiberg, A.S. Halsted, Appl. Opt. 8 (1969) 335

– 3 –

Generation of Ultrashort Laser Pulses

Lasers are divided into *continuous wave lasers* (*cw*) and *pulsed lasers*. In *cw* lasers, the intensity of the emitted light is constant as a function of time. In the *cw* regime, the gain is equal to losses after the round-trip time in the resonator. The value of B in eq. (1.25), for which $G^{(2)} = 1$, is called the *threshold gain*, B_t. When the gain achieved after the round-trip time in the resonator, $G^{(2)} > 1$, every subsequent passage leads to further amplification of the light emitted by the laser. This condition is required for pulsed lasers.

The temporal pulse duration in *cw* lasers goes to infinity. In pulsed lasers, the energy is released in the form of a short pulse of light. Initially, in 1960, laser pulses were around 10 ms long with a peak power of kilowatts ($1\ \mathrm{kW} = 10^3\ \mathrm{W}$). Shortly after 1960, the *Q-switching technique* reduced the pulse duration by a factor of 10^4 to nanoseconds ($1\ \mathrm{ns} = 10^{-9}\ \mathrm{s}$), with peak powers of megawatts ($1\ \mathrm{MW} = 10^6\ \mathrm{W}$). Unfortunately, we cannot produce pulses shorter than 10 ns by this method, because of the required pulse build-up time. The *cavity dumping* technique may reduce the pulse duration to 1–2 ns. These pulses are still too long for monitoring the dynamics in molecular systems because the fundamental chemical processes of life, such as proton transfer, electron transfer, photosynthetic reactions, or the folding of protein molecules, occur on the pico- and femtosecond ($1\ \mathrm{ps} = 10^{-12}\ \mathrm{s}$, $1\ \mathrm{fs} = 10^{-15}\ \mathrm{s}$) time-scale.

In 1964, the development of the *modelocking technique* reduced the pulse duration to picoseconds and pushed the peak power up to gigawatts ($1\ \mathrm{GW} = 10^9\ \mathrm{W}$). Shortly thereafter, pulses of just a few femtoseconds were produced using *colliding-pulse modelocking*. Until about a decade ago, dye lasers were used exclusively to generate ultrashort pulses. Now, they have been replaced by the more convenient technology of solid-state lasers and fiber lasers. The lasers working in the modelocking regime represent a special group of continuous wave lasers.

The breakthrough came in 1987 when the current methods of compressing laser pulses were first developed. The idea of *chirped pulse amplification* (CPA), combined with the employment of a solid-state material almost ideal for amplification—Ti:sapphire—led to the production of sub-5-fs pulses at nanojoule energies ($1\ \mathrm{nJ} = 10^{-9}\ \mathrm{J}$) that can be amplified to millijoules ($1\ \mathrm{mJ} = 10^{-3}\ \mathrm{J}$) with peak powers of terawatts ($1\ \mathrm{TW} = 10^{12}\ \mathrm{W}$). The commercial introduction in 1990 of the solid-state Ti:sapphire lasers, which produce femtosecond pulses through *Kerr-lens modelocking*, began a revolution in the field of ultrafast research and applications. The past decade has brought further spectacular progress in the development of ultrashort, powerful laser pulses. The femtosecond pulses have been superseded by attosecond pulses (1 as $=10^{-18}$ s), and peak powers of petawatts (1 PW $=10^{15}$ W) can be achieved. Femtosecond pulses have durations with the period of molecular oscillations whereas attoseconds are comparable with the periods of electrons revolving in atomic orbitals. A femtosecond pulse's interaction with an electronic wave packet under well-defined conditions gives rise to the generation of higher-harmonics, up to the soft X-ray region, with pulse durations reaching the attosecond region. Clearly, the limits have not been reached yet.

The progress in the development of ultrashort, powerful laser pulses has opened up an exciting area of potential applications such as energy production with laser beams used to ignite a pellet of fusion fuel, or experiments in basic physics to mimic the conditions within the center of stars. It is possible that ultrashort-pulse lasers will help investigations of the processes governing the evolution of stars and their explosion into Supernovae.

As we have learned from this Introduction, modelocking is an extremely important technique for generating ultrashort pulses. In the next section, we will explain what modelocking means, and how ultrafast pulses can be generated.

3.1. MODELOCKING. RELATIONSHIP BETWEEN LINEWIDTH OF STIMULATED EMISSION AND PULSE DURATION

In a free-running regime as discussed so far, lasers produce waves with a mixture of transverse and longitudinal modes, with a random mode-to-mode phase relationship that changes with time. This is not surprising, because coherence is only a single-mode feature. Each mode oscillates independently of the other modes. The intensity of the signal in the free multimode regime that results from the interference of longitudinal modes with random-phase relationships is a chaotic sequence of fluctuations that looks like the characteristics of thermal noises (Fig. 3.1a). The free-running laser output is a time-averaged statistical mean value of the intensity.

Suppose that it is possible to achieve a situation where the phases of the longitudinal modes are forced to maintain a fixed phase-relationship to each other. Such a laser is said to be *modelocked* or *phase-locked*. How to achieve this is different matter, which we will discuss later. We will show that, in this case, the laser output shows a periodic repetition of a wave-packet resulting from the constructive interference of longitudinal modes (Fig. 3.1b). The sequence of regular pulses occurs with the period $T=2L/c$ whereas the single pulse's temporal duration, t_p, is given by

$$t_p=\frac{T}{N}=\frac{2L}{cN}, \tag{3.1}$$

where N is the number of modes generated in an optical resonator (Fig. 3.2), L is the length of the resonator, and c is the speed of the light. Experimental techniques that will be discussed later stimulate the maintenance of a fixed phase-difference between modes, and leads to a laser work-regime called *modelocking*. Therefore, the modelocking results in a train of pulses with the repetition period, T, equal to the

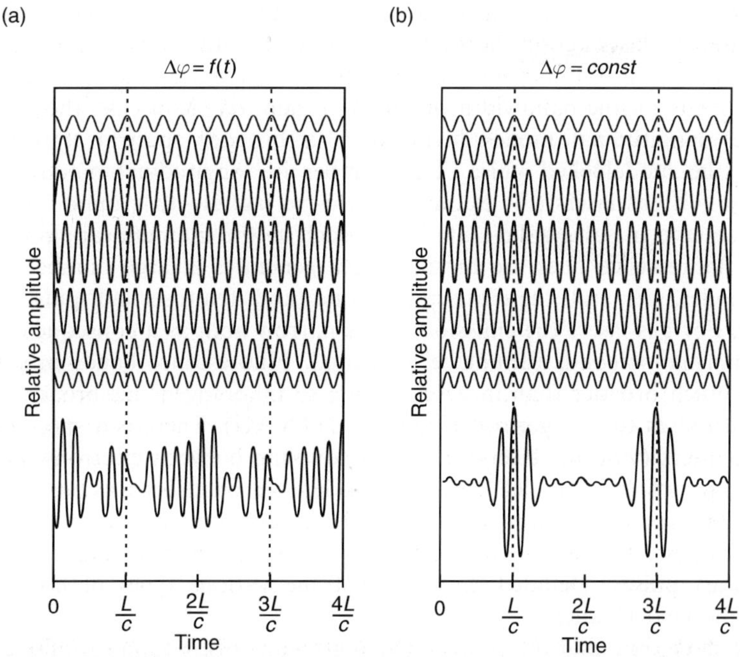

Fig. 3.1 (a) Time-evolution of the electric field in a free-running laser working in a multimode regime, (b) time-evolution of the electric field in a modelocked laser [1].

round-trip time in the optical resonator, and the pulse's temporal duration, t_p, equal to the round-trip time divided by the number of phase-locked modes, N.

As was shown earlier, in eq. (2.6), the number of modes, N, depends on the stimulated emission line width, $\delta\lambda$,

$$N = \frac{4L\delta\lambda}{\lambda_0^2}. \tag{3.2}$$

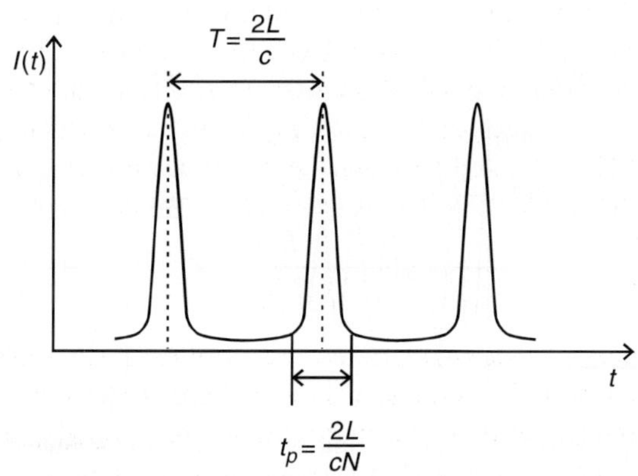

Fig. 3.2 Time-evolution of the intensity in a modelocked laser.

Equations (3.1) and (3.2) indicate that the spectral bandwidth of an active medium, and the laser action threshold, determine the duration of the modelocked pulse. The pulse duration depends on the number of longitudinal modes, N, which in turn depends on the bandwidth of the laser gain, $\delta\lambda$. As a rule, the greater the number of longitudinal modes involved in a broader spectral transition, the shorter is the modelocked pulse. The number of longitudinal modes can vary from a few—in gas lasers (for example in He-Ne lasers)—to around 10^4 or more in dye lasers and in some solid-state lasers, such as the titanium-sapphire laser. In dyes, the fluorescence lines are broad, which generates a large number of longitudinal modes, N, and therefore picosecond and femtosecond pulses can be generated. For gas lasers the emission line is narrow and, as a consequence, pulses shorter than nanoseconds cannot be generated. Fluorescence bands in solid-state lasers are much broader than in gases, owing to inhomogeneous broadening, and picosecond pulses can be generated (e.g., in Nd:YAG). There is a special class of solid-state lasers (vibronic lasers) in which coupling between electrons and vibrational degrees of freedom leads to a considerable band-broadening of spectral lines and, as a consequence, makes it possible to generate femtosecond pulses. The titanium–sapphire laser is the best candidate among the vibronic lasers for producing ultrafast pulses. Detailed discussion of the various types of lasers can be found in the next Chapter.

We will show now that in the modelocking regime one obtains a pulse sequence with the periodicity of $T = 2L/c$, with the duration of an individual pulse being $t_p = 2L/cN$. We shall assume for simplicity that the generated modes are plane waves, $E(t) = E_0 e^{i\omega t}$. This indicates that the spectral distribution of an individual longitudinal mode is described by the Dirac delta function $\delta(\omega - \omega_0)$ with infinitely narrow width. We will apply this approximation, which is not too bad if we recall one of the properties of the Fourier transform. The spectral line of width $\Delta\omega$ (Fig. 3.3a) corresponding to the damped signal in the time domain (Fig. 3.3b) measured in infinite time interval $(0, \infty)$ can be replaced by a non-damped signal in the finite time interval $(-\tau/2, +\tau/2)$ (Fig. 3.3c). Therefore, the plane wave is a reasonable approximation.

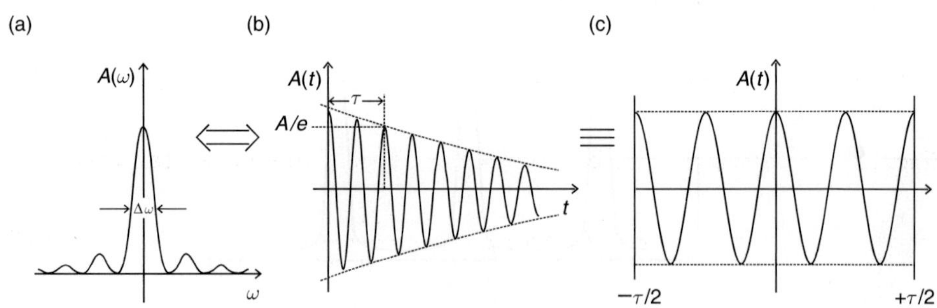

Fig. 3.3 Relationship between the spectral line width $\Delta\omega$ in the frequency domain (a) and the signal in the time domain (b). The signal (b) is equivalent to the signal (c). Explanation in the text.

The total electric field coming from $N = 2n + 1$ modes is represented by a sum

$$E(t) = \sum_{k=-n}^{n} E_0 \exp\{i[(\omega_0 + k\Delta\omega_q)t + k\Delta\varphi_q]\}, \tag{3.3}$$

where $\Delta\omega_q$ and $\Delta\varphi_q$ are respectively the frequency and the phase difference between the neighboring longitudinal modes. We now use the following relationships

$$\sum_{k=-n}^{n} e^{ik\alpha} = 2\sum_{k=0}^{n} \cos k\alpha - 1 \tag{3.4}$$

$$\sum_{k=-n}^{n} \cos k\alpha = \frac{\cos\frac{n\alpha}{2}\sin\frac{(n+1)\alpha}{2}}{\sin\frac{\alpha}{2}}. \tag{3.5}$$

Substituting (3.4) and (3.5) into (3.3), one obtains

$$\begin{aligned} E(t) &= E_0 \exp(i\omega_0 t) \sum_{k=-n}^{n} \exp[i(k\Delta\omega_q t + k\Delta\varphi_q)] \\ &= E_0 \exp(i\omega_0 t) \left[2\sum_{k=0}^{n} \cos(k\Delta\omega_q t + k\Delta\varphi_q) - 1 \right] \\ &= E_0 \exp(i\omega_0 t) \left[\frac{2\cos\left(n\frac{\Delta\omega_q t + \Delta\varphi_q}{2}\right) \sin\left[(n+1)\frac{\Delta\omega_q t + \Delta\varphi_q}{2}\right]}{\sin\frac{\Delta\omega_q t + \Delta\varphi_q}{2}} - 1 \right] \end{aligned} \tag{3.6}$$

By inserting, $\alpha = (\Delta\omega_q t + \Delta\varphi_q)$ into (3.6) one obtains

$$\begin{aligned} E &= E_0 \exp(i\omega_0 t) \left[\frac{2\cos\frac{n\alpha}{2}\sin\frac{(n+1)\alpha}{2} - \sin\frac{\alpha}{2}}{\sin\frac{\alpha}{2}} \right] \\ &= E_0 \exp(i\omega_0 t) \frac{2\cos\frac{n\alpha}{2}\left(\sin\frac{n\alpha}{2}\cdot\cos\frac{\alpha}{2} + \cos\frac{n\alpha}{2}\cdot\sin\frac{\alpha}{2}\right) - \sin\frac{\alpha}{2}}{\sin\frac{\alpha}{2}} \\ &= E_0 \exp(i\omega_0 t) \frac{\cos\frac{2n\alpha}{2}\sin\frac{\alpha}{2} + \sin\frac{2n\alpha}{2}\cos\frac{\alpha}{2}}{\sin\frac{\alpha}{2}} \\ &= E_0 \exp(i\omega_0 t) \frac{\sin\frac{(2n+1)\alpha}{2}}{\sin\frac{\alpha}{2}} \end{aligned} \tag{3.7}$$

Since $2n + 1 = N$ is equal to the number of longitudinal modes, one can write

$$E = E_0 \exp(i\omega_0 t) \frac{\sin\frac{N(\Delta\omega_q t + \Delta\varphi_q)}{2}}{\sin\frac{(\Delta\omega_q t + \Delta\varphi_q)}{2}}. \tag{3.8}$$

If the phase-difference between the neighboring longitudinal modes, $\Delta\varphi_q$, changes with time in a random way, the resultant electric field, E, in eq. (3.8) changes chaotically with time as in Fig. 3.1a. However, if the phase difference, $\Delta\varphi_q$, between modes is

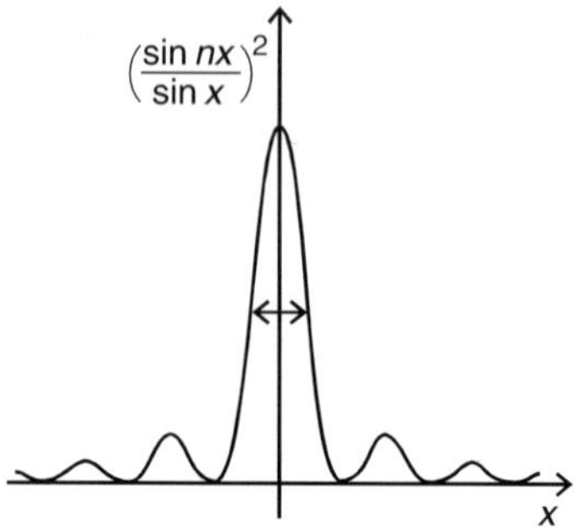

Fig. 3.4 Profile of a function $\left(\frac{\sin nx}{\sin x}\right)^2$.

constant, the total intensity of the electric field, E, arising as a result of the interference from N synchronised longitudinal modes, is an amplitude-modulated wave at a carrier frequency ω_0, equal to the central mode with the envelope expressed in the form

$$A(t) = E_0 \frac{\sin N(\Delta\omega_q t + \Delta\varphi_q)/2}{\sin(\Delta\omega_q t + \Delta\varphi_q)/2}. \tag{3.9}$$

The intensity, $I(t) = A^2(t)$, generated as a result of the interference between the N modes is a function of type $\left(\frac{\sin nx}{\sin x}\right)^2$, well known from diffraction theory, with the maximum at $x = 0$ illustrated in Fig. 3.4.

Since the function (3.9) is periodic, the radiation intensity generated as a result of the interference of N synchronised longitudinal modes is a repetition of pulses, periodic in time, as represented in Fig. 3.5.

The result derived in eqn. (3.9) shows that the laser emission is a sequence of regular pulses with temporal intervals of T, if the phase difference $\Delta\varphi_q$ between the neighboring modes is constant. The temporal intervals, T, between the pulses can be

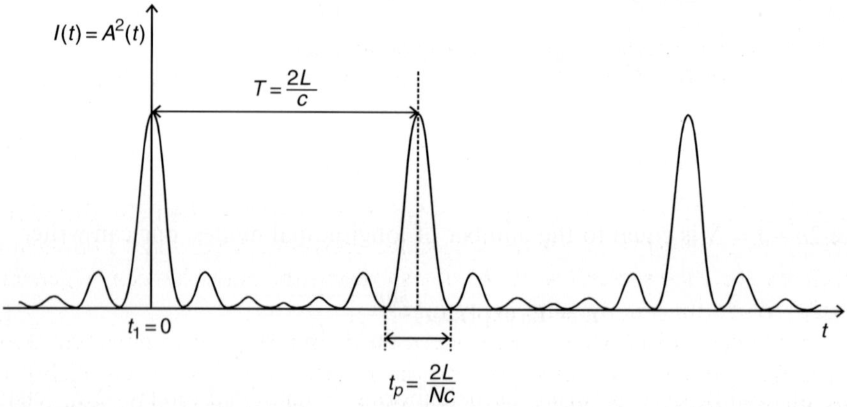

Fig. 3.5 Diagram of radiation-intensity dependence generated as a result of N longitudinal modes' interference, as a function of time.

calculated easily from eq. (3.9). We simply have to find the distance between the two subsequent largest maxima in Fig. 3.5.

From eq. (3.9), the first maximum, at time t_1, occurs when

$$\Delta\omega_q t_1 + \Delta\varphi_q = 0. \tag{3.10}$$

The next maximum, at time t_2, has to fulfil the condition

$$\Delta\omega_q t_2 + \Delta\varphi_q = 2\pi. \tag{3.11}$$

Subtracting the equations (3.11) and (3.10), we obtain $\Delta\omega_q T = \Delta\omega_q(t_2 - t_1) = 2\pi$. Therefore, the interval T between the modelocked pulses is

$$T = \frac{2\pi}{\Delta\omega_q} = \frac{2\pi}{2\pi\Delta\nu_q} = \frac{2L}{c}. \tag{3.12}$$

The equation (3.12) employs the relationship (2.3) derived in Chapter 2 for the frequency difference between the neighboring modes $\Delta\nu_q$, which is $\frac{c}{2L}$. In a similar way, we can estimate a single-pulse duration, t_p. One can see from Figure 3.5 that it corresponds approximately to the distance between the first two minima around the "large" maximum. The minimum occurs when the numerator of eq. (3.9) is equal to zero

$$\sin N(\Delta\omega_q t + \Delta\varphi_q)/2 = 0, \tag{3.13}$$

corresponding to

$$N(\Delta\omega_q t_1 + \Delta\varphi_q)/2 = 0, \tag{3.14}$$

and

$$N(\Delta\omega_q t_2 + \Delta\varphi_q)/2 = \pi. \tag{3.15}$$

Thus, the single-pulse duration, t_p, is

$$t_p = t_2 - t_1 = \frac{2\pi}{N\Delta\omega_q} = \frac{2L}{Nc}. \tag{3.16}$$

Therefore, we have proved that the relationship (3.1) is valid.

By inserting eqn. (3.2) into (3.16), one obtains the pulse duration, t_p, in another form

$$t_p = t_2 - t_1 = \frac{\lambda_0^2}{2c\delta\lambda}. \tag{3.17}$$

Equation (3.17) is a very important relationship between the pulse duration, t_p, and the gain bandwidth, $\delta\lambda$ for the stimulated emission. According to this relationship, the broader the spectral width $\delta\lambda$, the shorter is the pulse that can be generated. We will refer to this relationship many times.

The relationship (3.17) is a consequence of the relationship between the time domain and the frequency domain described by the Fourier transform in eq. (2.14), as is discussed in Chapter 2. It simply illustrates the Heisenberg uncertainty principle

$$\Delta t\Delta E \geq h/2\pi, \tag{3.18}$$

where Δt denotes the uncertainty of the time, which may be interpreted as a pulse duration, t_p, and $\Delta E = \hbar\Delta\omega$ defines the uncertainty of the energy corresponding to the width of the spectral band.

The magnitude of the product $\Delta t \cdot \Delta E$ depends on a temporal pulse shape. We assume that the pulse shape is described by a Gaussian function

$$E(t) = \frac{E_0}{\tau} \exp\left(-\frac{t^2}{2\tau^2}\right). \tag{3.19}$$

The frequency spectrum, $E(\omega)$, can be obtained from the Fourier transform

$$E(\omega) = \frac{1}{2\pi}\int_{-\infty}^{\infty} E(t)e^{-i\omega t}dt = \frac{E_0}{2\pi}\exp\left[-\frac{\tau^2}{2}(\omega-\omega_0)^2\right]. \tag{3.20}$$

This indicates that the shape of the spectral band in the frequency domain is also described by a Gaussian distribution. The full width at half height (FWHH) of the temporal pulse profile $E(t)$ given by (3.19) is

$$\Delta t_{FWHH} = 2\tau(\ln 2)^{1/2}, \tag{3.21}$$

and the FWHH of the spectral profile, $E(\omega)$, in the frequency domain given by eq. (3.20) is

$$\Delta\omega_{FWHH}/2\pi = \Delta\nu_{FWHH} = (\ln 2)^{1/2}/\pi\tau. \tag{3.22}$$

Thus, for the Gaussian profile the product $\Delta t \cdot \Delta E$ is equal to

$$\Delta t_{FWHH} \cdot \Delta\nu_{FWHH} = 0.441. \tag{3.23}$$

For other shapes of temporal profiles, this product is different from 0.441. Table 3.1 shows the values of the product for the most common pulse shapes.

Table 3.1

The product of time and spectral widths, $\Delta t_{\mathrm{FWHH}}\Delta\nu_{\mathrm{FWHH}}$, for different shapes of temporal pulses, $I(t)$

Function	$I(t)$	$\Delta t_{FWHH}\Delta\nu_{FWHH}$
Square	$I(t) = 1;\ \lvert t\rvert \le t_p/2$ $I(t) = 0;\ \lvert t\rvert > t_p/2$	1.000
Diffractive	$I(t) = \dfrac{\sin^2\left(\frac{t}{\Delta t_{FWHH}}\right)}{\left(\frac{t}{\Delta t_{FWHH}}\right)}$	0.886
Gaussian	$I(t) = \exp(-(4\ln 2)t^2/2\Delta t_{FWHH}^2)$	0.441
Hyperbolic Secant	$I(t) = \sec h^2\left(\frac{1.76t}{\Delta t_{FWHH}}\right)$	0.315
Lorentzian	$I(t) = \dfrac{1}{1+\left(\frac{4t^2}{\Delta t_{FWHH}^2}\right)}$	0.221
Exponential	$I(t) = \exp\left(\frac{-(\ln 2)t}{\Delta t_{FWHH}}\right)$	0.142

The relationship derived in eq. (3.23) corresponds to an ideal situation of a perfectly modelocked laser with a pulse called the *Fourier-transform limited pulse*. Such a pulse is the shortest pulse, of Δt_{FWHH}, that can be generated for a given gain-spectrum, $\Delta\nu_{FWHH}$. In practice, such pulses are seldom produced. The uncertainty relationship (3.23) holds only when the individual longitudinal modes are perfectly synchronised with each other, or in other words, when the spectral phase is a linear function of frequency, as we assumed in eq. (3.3)

$$E(\omega) = A(\omega)e^{i\Phi(\omega)}$$
$$\Phi(\omega) = \Phi_0 + \Phi_1(\omega - \omega_0). \tag{3.24}$$

It is crucial for perfect modelocking that all frequency components experience the same round-trip cavity-time, which is ensured by the phase linearity in eq. (3.24). Owing to material dispersion, each frequency component travels with a different velocity (the so-called *group velocity*, which will be discussed in Chapter 5), and the spectral phase is usually more complicated than linear

$$\Phi(\omega) = \sum_0^{\infty} \frac{1}{n!} \frac{d^n\phi}{d\omega^n}\bigg|_{\omega_0} (\omega - \omega_0)^n. \tag{3.25}$$

Owing to the quadratic term in the phase, each frequency component that comprises the spectrum of the pulse experiences a delay which is linearly proportional to the offset from the central frequency, ω_0. The pulse is said to be "*linearly chirped*". In this case, a Gaussian pulse for an ideal modelocking presented in Fig. 3.6 is replaced by a Gaussian pulse linearly chirped (Fig. 3.7). The linear chirp can be negative or positive (what this means exactly, we will discuss in Chapter 7). One can see from Fig. 3.7 that for the positive chirp, red components travel faster than blue ones, in contrast to the negative chirp.

To produce pulses as short as possible, dispersion in the cavity must be compensated for by adding optical elements—typically pairs of prisms or gratings and, especially, coated mirrors or a length of optical fiber. We will discuss these methods in Chapter 5.

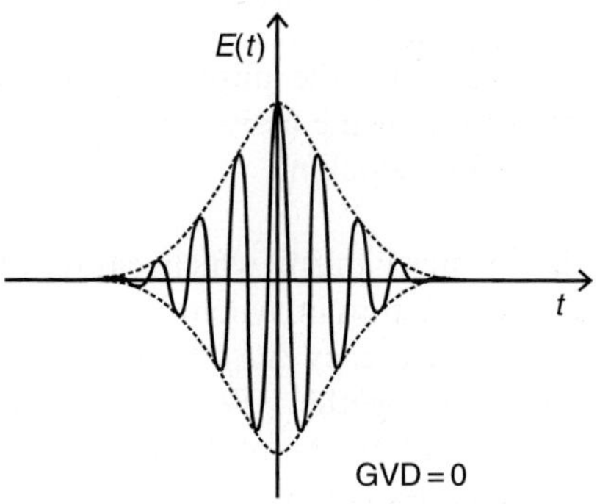

Fig. 3.6 Illustration of a Gaussian pulse for perfect modelocking (zero chirped).

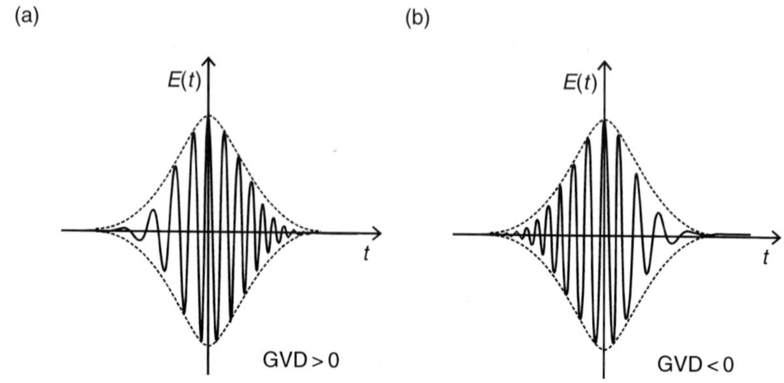

Fig. 3.7 Illustration of a Gaussian pulse with positive (a), and negative (b), chirp.

3.2. METHODS OF MODELOCKING. ACTIVE AND PASSIVE MODELOCKING

The question arises of how to achieve the modelocking, or in other words, how to create a situation with a fixed mode-to-mode phase-relationship between the neighboring modes. There are many different ways of modelocking but the principle is always the same—the periodic modulation of the optical resonator parameters (amplitude or frequency) with a frequency equal to the difference of frequencies between the neighboring longitudinal modes, $\Delta\omega_q$.

Generally, the methods of modelocking can be classified into *active modelocking* and *passive modelocking*. A special case of passive modelocking is the *self-modelocking* which occurs spontaneously in an active medium as a result of self-focusing.

The modulation of the optical resonator parameters with the frequency $\Delta\omega_q$ can be obtained in a variety of methods including:

a) acousto-optic devices which produce a sound-wave, modulating the laser beam's intensity propagating through a resonator;
b) electro-optical modulators driven at exactly the frequency separation of the longitudinal modes, $\Delta\omega_q$;
c) the *saturable absorbers* modulating the amplification factor of an active medium. The first two methods belong to the active modelocking methods, whereas the last represents the passive modelocking.

We now ask, "What is the mechanism which causes the randomly oscillating longitudinal modes to begin oscillating in synchronised phases, under the influence of the modulating factor, at the frequency $\Delta\omega_q$?" This can be achieved only when the longitudinal modes are coupled together. When we modulate the amplitude or frequency of a given longitudinal mode of frequency ω_0, with the modulation frequency Ω, an additional radiation component appears at $\omega_0 \pm n\Omega$. If the modulation frequency Ω is equal to the frequency-separation, $\Delta\omega_q$, of the longitudinal modes, these additional components overlap with the neighboring modes, causing coupling of the modes and stimulating oscillations in the same phase.

Here, we will describe the basic aspects of frequency- or amplitude modulation. In the first method, (a), *an acousto-optic transducer* generates a sound wave that modulates the amplitude of the laser beam in the optical resonator. Understanding of the mechanisms governing the interactions between light and sound waves is very important, since the acousto-optic devices are often used in laser technologies—not only for modelocking, but also in pulse-selection (*cavity dumping*) and in the *Q*-switching amplification.

A brief description of the interactions between light and sound waves is given below. We follow an excellent tutorial discussion in ref. [1]. A more rigorous discussion of these phenomena can be found in ref. [2].

If a transducer emitting ultrasonic waves at frequency Ω in the range of megahertz is placed in a glass of water (Fig. 3.8) illuminated with a laser beam of frequency ω, one notices that the light passing through the glass splits into several beams. At each side of the fundamental beam, which is unaffected in frequency ω and direction, one observes side beams having frequencies $\omega \pm n\Omega$. This phenomenon is named the *Debye and Sears effect*, after the authors who first described it in 1932 [3]. The Debye and Sears effect is similar to light diffraction by a slit. The difference is that in diffraction by a slit all the side beams have the same frequency, ω as the incident beam. Because the sound wave is a longitudinal wave, and its propagation occurs by creating regions of different density (Fig. 3.8), the analogy to diffraction is not surprising, because the regions of compression and dilation generated by the sound wave may remind us of a diffraction grating. Indeed, the regions of dilation can be treated as the slits through which more light passes than through the regions of greater density. However, why do the frequencies $\omega \pm \Omega$, $\omega \pm 2\Omega$, $\omega \pm 3\Omega n$ appear? We may imagine that light of frequency ω arrives at a medium characterised by a refractive index n_1 (Fig. 3.9). If the refractive index n_1 is larger than that of the

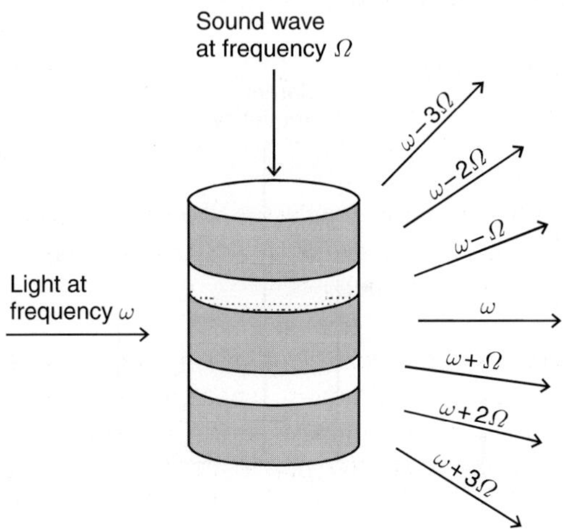

Fig. 3.8 Illustration of light and sound waves interaction.

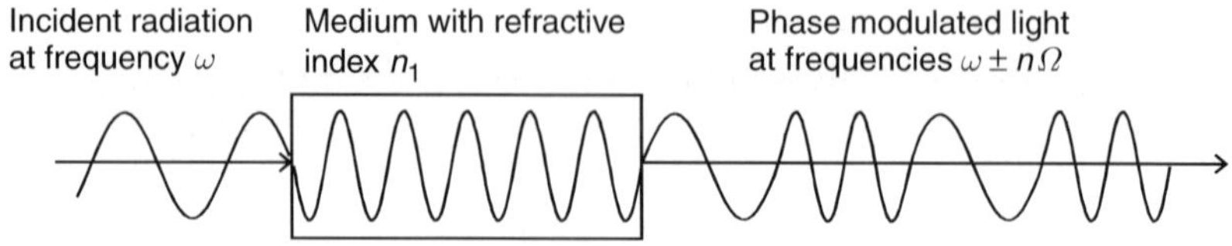

Fig. 3.9 Light modulation by periodic changes of the refractive index, n_1.

environment, n_0, the light in the medium travels n_1/n_0 times slower (since $\lambda\nu = c/n$). Let us assume that we have found a way of modulating the refractive index, n_1, with frequency Ω. This modulation causes the light in the medium to propagate faster or slower, and the output light from the medium is also modulated. The output light is characterised by the carrier frequency, ω, of the incident light and a side frequency of Ω leading to the appearance of additional components at frequencies of $\omega \pm n\Omega$ (Fig. 3.10).

The longer the optical path, l, in the material, the greater are the amplitudes of the sidebands at the frequencies $\omega \pm n\Omega$. The sidebands' amplification is reached at the expense of the amplitude of the fundamental beam at the carrier frequency, ω. The optical pathlength, l, is the parameter defining when the Debye–Sears effect can occur. We can distinguish two limiting cases

$$l \ll \frac{\Lambda^2}{2\pi\lambda} \tag{3.26}$$

and

$$l \gg \frac{\Lambda^2}{2\pi\lambda}, \tag{3.27}$$

where λ is the optical wavelength, and Λ is the length of an acoustic wave. The relationship (3.26) defines the critical length of the optical path for which the Debye–Sears effect can be observed. The relationship (3.26) characterises the conditions required for modelocking with *acousto-optic devices*, and is called the

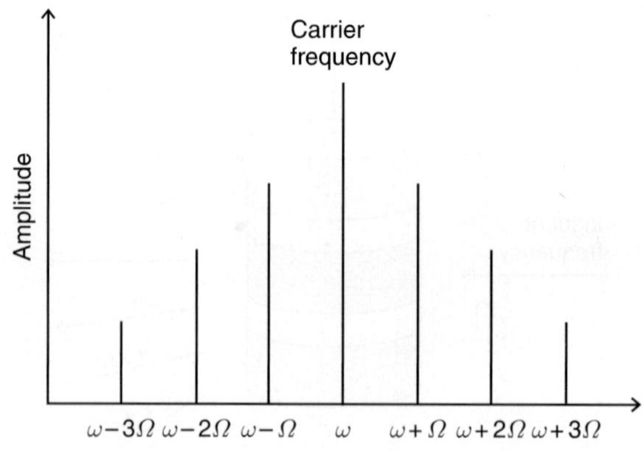

Fig. 3.10 Spectral distribution achieved by periodical modulation of the refractive index at frequency Ω.

Raman–Nath regime after the authors who derived it. The relationship (3.27) is employed in another acousto-optic configuration—a device for pulse selection called *the cavity dumper*, where it defines the conditions for the *Bragg reflection*.

The simplest way to modulate the refractive index n_1 is to make a periodic change of a medium's density, which can be achieved by passing the acoustic wave through the medium. The acoustic wave then creates regions of compression and dilation at its frequency Ω. In real acousto-optic devices, a standing acoustic wave is generated instead of a travelling wave whose forefront moves downward as in Fig. 3.8. The standing wave shown in Fig. 3.11 remains in place instead of moving down the column, and the refractive index, n_1, at each place in the column (e.g., the dashed line in Fig. 3.11) changes sinusoidally with the frequency Ω. Twice during the cycle the density is distributed uniformly along the whole column (3.11b and 3.11d), and twice it achieves a maximum at which the refractive index, n_1, is largest (3.11a and 3.11e), as well as once when it achieves the minimum density at which the refractive index, n_1, is the smallest (3.11c). Thus, twice during the cycle $T = \frac{1}{\Omega}$ when the density is distributed uniformly, the incident beam passes unaffected and the frequency of the transmitted beam is equal to ω, and the radiation amplitude is equal to the amplitude of the incident light. At other times diffraction occurs, leading to the appearance of additional bands at $\omega \pm n\Omega$, at the expense of weakening the amplitude of the carrier wave at frequency ω. Now we understand why an acousto-optical transducer modulates the amplitude of the light in an optical resonator. If this modulation is held at the frequency equal to the difference between the longitudinal modes, $\Delta\omega_q = c/2L$, the Debye–Sears effect leads to modelocking.

In practical applications an acousto-optic modulator consists of a small fused-silica (SiO_2) element (prism or plate) placed close to the optical resonator mirror (Fig. 3.12). The prism is used in multimode lasers, e.g., in argon lasers for wavelength selection.

The piezoelectric transducer at one end of a prism or a plate generates an acoustic wave of frequency $\frac{c}{2L}$. The end walls of the prism are polished to permit acoustic

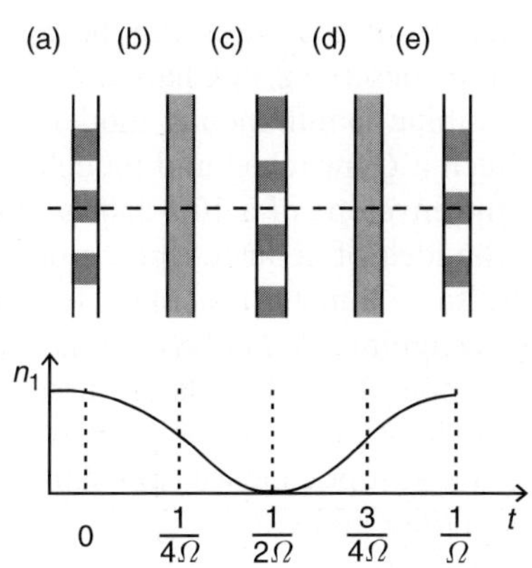

Fig. 3.11 Illustration of periodic changes of the refractive index by changes in the density of the medium caused by a soundwave. [1]

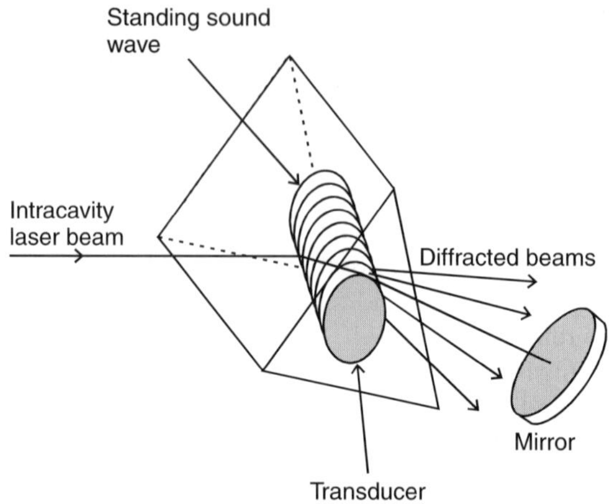

Fig. 3.12 Model of piezoelectric transducer.

resonance to produce the standing acoustic wave inside. A laser beam inside the optical resonator passes through the region of formation of the standing acoustic wave, interacting with it in the manner described above. As a consequence of this interaction, the laser beam with frequency ω is periodically modulated at the frequency $\Omega = \frac{c}{2L}$ by losses coming from the sidebands at frequency $\omega \pm n\Omega$. Only the axial beam participates in the laser action: the sidebands which are deflected from the main axis will be suppressed, since the length of the optical path for the sidebands is different from L at which the condition $\frac{n\lambda}{2} = L$ is fulfilled.

Traditionally, the acousto-optic modulation is used in flash-lamp pumped solid-state lasers such as Nd:YAG lasers. Recently, acousto-optic modulation has been utilized for Q-switching and modelocking in *diode-pumped solid-state* (*DPSS*) *lasers*.

A continuous-wave actively modelocked laser produces a train of pulses at a repetition rate in the range of 80–250 MHz and energy of a few nJ. If more energy is required, a pulse selected from the train can be amplified in a regenerative amplifier to reach a few mJ, as described in Chapter 6. If a more powerful pulse is needed, techniques that combine simultaneous modelocking and Q-switching or cavity dumping are used. Such a Q-switched and modelocked laser emits a burst of modelocked pulses within the envelope of a 100–250 ns Q-switch pulse.

We have just presented the idea of acousto-optic modulation. However, electro-optic devices can serve the same function as acousto-optic modulators both for active modelocking and Q-switching. A *Pockels cell* is a particular example of an electro-optic device. For example, Pockels cells are used to select and retain high-peak-power pulses from a modelocked Ti:sapphire laser for *chirped pulse amplification* (*CPA*). We will discus electro-optic devices in Chapter 6, where we will explain the idea of *a regenerative amplifier* and CPA.

Another way of modelocking is to use passive modelocking and *saturable dye absorbers*. There are various designs of passive modelocking, but a dye inside the resonator is a major requirement. One of many possible configurations is presented

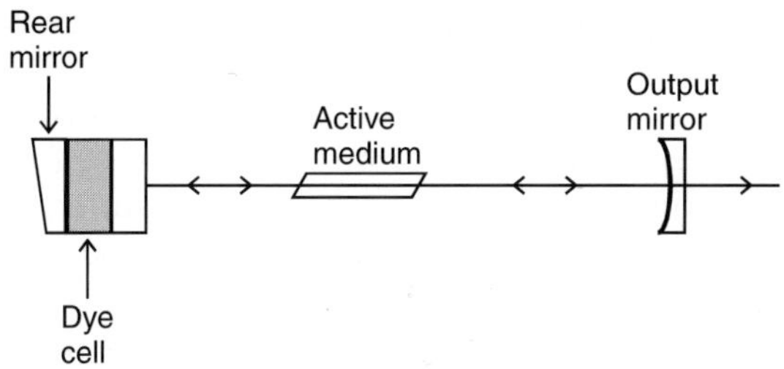

Fig. 3.13 The passive modelocking achieved by the saturable absorbers method.

in Fig. 3.13. In this configuration a dye cell and the rear mirror are combined to reduce the number of reflective surfaces in the laser cavity, and to minimise unwanted losses.

Let us assume that the absorbing dye in a cell is characterised by the energy levels E_1 and E_2 with $E_2 - E_1 = \hbar\omega$, where ω is the frequency of one of many longitudinal modes in the optical cavity. The lifetime of the absorbing dye molecules in the excited state is taken to be τ. If the lifetime is of the order of magnitude of the cavity round-trip time $T = \frac{1}{\Delta\nu} = \frac{2L}{c}$, i.e., a few nanoseconds in typical resonators, the dye molecules act like a passive Q-switching (see Section 3.3). If the lifetime is comparable to the pulse duration of a modelocked pulse, i.e., a few picoseconds, simultaneous modelocking and Q-switching can occur.

We will show that the dye in a cell plays the role of a filter (or a natural Q-switch shutter). Indeed, light in the optical resonator arriving at the cell-mirror promotes some molecules from the lower level, E_1, to the upper level, E_2, causing losses in the light intensity as a result of absorption by the dye. Initially, just at the beginning of pumping, the laser gain barely overcomes the losses of the saturable dye. In the early stage of pulse generation, the longitudinal modes are not synchronised in phase, and the laser output represents a chaotic sequence of fluctuations. As a result, both the amplification and dye absorption are not very efficient. As the pump process continues to increase the intensity above a threshold, light-amplification in the resonator approaches values of the saturation intensity in the dye. The gain in the laser medium is still linear, but the absorption of the dye becomes non-linear. With absorption of light at the high intensity the substance undergoes saturation (bleaching), so the condition $N_1 = N_2$ is fulfilled (where N_1 and N_2 indicate the number of molecules at the levels E_1 and E_2). The dye in the cell becomes transparent to the laser beam, which can arrive at the reflective rear mirror and back to the active medium, which in turn causes quick gain amplification. Now the intensity is sufficiently high, and the amplification in the medium becomes non-linear. The dye molecules return to the ground state, E_1, after time τ, and the process of light absorption is repeated. Therefore, the transmission in the cavity is modulated by successive passages of the high-intensity pulses resulting in a modelocked pulse train appearing in the laser output. Finally, the population inversion is depleted, and the pulse decays.

To summarize we may say that the mechanism of the passive modelocking with the saturable dyes consists of three main steps: 1) linear amplification and linear dye absorption; 2) non-linear absorption in the dye; 3) non-linear amplification when the dye is entirely bleached.

We have described only the most basic aspects of passive modelocking with saturable absorbers. Deeper treatments of this subject can be found in refs. [4–8]

The passive modelocking with saturable dye absorbers suffers from many inherent shortcomings. The laser output obtained with this method can be unpredictable unless the resonator alignment, pumping intensity, and dye concentration are carefully adjusted. Also, the mixing, handling, and maintaining the proper dye concentration can be cumbersome.

We have shown that saturable dyes can be applied for passive modelocking only when a dye has a lifetime comparable to the duration of modelocked pulses. In practice, this means that the method can be employed only for modelocking of picosecond pulses. If we wish to use this method for shorter, femtosecond pulses, we need a faster "shutter" than a saturable dye. In recent years, the method has been replaced by continuous-wave passive modelocking in solid-state lasers by using, *non-resonant Kerr effect—Kerr-lens modelocking* (KLM), or other passive techniques such as *saturable Bragg reflectors* (*SBR*) [9–11].

As long ago as the 1980s, engineers realised that a semiconductor quantum well, which will be described in Chapter 4, could play the role of a saturable absorber. A typical saturable Bragg reflector consists of alternate layers of high and low-index semiconductor materials, which also act as saturable absorber layers (Fig. 3.14).

In one particular configuration, the substrate material is GaAs, with alternate layers of AlAs and AlGaAs forming a multilayer Bragg mirror (Fig 3.14). Neither AlAs nor AlGaAs absorbs at 800 nm. A thin AlGaAs layer of a few microns' thickness is buried in the topmost layer of this stack, acting as a quantum well with a strong absorption at 800 nm. At low laser intensities, a typical SBR has a reflectance of 95%, whereas under modelocked intensities the reflectance rises to

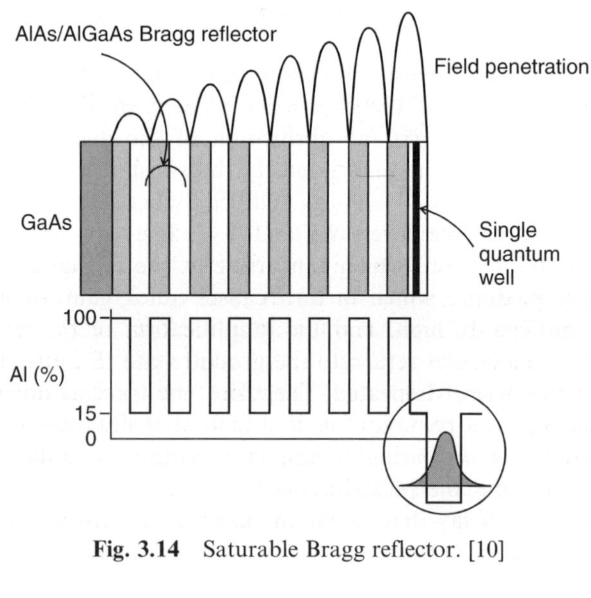

Fig. 3.14 Saturable Bragg reflector. [10]

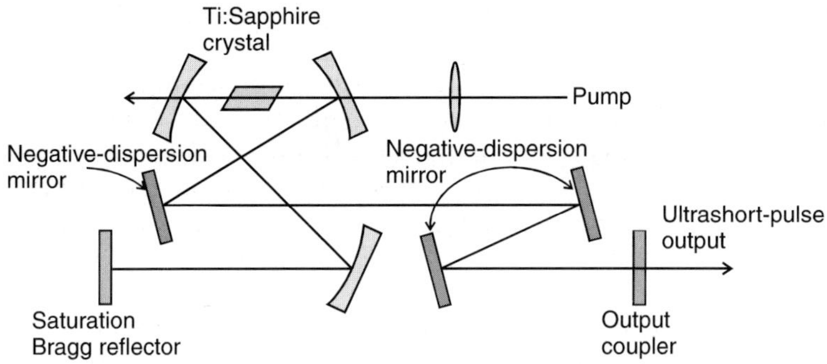

Fig. 3.15 Laser design in <30-fs Ti:sapphire that incorporates a saturable Bragg reflector [11].

almost 99%. Because of the multipass nature of a *cw* laser, this 4% change is more than sufficient to induce strong modelocking. Because of the absorption at 800 nm the saturable Bragg reflector has been applied in commercial Ti:sapphire laser technology in recent years. The saturable Bragg reflector method provides reliable, easy-to-use modelocked lasers for both laboratory and industrial applications. Figure 3.15 shows one of many commercial designs in which the Ti:sapphire oscillator and a solid-state pump laser are packaged in a single, compact sealed box.

We will now discuss the most common method of modelocking–*Kerr-lens modelocking* (KLM). This method requires no additional passive or active elements to modelock, and is employed in most solid-state lasers. A combination of a hard aperture and the Kerr effect leads to amplitude modulation of the resonator modes, with the frequency corresponding to the double round-trip time which is required to achieve modelocking. The Kerr effect, which has been known for a long time in nonlinear optics, implies that a refractive index is a function of a light intensity, I,

$$n = n_0 + n_2 I. \tag{3.28}$$

It indicates that the Kerr effect leads to the intensity-dependent variation of the laser beam profile. Indeed, for a Gaussian beam profile in the transverse direction the spatial index-distribution can be written as

$$n(r) = n_0 + n_2 I(r) \tag{3.29}$$

where $I(r)$ is given by Gaussian distribution $I(r) = \exp(-\rho r^2)$.

Figure 3.16 shows the refractive-index distribution along the x-axis for the Gaussian beam propagating along the z-axis. One can see that the index modification in an active medium follows the intensity of the laser beam. For a positive term n_2, the index has a maximum at $x = 0$ for the centre of the Gaussian beam, and is much smaller at the wings. Therefore, the refraction index is not homogeneously distributed in the Kerr medium, and corresponds to a situation as by inserting an additional material in a shape of a Gaussian lens into an optical resonator. The refractive-index lens formed in the Kerr medium focuses the laser beam towards a centre, as illustrated in Fig. 3.17. If we introduce an aperture into a resonator, it begins to act as a selective "shutter".

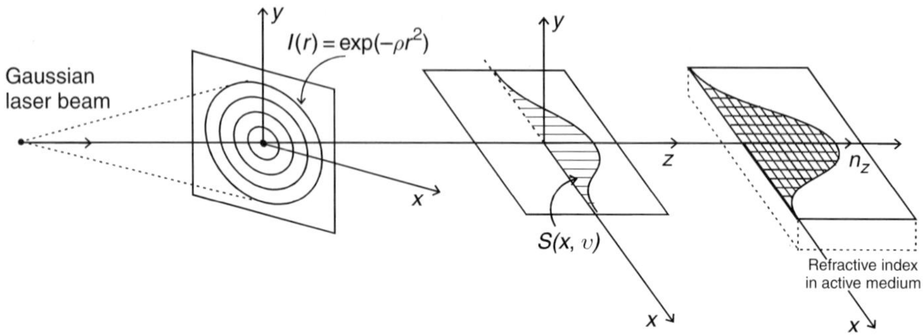

Fig. 3.16 Illustration of intensity-dependent refractive-index lens in an active medium.

This preferentially induces more loss at the edge of the beam, which is still a continuous wave, allowing the pulsed central mode to monopolise the laser gain. The modes of higher intensity are transmitted through the aperture because of the smaller size (due to the stronger Kerr lens focusing), whereas the modes of lower intensity cannot pass to the mirror M2 through the aperture, and they are lost. The high-intensity beam is reflected from the mirror M2 with some losses at the aperture, and passes through the active medium again, where is amplified. The loss–amplification process is repeated during every round-trip, leading to the amplitude modulation of the modes in the resonator, and during the modelocking.

The "shape" of the lens changes during the propagation of a pulse through a material and with the intra-cavity intensity. It can be shown, [12] that a focal length, f, is governed by the following expression

$$f = \frac{w^2}{4n_2 I_0 L}, \tag{3.30}$$

where w is the beam waist, n_2 is the non-linear index in eq. (3.29), I_0 is the peak intensity, and L is the length of the active medium. This kind of modelocking is called, *self-modelocking*, because the Kerr-lens medium can be the laser crystal itself. Actually, the self-modelocking does not require any additional passive or active elements in the resonator. Even the hard aperture for the mode selection on the size basis is not necessary, because the natural aperture is formed by the gain profile within the lasing material. For details, the reader is referred to ref. [13]. The KLM

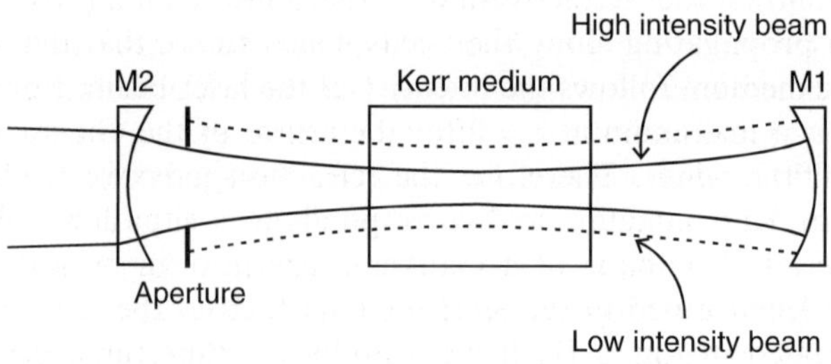

Fig. 3.17 Kerr lens amplitude modulation in an optical resonator.

effect has the result that in practically all solid-state lasers (Cr:YAG, Pr:YLF, Ti:sapphire) the modelocking is generated spontaneously without any additional modulating devices, because the active medium itself plays a role of the modulator. This kind of modelocking is commonly employed in Ti:sapphire (Ti:Al_2O_3) oscillators for the generation of femtosecond pulses of excellent reliability and great stability from pulse to pulse. They have been available commercially since 1990. However, in order to produce pulses which are stable and repeatable, with a strictly defined shape, one should apply devices controlling the *group velocity dispersion*—the *GVD*. The GVD effect will be discussed in Chapter 5.6.

It often happens that picosecond and femtosecond lasers using exclusively the KLM effect can become unstable owing to changes of temperature in the environment, vibrations in the laboratory, and other uncontrolled factors. Therefore, some manufacturers choose a method combining KLM and the active modelocking by an acousto-optic device. This type of modelocking is called, *regenerative modelocking*. When a solid-state laser begins to operate in the continuous work (*cw*) regime, the longitudinal modes are produced at frequencies that differ by $\Delta\nu = c/2L$. The modes are not well coupled at the beginning, and the phase difference between them changes chaotically. However, a small number of modes is partially correlated and the frequency $\Delta\nu = c/2L$ begins to modulate the beam intensity at the initial step of laser emission. This modulation frequency is recorded by a photodiode, amplified, and sent to an acousto-optic modulator. The modulator begins to modulate an active medium at the frequency that had been received from the photodiode. This solution removes the main limitation of the active modelocking which depends on the resonator cavity length, L. In the regenerative modelocking method, the signal is sent automatically to the acousto-optic transducer to change the frequency of modulation when the length of the resonator is changed a little for any reason. A detailed description of the regenerative modelocking can be found in ref. [14] and an advanced discussion of modelocking can be found in excellent tutorial presentations in refs. [13–19].

3.3. *Q*-SWITCHING

The peak power of a laser depends both on the pulse's duration and its energy. The shorter the duration of the pulse, and the higher its energy, the higher is the peak power. For continuous-wave modelocking, the ultimate limit of pulse duration for a Ti:sapphire laser is about 3 fs, with the typical energy of a single pulse being a few nJ, giving a peak power in the range of MW. Chirped-pulse amplification techniques, which will be discussed in Chapter 6, may help to achieve peak powers greater than TW in typical commercial configurations. However, the average power in such systems is low—typically about 1 W. In commercial systems, with a repetition of 1 kHz, the amplified pulse energy is 1mJ and the average power is 1 W.

Not all applications require cutting-edge performance, and the ultrafast Ti:sapphire systems are still complex and expensive devices. Another technique, called *Q-switching*, is employed to generate short pulses: this does not permit the generation of femtosecond pulses, but is very useful for the generation of picosecond or nanosecond pulses of high energy. The most powerful laser that employs the

Q-switching technique (Nd:glass) has achieved the peak power greater than a petawatt.

The Q-switching method takes its name from the resonator quality factor Q discussed in Chapter 2.2. The quality factor, Q is defined as the ratio of the energy stored in the cavity to the energy loss per cycle. We will show that by a fast change of Q we can force a laser to produce pulses. The pulses produced with the Q-switching technique are longer (picosecond, nanoseconds) than those obtained with modelocking (femtoseconds), but they have much higher energy. For typical femtosecond modelocked lasers the energy of a single pulse is around a few nJ at the high repetition rate of 76–82 MHz, whereas for the Q-switched pulses the typical energy of a single pulse is a few mJ at a repetition of a few Hz.

In Q-switching, the energy is stored in the optical cavity, with the population inversion building up until the Q-switch is activated. Once the Q-switch is activated, the stored energy is released in a single pulse. The higher the quality factor, the lower are the losses, and the more energy can be stored inside the cavity. In the Q-switched lasers, the energy obtained from the population inversion by pumping (usually flash lamps) is stored in the active medium. Although the stored energy is far above the threshold for lasing action, the action does not start, being prevented by introducing controlled losses to the resonator (low Q). So, the gain in the resonator is high, but the cavity losses are also high and the laser does not lase. The energy may be stored in the upper level as long as the pumping pulse from the flash lamp builds up and the relaxation processes do not drop the molecules back to the ground state and destroy the population inversion. This time is of the order of the lifetime of the upper state. Once the Q-switch is activated, and the high quality factor Q is restored, the lasing suddenly starts and the stored energy is released in a single short pulse. The peak power of such a pulse is extraordinarily high.

The mechanism of generation of a Q-switched pulse is illustrated in Fig. 3.18, and the theory of the Q-switch is given in refs. [20–22]. Here we only present the equation derived in ref. [20] for the pulse duration in a three-level system for rapidly Q-switched lasers;

$$\Delta t_p = \tau_c \frac{n_1 - n_0}{n_1 - n_t\left[1 + \ln\left(\frac{n_1}{n_t}\right)\right]} \tag{3.31}$$

where τ_c is the photon lifetime, n_1, n_t, n_0 are respectively the initial, the threshold value, and the final population inversion densities.

Now, when we understand the major steps in the mechanism of generation of Q-switched pulses we need to ask how to control the resonator quality-factor, Q, and how to switch rapidly between the low and high values of Q. There are several methods, including acousto-optic, electro-optic, mechanical, and dye switches. The idea of the acousto-optic modulator was explained in Chapter 3.2. We presented the acousto-optic modulator employed in the active modelocking and showed that the active modelocker usually works in the Raman–Nath regime (eq. 3.26), in contrast to the Bragg regime (eq. 3.27) that is employed in the cavity dumper and in the Q-switching modulators. The main difference between these regimes is presented in Fig. 3.19.

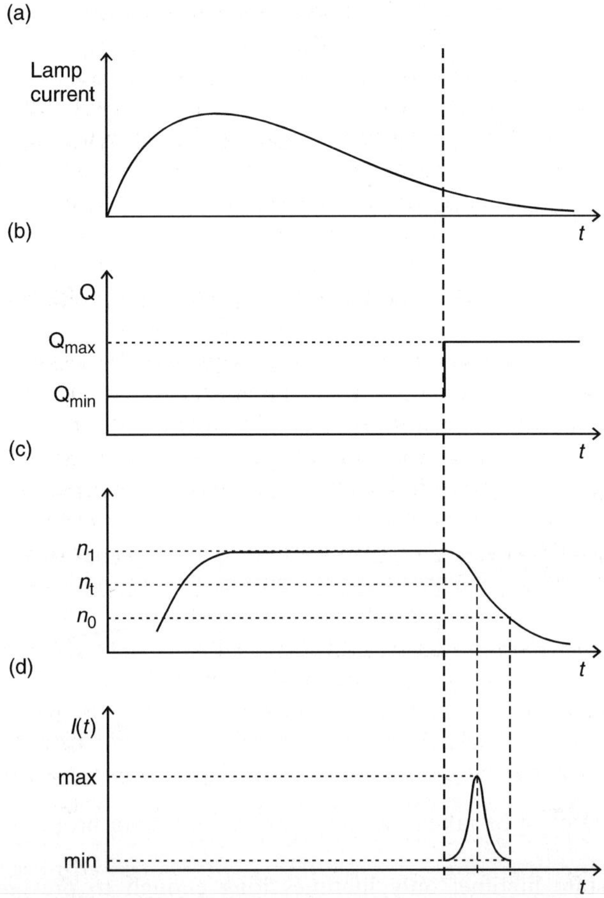

Fig. 3.18 Mechanism of generation of a Q-switched pulse: (a) pumping, (b) Q-switching, (c) energy storage in a three-level system, (d) pulse generation.

Briefly, an acousto-optic switch consists of a block of optical material (quartz, fused silica SiO_2, flint glass, tellurium dioxide) that is transparent to the laser beam. A piezoelectric transducer, usually a crystal such as lithium niobate, is bonded to one side of the block by epoxy or vacuum metallic bonding. The radiofrequency (RF) driver generates in the transducer the acoustical wave that propagates through the medium. The radiation inside the resonator interacts with the sound wave leading to diffraction of the incident beam. Compared to the Raman–Nath regime presented in Chapter 3.2, the frequency of the acoustic wave is higher, the interaction path is lengthened, and higher-order diffraction is eliminated. Only zero and first order rays are not suppressed. The diffracted beam reduces the quality of the resonator, Q, allowing the energy to build up and store inside the resonator without lasing. When the sound-wave stops travelling (the transducer is switched off) the radiation beam is no longer diffracted (high Q), the laser begins to lase, and the energy is released from the resonator in a single pulse.

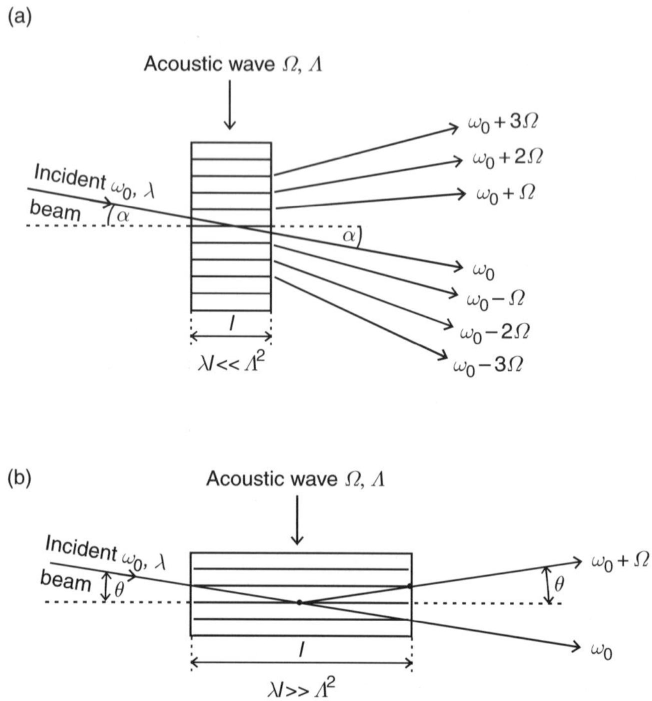

Fig. 3.19 Limit cases of acousto-optic devices: (a) Raman–Nath regime; (b) Bragg regime.

Several criteria must be taken into account in choosing a proper *Q*-switch:

1. the upper-state lifetime; only lifetimes long enough to prevent spontaneous energy emission can be *Q*-switched;
2. the gain parameter; if the gain is high, the diffracted light may not be able to prevent a feedback in the cavity leading to laser lasing;
3. the storage capacity, which denotes how much power the *Q*-switch will have to accommodate.

The *Q*-switching is employed in flash lamp-pumped solid-state lasers and diode-pumped solid-state lasers such as Nd:YAG, Nd:YVO_4, Nd:YLF, as well as ruby and Nd:glass.

3.4. CAVITY DUMPING

Cavity dumping is not a technique for generation of ultrashort pulses. It is usually used to increase the pulse energy or change the repetition rate. We will discuss cavity dumping in this Chapter because we want to compare it with the *Q*-switching technique, and to illustrate similarities and important distinctions.

The author's experience is that beginners find it difficult to understand the difference between Q-switching and cavity dumping. In both cases, energy is stored in the resonator cavity—often with the help of the acousto-optic devices working in the same Bragg regime. What distinguishes and provides the specificity of these methods? In the Q-switching regime, the energy is "stored" in the population inversion—in the amplifying medium. During energy storing, shortly before the energy is released from the cavity, the laser does not lase, because the cavity is kept below the threshold conditions. Although the gain in the active medium is high, the cavity losses (low Q) are also high, preventing lasing action.

In contrast, in the cavity dumping mode the cavity is not kept below the threshold conditions, and the laser lases all the time (both when it emits pulses or does not), because the energy is stored in the optical radiation energy inside the cavity, and not only in the population inversion. The only threshold that must be kept is the damage threshold for the optical elements inside the cavity. Cavity dumping can be employed in any dye laser or solid-state laser. The cavity dumping can be employed in *cw*-pumped lasers, flash-lamp pumped lasers, and lasers pumped with modelocked lasers.

Cavity dumping, like Q-switching, can significantly increase the pulse energy of a modelocked laser. In contrast to Q-switching, which produces a burst of modelocked pulses within the envelope of a 100–200 ns Q-switch pulse, the cavity-dumped laser produces a single modelocked pulse (Fig. 3.20).

The other function of cavity dumpers is to change the repetition rate. Cavity dumping of continuously pumped lasers is a way to obtain pulses of higher repetitions (from kHz to MHz) than those available by Q-switching. For example, repetition rates from 125 kHz to several MHz were achieved with cavity dumping for Nd:YAG lasers pumped by *cw*-sources [23, 24]. In contrast, cavity-dumped dye lasers, pumped with Q-switched and modelocked pulses from Nd:YAG lasers at 76 MHz repetition, can change the high repetition to lower repetition of a few hundreds of kHz. To help the understanding of cavity dumping we will

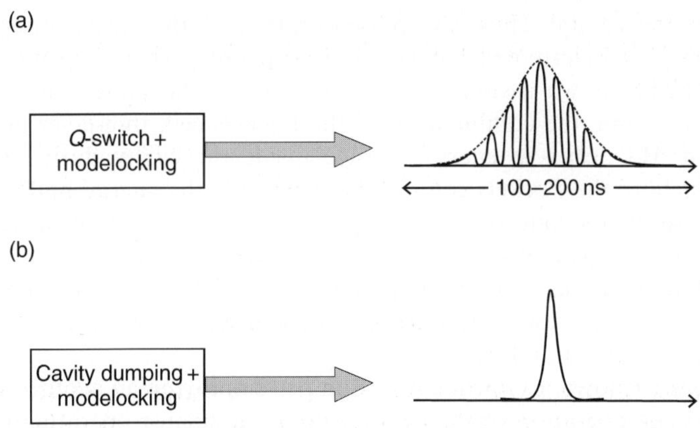

Fig. 3.20 (a) Q-switching produces a burst of modelocked pulses within the envelope of a 100–200 ns Q-switch pulse; (b) cavity dumped laser produces a single modelocked pulse.

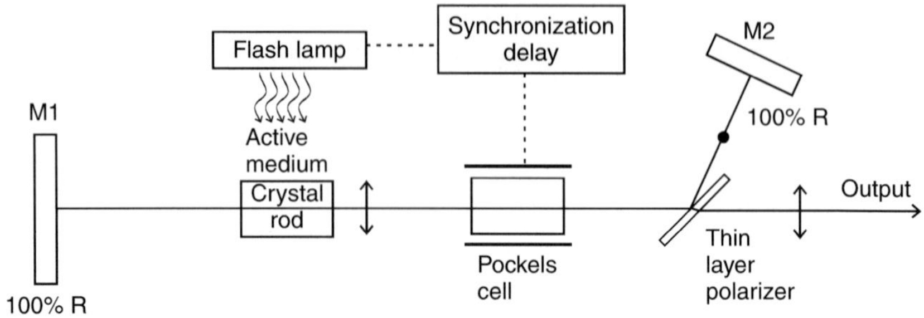

Fig. 3.21 Optical layout for cavity dumping. The arrow ↕ illustrates the polarization in the plane of the drawing; ● – polarization perpendicular to that plane.

explain it for a simple configuration based on an electro-optic device (Pockels cell). Later we will discuss the cavity dumping with an acousto-optic device for mode-locked lasers.

The typical optical layout for cavity dumping is presented in Fig. 3.21.

The flash lamp is fired at $t=0$, and its intensity begins to increase, producing fluorescence in the active medium. The horizontally polarized fluorescence (in the plane of the drawing in Fig. 3.21) passes through the thin layer polarizer as lost output light. Upon reaching the maximum lamp-current ($t_1 \sim 0.8$ ms) and peak-energy storage (and maximum population inversion) in the crystal, the Pockels cell ($\lambda/2$) is switched on at t_1, changing the polarization of the light to the vertical plane. The resulting vertically polarized light is reflected, and not transmitted, by the thin-layer polarizer, and reflected by the 100% R mirror, M2. Therefore, the beam is kept inside the cavity, leading to energy storage. When the power in the cavity reaches its peak value at t_2 ($\Delta t \sim t_2 - t_1 \cong 60$ ns for the ruby laser [25]) the Pockels cell is switched off and the polarization returns to the horizontal. The energy stored in the cavity can now be released through the thin-layer polarizer as an output pulse. This takes roughly the round-trip time, which is required to completely drain the energy from the cavity. Thus, the pulse-duration of the cavity-dumped pulse is almost completely determined by the round-trip time, which depends on the resonator geometry. If we assume a 1 m long cavity, the pulse duration, $t_p = 2L/c \approx 6.7$ ns. Therefore, the combination of the Pockels cell, thin-layer polarizer, and 100% mirror M2, leads to energy storage inside the cavity during the time Δt between switching the Pockels cell on and off, when the energy builds up. Within this time of $\sim$60 ns the light passes through the resonator about 60 ns$/t_p \approx 12$ times. Without cavity dumping, it would be released every round-trip time.

We shall now discuss cavity dumping for modelocked lasers employing an acousto-optic device to store the energy inside a cavity. We follow the excellent explanation presented in ref. [1].

In Fig. 3.22 is shown a typical cavity dumper employing an acousto-optic device (Bragg cell). The operation of the acousto-optic modulator operating in the Bragg regime is explained above (Fig. 3.19b). This configuration is often used in dye lasers pumped by the Q-switched, modelocked Nd:YAG lasers.

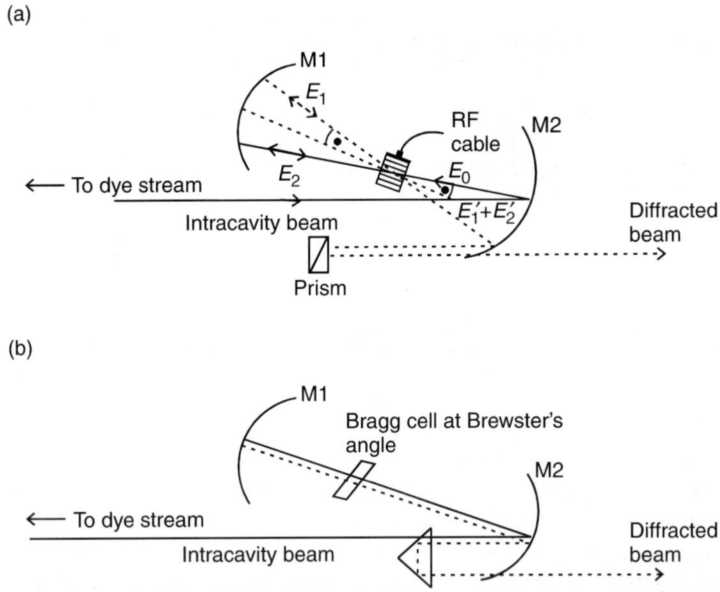

Fig. 3.22 The configuration of a cavity dumper: (a) top view, (b) side view.

In the vertical plane (side view) the Bragg cell is oriented at Brewster's angle, to minimise reflective losses and favour lasing of vertically polarized light. In the horizontal plane (top view) the Bragg cell is oriented at the Bragg angle θ ($\sim$2.3° from the normal for $\lambda = 600$ nm, and acoustic frequency $\Omega = 779$ MHz). The incident radiation (E_0) is split into one diffracted beam (E_1) and one directly transmitted beam, after crossing the modulator (E_2). The two beams are sent back upon themselves, since the acousto-optic cell is placed at the centre of curvature of the spherical mirror M1. They cross for a second time in the modulator. Part of the reflected light is sent back in the incident direction (E_0) to the resonator cavity, and the rest corresponding to the directly transmitted beam ($E_1 \rightarrow E_1'$) plus the diffracted part of E_2 (E_2'), is sent out of the cavity ($E_1' + E_2'$). We now calculate the output field ($E_1' + E_2'$).

The incident field E_0 is described by;

$$E_0 = A_0 \exp(i\omega t), \tag{3.32}$$

where ω is the frequency of the incident laser beam. The acoustic field in the Bragg cell is written as;

$$E_s = A_s \exp i(\Omega t + \phi_s), \tag{3.33}$$

where Ω and ϕ_s are respectively the frequency and the phase of the sound wave. The two fields (E_1 and E_2) after the first passage of the Bragg cell are given by

$$E_1 = E_0\sqrt{\eta}\exp[i(\omega - \Omega)t]\exp[i(\pi/2 - \phi_s)]\exp(i\phi_1) \tag{3.34}$$

$$E_2 = E_0\sqrt{1-\eta}\exp[i(\omega t - \phi_2)]. \tag{3.35}$$

The terms ϕ_1 and ϕ_2 are the phase changes for the two beams travelling through the Bragg cell at the exit plane of the cell, and are functions of the geometry of the Bragg cell. The additional phase $\pi/2$ originates from the fact that the electric field is a transverse wave, in contrast to the longitudinal acoustic wave. Thus, the interaction between them is shifted in phase by $\pi/2$. The term η characterises losses by diffraction in one passage across the Bragg cell

$$\eta = \frac{I_{diff}}{I_{in}}, \tag{3.36}$$

where I_{diff} and I_{in} are respectively the intensity of the diffracted- and the incident light. Since the intensity I is proportional to the square of the electric field E, the terms $\sqrt{\eta}$ and $\sqrt{1-\eta}$ appear in eqs. (3.34) and (3.35). The parameter η characterizing the diffraction efficiency depends on the acoustic power, the geometry of the acousto-optic device, and characteristics of the material. The fields after the second crossing of the Bragg cell are written as

$$E_1' = E_0\sqrt{\eta}\sqrt{1-\eta}\exp[i(\omega - \Omega)t]\exp[i(\pi/2 - \phi_s)]\exp[i(\phi_1 + \phi_1')] \tag{3.37}$$

$$E_2' = E_0\sqrt{\eta}\sqrt{1-\eta}\exp[i(\omega - \Omega)t]\exp[i(\pi/2 - \phi_s)]\exp[i(\phi_2 + \phi_2')] \tag{3.38}$$

where ϕ_1' and ϕ_2' are the phase changes during the round trip. The resulting field $(E_1' + E_2')$ that is sent out of the cavity can be written as

$$\begin{aligned} E_{out} &= E_1' + E_2' \\ &= 2E_0\sqrt{\eta}\sqrt{1-\eta}\exp[i(\omega t + \pi/2)]\{\exp[-i(\Omega t + \phi_s - \phi_1)]\exp[i(\Omega t + \phi_s - \phi_2)]\} \end{aligned} \tag{3.39}$$

where $\phi_1 = \phi_1 + \phi_1'$; $\phi_2 = \phi_2 + \phi_2'$. By combining these equations we obtain

$$E_{out} = 2E_0\sqrt{\eta}\sqrt{1-\eta}\exp[i(\omega t + \pi/2)]\exp\left(i\frac{\phi_1 + \phi_2}{2}\right)[\cos(\Omega t + \phi_s + \phi_2 - \phi_1)] \tag{3.40}$$

and the corresponding intensity,

$$I_{out} = |E_{out}|^2 = 4E_0^2\eta(1-\eta)[1 + \cos 2(\Omega t + \phi_s + \phi_1)]; \tag{3.41}$$

where $\phi = \frac{\phi_2 - \phi_1}{2}$.

The expression (3.41) shows that the diffracted intensity, I_{out}, is modulated according to twice the acoustic wave frequency Ω. It has maximum values for times t given by

$$\Omega t + \phi_s + \phi = k \tag{3.42}$$

where k is an integer, and a minimum for t given by

$$\Omega t + \phi_s + \phi = k + \frac{1}{2}. \tag{3.43}$$

The result obtained for the output intensity, I_{out}, simply says that the two diffracted beams $E_1' + E_2'$ can interfere with each other either, constructively or destructively, depending on the phase-relationship between the acoustic wave ϕ_S and the light beam in the laser ϕ. Additionally, the term $4\eta(1-\eta)$ in eq. (3.41) explains why the double passing across the Bragg cell is preferred. If we assume $\eta = 0.5$ for a single-pass intensity diffraction, the double pass gives 100% efficiency. Therefore, it is possible to time the acoustic pulse in the Bragg cell relative to the phase of the laser pulse such that either a maximum intensity is deflected out of the beam, or essentially no intensity at all—destructive interference prevents the light from being deflected. This method is often called "*integer plus* $\frac{1}{2}$ *timing*", and can be used to lower the repetition of the dye laser pumped by the high-repetition modelocked lasers. This can be done simply by properly choice of the acoustic frequency with respect to the modelocked laser repetition. If the acoustic frequency is chosen in such a way that dividing it by the modelocking frequency yields an integer k plus $\frac{1}{2}$, every

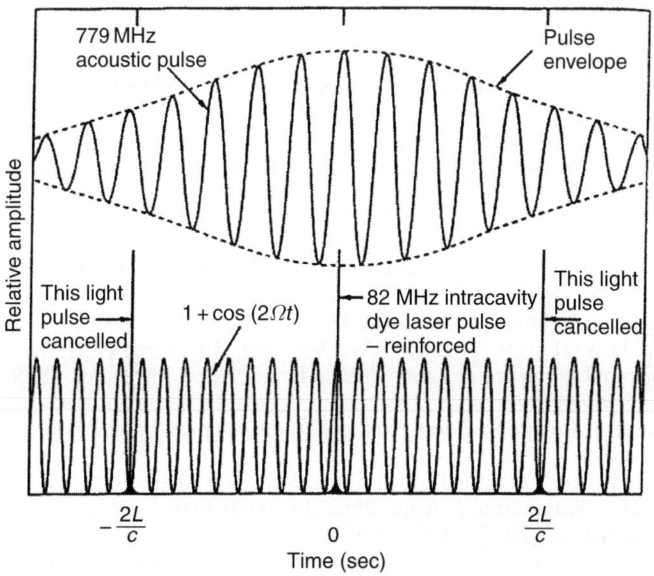

Fig. 3.23 A diagram illustrating *integer plus* $\frac{1}{2}$ *timing* in the cavity dumper. Three adjacent mode-locked dye laser pulses are illustrated in the bottom part of the Figure. Since the repetition rate of the laser is 82 MHz, they are separated in time by $2L/c$, or about 12.2 ns, where L is the optical cavity length of the dye laser. The acoustic pulse, which is sent through the Bragg cell by the transducer, is illustrated in the top part of the Figure. Its frequency is chosen to be 779 MHz which, when divided by the repetition rate, yields $9\frac{1}{2}$—an integer plus $\frac{1}{2}$. Owing to the double-pass configuration in the cavity dumper, the dye laser output consists of two light beams, one shifted to higher frequency by the acoustic frequency, and one shifted down. They thus interfere with one another in a manner which modulates the output at a frequency twice that of the acoustic wave. This modulation function is shown in the bottom part of the Figure. It shows how the output of the laser would be modulated if the laser were operated with continuous light, rather than being modelocked. Depending on the time of arrival of the intra-cavity dye laser pulse relative to that of the acoustic pulse, the output pulse is either reinforced or cancelled. Reproduced from ref. [1]

k-modelocked pulse will be sent out with a maximum intensity; the other pulses will not be sent out, because their intensity will be zero. Figure 3.23 illustrates, integer plus $\frac{1}{2}$ timing in the cavity dumper [1].

REFERENCES

1. E.W. Small, *Topics in Fluorescence Spectroscopy*, vol. 1, J.R. Lakowicz, ed., Plenum Press, New York (1991)
2. M. Born, E. Wolf, *Principles of Optics*, Pergamon Press, Oxford (1965)
3. P. Debay, F.W. Sears, *On the Scattering of Light by Supersonic Waves*, Proc. Natl. Acad. Sci. USA 18 (1932) 409–414
4. H.W. Mocker, R.J. Collins, *Mode Competition and Self-locking Effects in a Q-switched Ruby Laser*, Appl. Phys. Lett. 7 (1965) 270–272
5. A.J. DeMaria, D.A. Stetser, H. Heynau, Appl. Phys. Lett. 8 (1966) 174–176
6. D. von der Linde, Appl. Phys. Lett. 2 (1973) 281
7. G. Girard, M. Michon, IEEE J. QE-9 (1973) 979
8. D. von der Linde, K.F. Rodgers, IEEE J. QE-9 (1973) 960
9. U. Keller, D.A.B. Miller, G.D. Boyd, T.H. Chiu, J.F. Ferguson, M.T. Asom, Opt. Lett. 17 (1992) 505
10. S. Tsuda et al., IEEE. J., Sel. Topics in Quant. Elect. 2 (1996) 454
11. B. Craig, A. Krueger, *Saturable Bragg Reflectors Simplify Modelocking*, Laser Focus World 36 (8) (August, 2000) 227–228
12. M. Piché, Opt. Commun. 86 (1991) 156
13. F. Krausz, M.E. Fermann, T. Brabec, P.F. Curley, M. Hofer, M.H. Ober, C. Spielmann, E. Winter, A.J. Schmidt, IEEE J. QE-28 (1992) 2097
14. J.D. Kafka, M.L. Watts, J.W.J. Pieterse, *Picosecond and Femtosecond Pulse Generation in a Generatively Mode-Locked Ti:Sapphire Laser*, IEEE J. Quant. Electron. 28 (1992) 2151
15. W. Koechner, *Solid-State Laser Engineering*, 5th Edition, vol. 1, Springer-Verlag (1999)
16. H.A. Haus, J.G. Fujimoto, E.P. Ippen, IEEE J. QE-28 (1992) 2086
17. E.P. Ippen, H.A. Haus, L.Y. Liu, J. Opt. Soc. Am. B 6 (1989) 1736
18. S.L. Shapiro (ed.), *Ultrashort Laser Pulses*, Topics Appl. Phys., vol. 18, Springer, Berlin, Heidelberg (1977)
19. W. Kaiser (ed.), *Ultrashort Laser Pulses*, 2nd edn., Topics Appl. Phys., vol. 60, Springer, Berlin, Heidelberg (1993)
20. W.G. Wagner, B.A. Lengyel, J. Appl. Phys. 34 (1963) 2040
21. R.B. Kay, G.S. Waldmann, J. Appl. Phys. 36 (1965) 1319
22. J.J. Degnan, IEEE J. QE-25 (1989) 214
23. D. Maydan, R.B. Chesler, *Q-Switching and Cavity Dumping of Nd:YalG Lasers*, J. Appl. Phys. 42(3) (1971) 1031–1034
24. R.B. Chesler, D. Maydan, *Calculation of Nd:YalG Cavity Dumping*, J. Appl. Phys. 42(3) (1971) 1028–1030
25. I.F. Balashov, V.A. Berenberg, V.V. Blagoveshchenskii, Sov. Phys. 14 (1965) 692

– 4 –

Lasers

Lasers can be classified according to the active medium or the wavelength of the emitted radiation. The most general division based on the first criterion consists of gas lasers, liquid lasers and solid-state lasers. Gas lasers include all lasers in which the gain medium is a gas. The best-known gas lasers are:

a) *CO_2 laser* (10.6 μm) (near infrared),
b) *CO laser* (5–6.5 μm) (near infrared),
c) *N_2O laser* (10.6 μm),
d) *molecular gas lasers* (CH_3OH, $C_2H_2F_2$, CH_3F) (40 μm–1 mm) (far infrared),
e) *chemical lasers* (in which one of the reaction products is: I, HF, HCl, HBr, CO, CO_2), near infrared (1.3–11 μm),
f) *ion argon laser* (a few lines in the 476.5–514.5 nm range) (visible region),
g) *krypton laser* (503.9–752.5 nm) (visible region),
h) *helium–neon laser* (632.8 nm) (visible region),
i) *excimer lasers*, such as: ArF, XeCl, KrF, emitting radiation at the wavelengths 193, 248, 308 nm (UV),
j) *nitrogen laser N_2* (337 nm) (UV).

Liquid lasers include all lasers in which the gain medium is a liquid. Dye lasers emitting in the visible region are the liquid lasers that are used most often. Their outputs span the spectrum from the near-UV to the near-IR, depending on the dye used.

Solid-state lasers include all optically pumped lasers in which the gain medium is a solid. The best-known solid-state lasers are the:

a) *ruby laser* (694.3 nm) (visible region),
b) *Nd:YAG* laser with neodymium-doped *yttrium–aluminum garnet* matrix (1064 nm)
c) *Titanium–sapphire laser* (670–1070 nm) and other solid-state tunable vibronic lasers,
d) rare-earth elements other than neodymium (holmium, erbium, thulium) lasers, such as *Ho:YAG*, *Er:glass*, *Er:YAG*, *Tm:YAG*, emitting at about 2 μm depending on the matrix and doped material used.

Semiconductor lasers do not belong to the category of solid-state lasers although the gain-medium is also a solid. Conventionally, they form a separate category because diode lasers are usually electrically pumped, and different physical processes are responsible for the emission that occurs over a broad spectral range of the visible and infrared region.

The classification based on the wavelength of the emitted radiation is related to the nature of the quantum transitions between which the laser action occurs. The lasers emitting far-infrared radiation operate on population inversion between the rotational levels. The rotational gas lasers employ substances such as CH_3OH, $C_2H_2F_2$, or CH_3F. Lasers emitting in the infrared range operate on transitions between vibrational levels; the CO_2 laser is the most representative of this group. Lasers emitting in the visible and ultra-violet range employ transitions between electronic levels: the most representative ones are argon, helium, neon, krypton, nitrogen, dye lasers, solid-state, and excimer lasers.

4.1. RUBY LASER

Theodore Maiman constructed the world's first laser from a ruby crystal. Since that first ruby laser, researchers have discovered many other materials for use as the gain medium, but the oldest laser still finds some applications. The large energy pulses and the red visible light radiation (694.3 nm) of the ruby laser still find applications in holography and dermatology. The ruby laser is utilized in dermatology to remove tattoos and pigment defects of the skin. The ruby laser is a pulsed laser of low repetition rate—the repetition rate being the number of pulses that are sent by a laser per 1 second. The pulse-duration is of the order of milliseconds, with an energy of 1 J, and an average pulse power of the order of kilowatts. The active medium of a ruby laser is a rod made of a synthetic sapphire (Al_2O_3) doped with chromium (0.01–0.5%)—having the same valence number (+3) as the aluminum. The electronic transitions in a ruby laser are presented in Fig. 4.1.

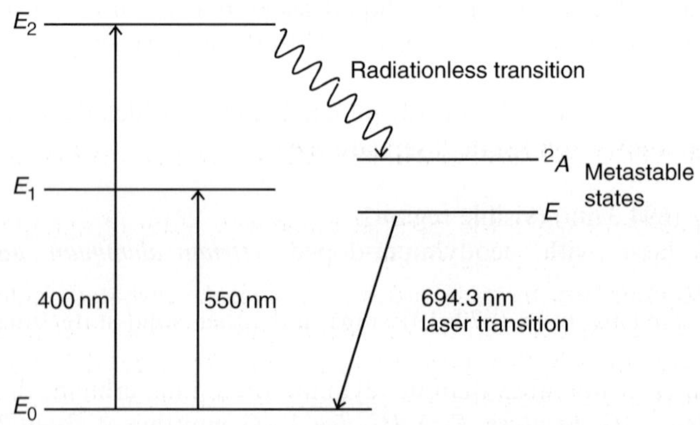

Fig. 4.1 Electronic transitions in ruby laser.

The ruby laser is pumped with a xenon flash lamp. When the chromium ions, Cr^{+3}, in a ruby crystal absorb photons of visible light at 400 nm or 550 nm, some electrons jump from their ground-state, E_0, to the excited states E_1 or E_2. The electrons excited to these states almost immediately (ca. 100 ns) dissipate their excess energy to the surrounding crystal lattice. As a result of the radiationless transitions the electrons jump to one of two closely spaced metastable states, E or ${}^2A$, where they stay for the relatively long time of 3 ms at room temperature. This time is long enough to achieve population inversion. Stimulated emission occurs on the $E \rightarrow E_0$ transition, generating light at a wavelength of 694.3 nm. Because the whole cycle of excitation, relaxation, and stimulated emission, involves transitions between three energy levels, the ruby laser is known as a *three-level laser*. The three-level lasers are relatively inefficient because the laser transition terminates in the ground state, and large pumping energies are required to achieve population-inversion. In the three-level lasers, more than half the atoms have to be transferred to the excited state to create the population inversion between the metastable and the ground state. The energy produced in the active medium as a consequence of radiationless transitions heats the ruby crystal, limiting the repetition rate to several pulses during one second. This inconvenience of the three-level system is partially compensated for by a long lifetime of the metastable state. This long lifetime permits the ruby rod to store an amount of energy many times greater than that in the neodymium Nd:YAG rod under the same conditions and, as a consequence, to generate pulses of much larger energy.

Ruby lasers find wide application in holography, plasma diagnostics, and dermatology. In dermatological applications the *Q*-switched ruby lasers are utilized, generating pulses of energy of 2–3 J with a repetition rate of 0.5–1 Hz. Such an energy is sufficient to remove black, blue, and green pigments of skin tattoos, as well as stains caused by excess melanin. The laser beam destroys cells containing the pigment by inducing photochemical reactions, whereas the surrounding non-pigmented areas of skin do not absorb the light energy and experience only negligible damage. The lymphatic system then slowly removes dead cells during the following several months. The ruby laser does not remove red pigments since these do not absorb the red light at 694.3 nm. The ruby laser can also work in the modelocking regime, emitting pulses of 20–30 picosecond duration, with an energy of 1 mJ, and repetition rate of 20–30 Hz.

4.2. MOLECULAR GAS LASERS FROM THE INFRARED REGION

Molecular gas lasers emit in the infrared range as a result of quantum transitions between rotational or vibrational states.

The most popular laser from the infrared range is the carbon dioxide (CO_2) laser, which emits radiation between 8 and 18 μm, but the most useful wavelength is centered at 10.6 μm. Another popular laser, the CO laser, delivers radiation at shorter wavelengths (5–6.5 μm). To generate longer wavelengths (30 μm–1 mm) from the far infrared range, simple organic substances such as methanol are used as gain media. In gas lasers such as CO_2 or CO the emission is generated by rotational-vibrational transitions, in contrast to the gas lasers emitting in the far-infrared radiation where lasing occurs exclusively on rotational transitions.

4.2.1. Lasers Operating on Rotational Transitions

Lasers which emit light as a result of rotational transitions generate far-infrared radiation. The powers of the commercially available lasers operating on rotational transitions are in the range from milliwatts to 1 W. *Far-infrared lasers* are excited with radiation at about 10 μm from CO_2 or N_2O lasers. The optical pumping causes the molecules to be transferred from the rotational state characterized by the rotational quantum number J_0 in the ground vibrational state ($\upsilon = 0$), to the rotational state J in the excited vibrational state ($\upsilon > 0$), for a given vibrational mode (Fig. 4.2). The symbol υ indicates the vibrational quantum number. The laser action occurs between the two rotational levels $J \rightarrow J-1$ in the excited vibrational state. The pumping laser should deliver strongly monochromatic radiation in a narrow spectral range, to avoid filling neighboring rotational levels which lie close to each other.

The active media in the far infrared lasers are simple organic substances, e.g., alcohols and hydrocarbons. In commercial models, the optical resonator contains a gas reservoir which can be exchanged and replaced by another one, making it possible to obtain far-infrared radiation from 40 μm to 2000 μm. Since the rotational energy depends strongly on the molecular mass, isotopic substitution may provide additional lines. For example, CD_3OD emits at 41.0 μm, CH_3OD at 46.7 μm, and CH_3OH at 70.6 μm. Although commercially available, the far-infrared gas lasers are less popular than gas lasers such as CO_2, CO, N_2O, which emit shorter wavelengths in the near-infrared. This results from technical problems in the far-infrared range, related to propagation in waveguides, as well as requirements imposed on the optical elements.

The optical resonator of the far-infrared laser consists of two mirrors, Z_1 and Z_2, an active medium, and also a vacuum waveguide which leads the emitted waves (Fig. 4.3). The vacuum waveguide must be used because air-absorption at the wavelengths of

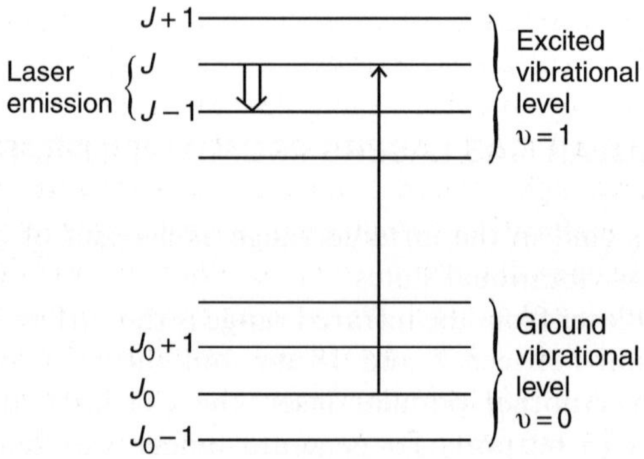

Fig. 4.2 Schematic diagram of quantum transitions in lasers which emit as a result of rotational transitions.

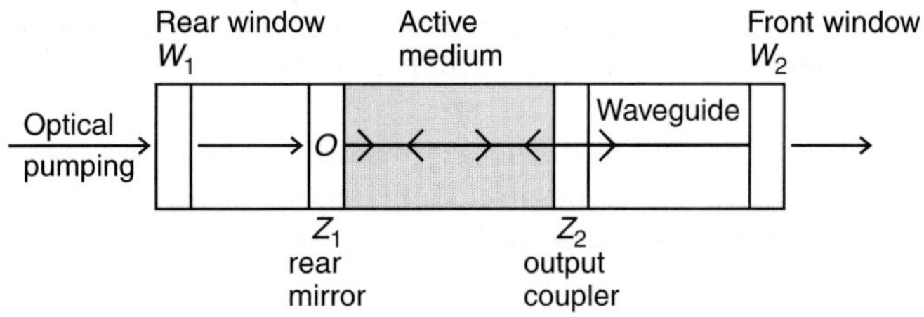

Fig. 4.3 Optical resonator of far infrared lasers on rotational transitions.

30 and 50 μm would effectively attenuate the light's propagation over longer distances. The optical resonator consists of the mirrors Z_1 and Z_2 between which the laser action occurs, as well as the windows W_1 and W_2 protecting the tube, under vacuum, from environmental influences. The optical elements inside the resonator have to obey several requirements, which are sometimes mutually exclusive:

1. The resonator mirrors Z_1 and Z_2 have to reflect radiation both from the range of 10 μm (the pumping radiation) and from the far infrared (the generated radiation). The reflection of the pumping radiation inside the resonator increases the feedback, owing to the increase of the optical path—which is very important, taking into account the small energies of the rotational transitions.
2. The rear window, W_1, has to transmit the pumping radiation, and is usually made of zinc selenide. There is a hole, O of 1–2 mm diameter in the rear mirror, Z_1, which allows the radiation at 10 μm to be introduced into the cavity.
3. The front window, W_2, has to transmit the far-infrared radiation generated in the resonator. Most materials are opaque to radiation at wavelengths longer than 50 μm. Only some materials—such as silicon, germanium, crystalline quartz, diamond, as well as polyethylene—can be used in the window W_2 as they are transparent for the far-infrared light.

4.2.2. Lasers Operating on Vibrational-Rotational Transitions: CO_2 and CO

Lasers operating on *vibrational-rotational transitions* generate radiation in the infrared range, both in the continuous and pulsed regimes. The main lines are at 10.6 μm and 9.6 μm (CO_2), and 5–6.5 μm (CO). The first *carbon dioxide* (CO_2) *laser* was invented by Kumar Patel from the Bell Telephone Laboratories in 1964. The CO_2 lasers can achieve high powers, from watts to kilowatts. Although CO_2 lasers have found many applications—including surgical procedures—their popular image is as powerful devices for cutting, drilling, or welding, as weapons for defense and security applications, with laser-beams aboard aircraft that can destroy ballistic missiles. In 1970, the United States Air Force built the Airborne Laser Laboratory with a 400 kW gas dynamic carbon dioxide laser housed in a militarized Boeing 707 airplane. Now, the defense and security programs have replaced carbon dioxide lasers by *chemical lasers*, but they are still symbols of the most powerful lasers.

The active medium in a CO_2 laser consists of carbon dioxide, nitrogen N_2, and helium He. The whole cycle of mechanisms involved in CO_2-laser operation is presented in Fig. 4.4. The CO_2 lasers are pumped by RF (radio frequency) or direct-current electrical discharge in the gas medium. The electrons produced as a result of ionization excite the bond-stretching mode of N_2 molecules, promoting them from the ground vibrational state ($\upsilon = 0$) to the first excited vibrational state ($\upsilon = 1$). Nitrogen has a very long-lived first-excited vibrational state that almost exactly matches the upper level of CO_2. Therefore, the process of transferring the excess energy via collisions to carbon dioxide is very efficient. The energy of the N_2 molecule in the first-excited vibrational state is practically the same as the energy of the first excited vibrational state (00^01) of the asymmetric stretching mode, ν_3, of the CO_2 molecule. The small energy gap ($\Delta E = 18\,\text{cm}^{-1}$) between the excited levels makes possible the energy-transfer from the N_2 molecule to the CO_2 molecule, then the return of the N_2 molecule to the ground vibrational state ($\upsilon = 0$), and the excitation of the mode ν_3 from the ground vibrational state, denoted in Fig. 4.4 as (00^00), to the excited vibrational state (00^01) in the CO_2 molecule. The main laser transitions in CO_2 occur between the excited states of the mode ν_3 (00^01) and the symmetric stretching mode ν_1 (10^00) (10.6 μm), or the bending mode ν_2 (01^10) (9.6 μm). Helium molecules do not take part directly in the excitation of CO_2 molecules but do play an important role in heat-transfer from the gas mixture to the tube walls, as well as facilitating the depopulation of the lower vibrational levels in CO_2—contributing in this way to maintenance of the population inversion. A volume-ratio of 1:2:3 for the CO_2:N_2:He mixture is typical.
In Fig. 4.4, the symbols such as (00^00), (01^10) denote the set of vibrational quantum numbers, υ, determining the energy of the quantum harmonic oscillator

$$E = h\nu\left(\upsilon + \frac{1}{2}\right). \tag{4.1}$$

The numbers in brackets denote the quantum vibrational numbers of the mode ν_1, ν_2, and ν_3, respectively. The upper indices define the vibrational level's degeneration. For

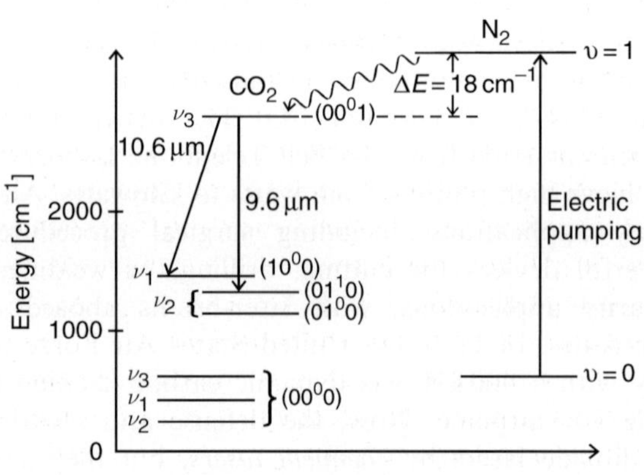

Fig. 4.4 Quantum transitions in the CO_2 laser.

example, the symbol (01^10) denotes the ground vibrational state for the mode ν_1, the first excited vibrational state for the mode ν_2 which is doubly degenerate, and the ground vibrational state for the mode ν_3.

While the fundamental principles of CO_2 lasers have remained unchanged for over 40 years, technological advances have enabled improved products to be offered in the low-power (<500 W) and high-power categories. The power of CO_2 lasers depends on their configuration. One can distinguish the following types of CO_2 lasers (Fig. 4.5):

a) *conventional laser with longitudinal direct-current discharge excitation,*
b) *waveguide laser with RF excitation,*
c) *longitudinal flowing-gas laser,*
d) *transverse flowing-gas laser,*
e) *transversely excited atmospheric lasers* (*TEA*).

The lasers of type (a) consist of a sealed tube filled with an active gas and equipped with electrodes to which a voltage is applied. The electrical discharge leads to dissociation of CO_2 into CO and O_2. To regenerate the CO_2, hydrogen or water is added to the active gas, or a nickel cathode is employed to catalyze the recombination reaction. The sealed-tube-CO_2 laser can work for thousands of hours. Typical powers of lasers of type (a) are in the range of 25–100 W.

The lasers of type (b) are pumped with an RF discharge. The vacuum-sealed tube is formed as a waveguide of several millimeters' dimension. The typical power of a waveguide laser with RF excitation is up to 50 W. Good beam quality and compact size, accompanied by large powers, make these lasers competitive with other configurations.

In the lasers of type (c), gas flows along the axis of the laser cavity. The active material is exchanged in a continuous way, enhancing the output power of the laser up to 800 W per meter of plasma tube at fast gas flow. As the output power of a CO_2 laser depends on the plasma tube's length, it can be enhanced considerably by the parallel alignment of several tubes (Fig. 4.6). The total power that can be achieved for fast-axial-flow configurations is from 500 W to 5 kW.

In lasers of type (d), the gas-flow is transverse to the laser-cavity's axis. The lasers with transverse flow can achieve even larger powers than type (c) (up to 50 kW).

In TEA lasers—type (e)—the pressures in the tubes are close to atmospheric pressure, and they are therefore called "atmospheric lasers". TEA lasers emit pulses of duration from nanoseconds to microseconds, and powers from 1 W to 100 W.

The optical materials used in lasers emitting radiation in the infrared range are obviously different than those used in the visible range. For example, materials such as germanium (Ge) or gallium arsenide (GaAs) are completely opaque in the visible range (VIS), while being transparent in the infrared range. Some materials, such as zinc selenide (ZnSe), are transparent in both spectral ranges. Typical materials transparent in the IR range are: NaCl, CsI, or (toxic) compounds of thallium (KRS-5). Metal mirrors (copper, molybdenum) are used in the IR range, owing to their small absorption (and large reflectivity) as well as their large heat capacity which enables removal of heat from the active medium.

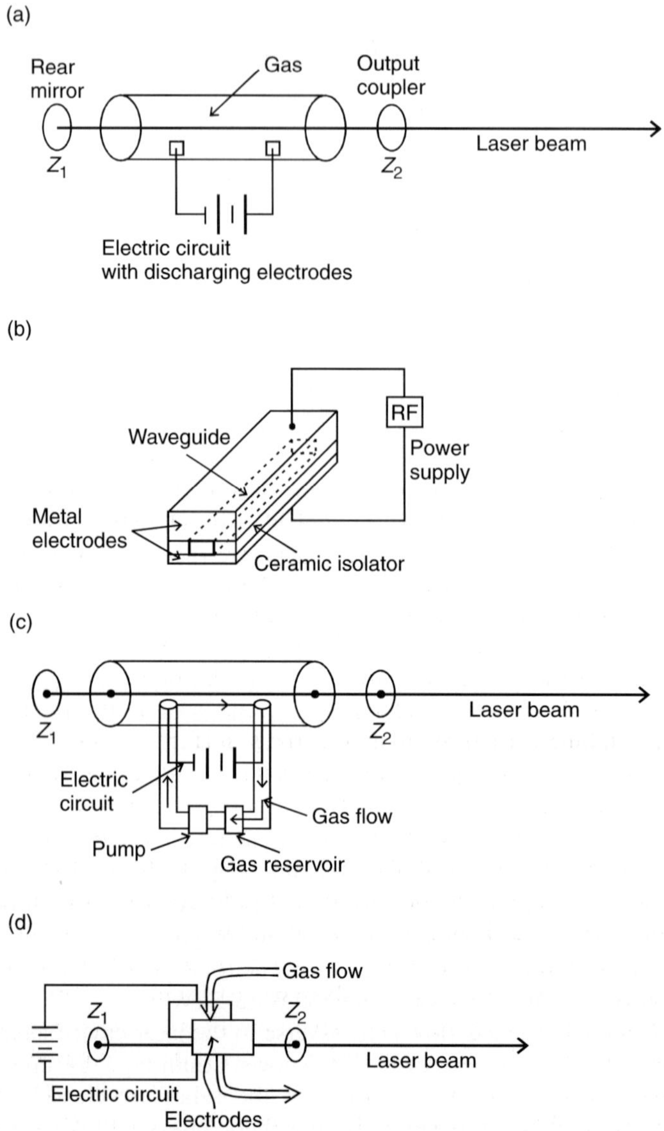

Fig. 4.5 Types of CO_2 lasers.

CO_2 lasers have long been used in industrial and medical applications, and show no signs of decline. Armies and defense agencies have noticed the high powers of CO_2 lasers as potential weapons for defense and security applications, such as destroying ballistic missiles. Moreover, CO_2 lasers have found application as lidars (laser radars). By irradiating the atmosphere with a CO_2 laser, remote IR spectra of airborne gases and pollutants can be obtained. The most common application for CO_2 lasers is in materials processing. Low-power CO_2 lasers of 50 W or less can serve to cut, mark, or

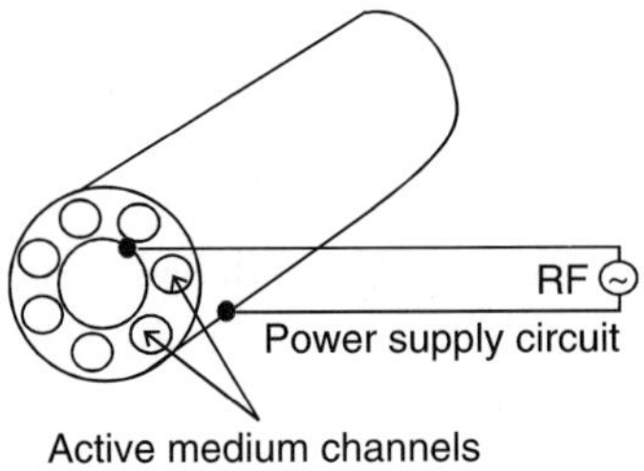

Fig. 4.6 CO_2 laser with an axial flow.

drill through plastics, rubber, ceramics and glass. They can also be used to mark a number of metals, including mild and stainless steels and titanium, because laser marks are more permanent than ink-jet printing. Although most metals show small absorption at 10 μm, titanium is an exception because it absorbs strongly in this region. Therefore, CO_2 lasers are particularly useful for cutting titanium. Low-power CO_2 lasers can be used to weld some thin metals and foils. Metal-marking and welding need powers between 50 and 200 W. Drilling with low-power CO_2 lasers of metals on printed circuit boards is increasingly used. Although copper reflects CO_2 radiation, new methods of oxidizing the copper surface of circuit boards have made it possible to drill through the copper. Precise drilling or cutting requires appropriate pulsing conditions, usually at 1–2 kHz for metals, and up to 100 kHz for non-metals or for welding.

A major industrial application of high-power CO_2 lasers is thermal etching, to change the crystalline structure of a material's surface in order to increase its durability and strength. High-power CO_2 lasers are also used to cut, drill, and weld steel and other less-reflective metals. In medicine, high-power CO_2 lasers have long been used as the scalpel in many surgical procedures, because the emission at 10 μm is absorbed quickly by water in flesh, leading to vaporization of the tissue, with a very fine cut-width and with the beneficial cauterizing effect essential in operations on blood-supplying tissue. However, recent developments in laser technology have demonstrated that other types of lasers are more suited for most medical applications. One exception to this trend which has been gaining interest is a cosmetic procedure of skin-smoothing and rejuvenation of severe acne and skin texture imperfections. Another vital medical application is in transmyocardial revascularization (TMR), designed to help people with severe coronary artery disease and angina to relieve chest pain caused by lack of oxygen to the heart. This procedure, which makes tiny holes in the heart to increase the flow of blood to oxygen-starved tissue, has received the US Food and Drug Administration's (FDA) clearance.

The *CO laser* was constructed soon after the CO_2 laser. A CO laser emits radiation at wavelengths in the range of 5–6.5 μm. Like the CO_2 laser, it is pumped by an electrical discharge, and its output power is high. It is produced both in a version of a sealed tube filled with an active gas, and as a flowing-gas laser. The flowing-gas laser is less safe because CO is odorless, and a content of 50 ppm in the atmosphere is already dangerous for health, while higher doses are deadly. The sealed-tube CO laser is much less dangerous, as the content of CO is small, and even when the tube is damaged the leakage to the atmosphere does not present a threat.

The CO lasers find applications in dentistry and gynecology. Moreover, with the development of the technology of flexible optical fibers, which are characterized by high transmission for this spectral range, CO lasers can be considered as tools in medicine and industry. As early as the 1960s, military projects showed interest in CO lasers because of their high powers, but technical obstacles—particularly in beam propagation, because of strong light absorption by the atmosphere in the range of 5–6.5 μm, has caused these investigations to decline.

4.3. CHEMICAL LASERS

Chemical lasers emit radiation in the far-infrared and infrared regions (1.3–11 μm). Chemical lasers emitting radiation in the visible range are of great interest for military technologies—for obvious reasons—although they are still in a less mature stage than the infrared chemical lasers.

So far we have described lasers that are optically or electrically pumped. The chemical lasers use another type of pumping—the energy released during chemical reaction. Table 4.1 shows the most typical chemical reactions utilized in chemical lasers. The reactions used most often are based on the generation of excited HF^* and DF^* molecules (Table 4.1). HF and DF lasers are commercially available chemical lasers which deliver radiation up to hundreds of watts.

The laser action in chemical lasers occurs between vibrational levels. Isotopic substitution by deuterium in the HF laser is also used to shift emission wavelengths from 2.6–3.4 μm to 3.5–4.4 μm. This shift is essential in military applications in which the laser beam travels over large distances, because it can allow strong absorption in the atmosphere. In commercially available HF/DF lasers the hydrogen fluoride is produced by an electric discharge in a tube, *via* a dissociation reaction of sulfur hexafluoride, SF_6. Oxygen is added to react with sulfur. Hydrogen, which is injected into the tube, reacts with fluoride, and generates the hydrogen fluoride HF^* in vibrationally excited states. The gas flows quickly through the resonator cavity, perpendicular to the axis (Fig. 4.7).

In military lasers, the gas flows with supersonic velocity, in contrast to commercially available lasers in which the velocity is much smaller than the velocity of the sound.

Table 4.1

Typical chemical lasers

Laser	Reaction	Radiation wavelength [μm]
CO_2	$DF^* + CO_2 \rightarrow CO_2 + DF$	10.0–11.0
CO	$CS + O \rightarrow CO^* + S$	4.9–5.8
HBr	$H + Br_2 \rightarrow HBr^* + Br$	4.0–4.7
DF	$F + D_2 \rightarrow DF^* + D$	3.5–4.2
HCl	$H + Cl_2 \rightarrow HCl^* + Cl$	3.5–4.1
HF	$F + H_2 \rightarrow HF^* + H$	2.6–3.3
I	$O_2^* + I_2 \rightarrow O_2 + 2I^*$	1.3

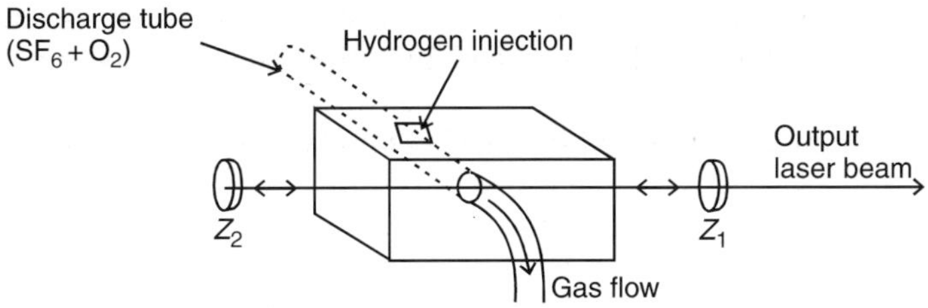

Fig. 4.7 Scheme of a chemical laser.

Another group of chemical lasers operates on transitions in atomic iodine using the reaction $O_2^* + I_2 \rightarrow O_2 + 2I^*$. Several reactions are possible to generate excited oxygen. For example, hydrogen peroxide is sprayed into chlorine gas, producing heat in an exothermic reaction that excites oxygen. The excited O_2^* molecules lead to excitation of atomic iodine, as I^* followed by population inversion and laser action.

There is no other group of lasers in which the interest is almost entirely dominated by military and defense applications. High-energy laser weapon projects date back to the 1970s. At that time the US Navy built a 2.2 MW deuterium fluoride chemical laser called MIRACL. In the 1980s the "star wars" idea shifted the strategic defense to study prospects for space-based laser battle stations, and large financial means were invested in the development of HF lasers. In the 1990s the military interest shifted to the 1.315 μm oxygen–iodine laser known as COIL. As early as 2005 the Airborne Laser program in the US plans to generate a megawatt beam that could destroy ballistic missiles several hundreds of kilometers away: the plan is to mount a chemical oxygen iodine laser on a Boeing 747 aircraft.

There is some interest in shifting applications of the chemical oxygen–iodine laser from military to civilian use. A high power COIL emitting at the wavelength 1.315 μm, which can easily be transmitted by standard fused-silica fiber optics, may soon be used for general industrial use such as cutting and welding. These lasers show promise for use as powerful tools for dismantling obsolete nuclear weapon production facilities and nuclear-power reactors.

4.4. SOLID-STATE LASERS

Solid-state lasers include all optically pumped lasers in which the gain medium is a solid. The first solid-state laser—the first laser ever—was the ruby laser described in chapter 4.1. Although semiconductor or diode lasers also have a solid gain medium, they were developed later, and historically the term "solid-state laser" refers to all optically pumped solid-state lasers and not to semiconductor lasers. The first design of a solid-state laser consists of a crystal doped with a metallic or metal-like element pumped by a flash lamp. Within the past few years, *diode-pumped solid-state lasers* (*DPSSL*) have emerged from the laboratory into everyday use in industrial and medical applications.

Solid-state lasers can be divided into two broad groups of non-tunable and tunable lasers. Most solid-state lasers belong to the class of four-level lasers, and they operate on electronic transitions, in contrast to the lasers described in the previous chapter. Solid-state lasers utilize an active medium consisting of a solid-state host—most often a single crystal—doped with about 1% of different species, such as neodymium ion or other ions such as holmium, erbium, or thulium.

4.4.1. Neodymium Laser and other Rare-Earth Lasers

In this chapter we discuss the group of non-tunable solid-state lasers. The active media of many solid-state lasers are the rare earth elements belonging to the lanthanide series. Those used most often are neodymium, holmium, erbium and thulium. Optical glass, and most of the crystals do not exhibit absorption in the visible or far-infrared range. However, when they are doped with the rare earth elements, they begin to show strong absorption and fluorescence. The energy levels of the rare earth ions involved in a laser operation create the four-level scheme shown in Fig. 4.8.

The neodymium lasers belong to the most important group of solid-state lasers. The electronic structure of the neodymium atom is: $1s^2\ 2s^2\ 2p^6\ 3s^2\ 3p^6\ 3d^{10}\ 4s^2\ 4p^6\ 4d^{10}\ 4f^4\ 5s^2\ 5p^6\ 5d^0\ 5f^0\ 6s^2$, and that of the Nd^{+3} ion is: $1s^2\ 2s^2\ 2p^6\ 3s^2\ 3p^6\ 3d^{10}\ 4s^2\ 4p^6\ 4d^{10}\ 4f^3\ 5s^2\ 5p^6$. The ground-state, E_0, of the Nd^{+3} ion is the 4f state, occupied by three electrons (total spin $S = \frac{3}{2}$, total orbital moment $L = 6$). The 4f shell is screened by the external shells 5s and 5p, which is why the energy-level structure of 4f depends only slightly on the host matrix. The neodymium ion can be added as an impurity to crystalline solids or to glasses. When added to a host of yttrium–aluminum garnet ($Y_3Al_5O_{12}$ – YAG), the active medium of the *Nd:YAG laser* is provided. The other popular host crystals are YLF ($YLiF_4$—yttrium lithium fluoride) and YVO_4 (yttrium orthovanadate). The neodymium ion can be added to a glassy matrix that is utilized as an active medium in the *Nd:glass laser*. Detailed properties of solid-state laser materials can be found in ref. [1].

All electronic transitions in the neodymium laser (and many other lasers based on the rare earth elements) occur within the 4f shell. For this shell the spin–orbital coupling is very strong, which causes the total momentum $\mathbf{J} = \mathbf{L} + \mathbf{S}$, and not its separate components $\mathbf{L}$ and $\mathbf{S}$, to undergo the quantization. Although limited by screening, interaction with the crystalline field surrounding the Nd^{+3} ion removes the shell's degeneration, splitting it into $2J + 1$ levels.

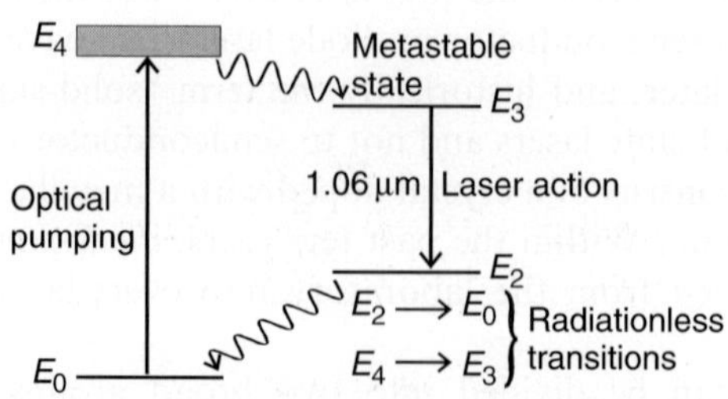

Fig. 4.8 Quantum transitions in the neodymium laser.

A scheme of the quantum transitions in the neodymium laser is presented in Fig. 4.8. The neodymium ions are pumped optically by a flash-lamp or a laser diode from the ground state, E_0, to the excited state, E_4, characterized by a short lifetime and not involved directly in a laser transition. Radiationless transitions promote the ion to the metastable state, E_3, where it stays for a longer time, of about 100 μs. The population inversion takes place between the states E_3 and E_2, and the laser action between these states produces radiation from the range of near-infrared at a wavelength of 1.064 μm. The lower laser state, E_2, relaxes quickly to the ground state, E_0. The neodymium laser can work in the continuous and pulsed regimes. The shortest pulses, of tens of picoseconds can be achieved in the modelocking regime combined with the Q-switching. Q-Switching without modelocking provides pulses of nanoseconds. The neodymium laser belongs to the group of non-tunable lasers. The typical construction of a lamp-pumped laser resonator is shown in Fig. 4.9.

Although the same triply-ionized neodymium ion is involved in the Nd:glass laser, with emission at slightly different wavelength, the Nd:YAG and Nd:glass lasers have little in common. The Nd:glass type is much more efficient at storing energy than is the Nd:YAG. Nd:glass lasers are used in high-energy applications provided in the pulse Q-switching regime. They can produce pulses of 100 J and more, in contrast to Nd:YAG lasers which are limited to about 1 J. On the other hand, Nd:YAG lasers are better for high average power applications, owing to the higher thermal conductivity of YAG than that of glass. A Nd:YAG lamp-pumped laser consisting of several Nd:YAG rods can produce an output of kilowatts or more.

There are many industrial applications of Nd:YAG lasers. Metals processing, including cutting, welding, or marking is almost always associated with Nd:YAG lasers. This kind of application usually needs large, lamp-pumped lasers. Spectroscopic research or surgical procedures usually need smaller, diode-pumped systems with frequency doubling. The neodymium lasers are utilized to pump another lasers or to pump the regenerative amplifiers. For example, the Ti:sapphire laser can be pumped by the second harmonic of the neodymium laser ($Nd:YVO_4$, *cw*, 5.5 W, 532 nm). The regenerative amplifier seeded by the Ti:sapphire laser (82 MHz, 80 fs, 12 nJ, modelocked) can be pumped with a neodymium laser (Nd:YLF, 1 kHz, 10 mJ, 250 ns, 527 nm, Q-switched). The specifications in brackets illustrate typical commercial parameters. Since the mid-1970s, Nd:YAG lasers were workhorses in laboratories around the world for pumping dye lasers. The advent of tunable solid-state laser systems and optical parametric oscillators has reduced this application significantly.

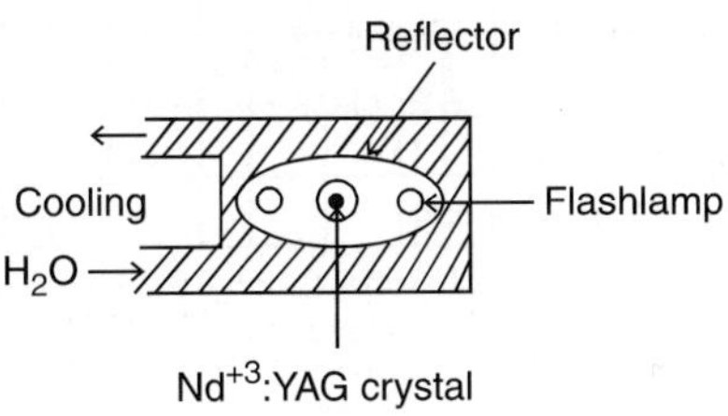

Fig. 4.9 Schematic section of a typical Nd:YAG laser.

The neodymium lasers have found many and diverse applications in medicine, although their applicability is limited by small absorption by water, the main component of human tissue, in the spectral range of 1 μm.

A schematic plot of the absorption in water as a function of radiation wavelength is given in Fig. 4.10. The absorption in water is determined here by the optical path-length at which a given radiation wavelength is entirely absorbed. The smaller is the optical path length, the larger is the absorption coefficient.
One can see that the water's absorption at the wavelength of 1.064 μm (Nd:YAG laser) is small, allowing a laser beam to penetrate more deeply into a tissue than, e.g., radiation at a wavelength of 2–3 μm (Er:glass, Ho:YAG), or 10 μm (CO_2). For these reasons, the Nd:YAG laser is chosen for medical applications which require deep penetration: it can cut comparatively tough tissue in operations on vascular tissue or for hair removal. Laser lithotripsy is another medical application. The radiation from Nd:YAG lasers can be propagated through optical fibers. Owing to the deep penetration of light of 1 μm wavelength into human tissue, this radiation is harmful to eyes. It limits lidar applications for remote monitoring, although the weak absorption in the atmosphere facilitates the light's propagation over long distances. Because of these drawbacks of 1 μm radiation, active media generating radiation in the range of 2–3 μm, corresponding to strong water absorption, were sought. Such radiation is not harmful to the eyes, and a strong absorption—preventing deep penetration—often expands the range of medical applications.

The lasers emitting radiation at wavelengths of 2–3 μm are the solid-state lasers that use YAG as the host, doped with holmium (Ho, 2.1 μm), thulium (Tm, 2 μm) or erbium (Er, 2.9 μm) ions. All these elements, like neodymium, belong to the rare earth series, and their ionization state is +3. The mechanism of the laser action in the rare-earth lasers is similar to that described earlier for the neodymium laser (Fig. 4.8). The lasers are pumped optically by flash lamps or diode lasers.

Wavelengths at about 3 μm are attractive because human tissue absorbs most of their energy, making it possible to ablate both hard and soft tissue. A drawback of these systems, however, has been the awkwardness of transmitting the 3 μm energy from the source laser to the tissue surface. Holmium lasers are used in endoscopies, and in the atmosphere, as well as in aviation to register the eddies and air currents in the neighborhood of control towers. They can also be part of standard navigational

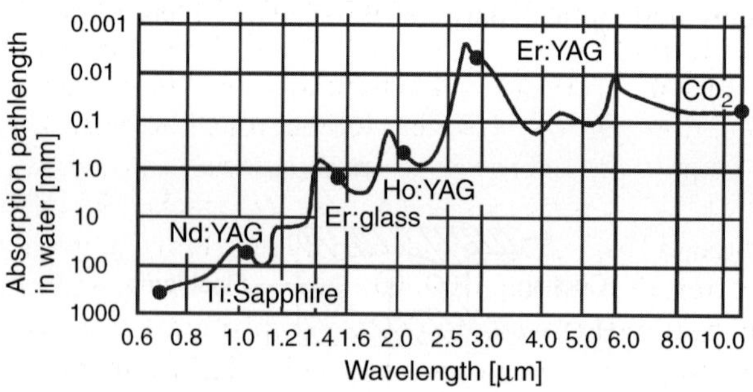

Fig. 4.10 Schematic diagram of absorption path-length in water versus the wavelength of the radiation.

equipment on commercial airliners to inform pilots about changes of wind direction and turbulence. The Er:glass laser is used in ponds and spinal operations, and removing hemorrhoids and polyps. Radiation emitted by erbium lasers penetrates less deeply than that from holmium lasers, but owing to their larger absorption coefficient they are used for cutting bones and drilling in dentistry. Erbium-doped YAG lasers have become accepted for a variety of medical applications, including dermatology, stomatology, and ophthalmology. The Er:glass medium, with a very long, thin glass fiber doped with erbium, is typically used as an amplifier in an erbium-doped fiber amplifier (EDFA). EDFA amplifiers are deployed widely in global communication systems, where they directly boost the signal carried on fiber-optic cables.

Historically, most visible lasers have been based on gas, dye, or semiconductor gain media. The advent of *diode-pumped solid-state lasers* (*DPSSL*) has led to the popularity of new visible or UV solid-state lasers based on second or higher harmonics. They provide an alternative to dye lasers, water-cooled argon-ion lasers, and to excimer lasers. The solid-state lasers are far simpler than their liquid dye counterparts and often expose the researcher to less toxicity. They often need no external cooling water, and many of the specifications, such as optical noise and pointing stability, are better than those for most argon-ion lasers. This is a result of the increase in power of the laser-diode pump sources, and more efficient use of frequency-conversion techniques. High-power laser-diode arrays are located in the laser power supply, and provide pumping of optical energy through a fiber-bundle coupled to the laser head. Some benefits of DPSSL technology are obvious—the lamp sources are high-voltage and high-temperature devices that require water-cooling and replacement every hundred hours. DPSSLs have lifetimes typically of thousands of hours, and small air-cooled devices may be battery-powered. High-power DPSS lasers, which are now available with kilowatt power outputs, are finding new mobile, airborne, industrial, and research applications. Nd:YAG is still the most commonly used DPSSL crystal in the frequency-doubled high-energy *Q*-switched lasers in industrial applications. Nd:YLF and Nd:YVO_4 crystals have been growing in popularity for about a decade in DPSS lasers. The exceptionally weak temperature-dependence of the refractive index of Nd:YLF crystals reduces the problems of thermal lensing at high pump and DPSSL powers. The long upper-state lifetime allows the storage of high energy in the laser transition, resulting in *Q*-switched pulses of high energy. A frequency-multiplied diode-pumped Nd:YLF laser can emit an average power of 15 W in the UV at 351 nm, or more than 30 W at 527 nm. They are usually used in DPSS lasers with repetition rates below 5 kHz. The Nd:YVO_4 crystal has an upper-state lifetime much shorter than Nd:YAG or Nd:YLF, which limits the energy that can be stored. Although Nd:YVO_4 lasers cannot produce as high an energy as competing *Q*-switched Nd:YAG and Nd:YLF lasers, the *Q*-switched Nd:YVO_4 lasers have a high average power output because of their short pulse duration (about 10 ns), and can work at repetition rates of 10 kHz or higher. The commercially available diode-pumped Nd:YVO_4 laser produces more than 10 W of *cw* output at 532 nm in a low-noise TEM_{00} mode. The peak-to-peak amplitude of noise is lower than that of a conventional argon-ion laser, and its beam-pointing and power stability are comparable to large-frame argon-ion lasers.

The major benefit from using Nd:YVO_4 crystals is its high gain and low threshold, which makes this material a good choice for low-power applications, such as hand-held pointers.

The frequency-tripled DPSSLs reach the ultraviolet range, and can be used for material processing, drilling holes for computer circuit-board manufacturing, image recording, and medical diagnostics. Commercial DPSSLs with an 4 W output at 355 nm have been available for a few years.

4.4.2. Solid-State Tunable Lasers (Vibronic Lasers)

Tunable lasers represent a very important class of solid-state lasers. The solid-state lasers discussed in the previous Chapter belong to the group of non-tunable lasers that can emit radiation at a few discrete wavelengths only. As applications have become increasingly sophisticated, the demand for more wavelengths has also increased. The tunable lasers are able to emit radiation in a wide spectral range by tuning continuously across a range of wavelengths. The solid-state tunable lasers can fulfil many of these needs. They are termed, *vibronic lasers*.

In practical terms, the phrase "*tunable solid–state laser*" is almost always synonymous with the Ti:sapphire laser which will be discussed in this Section. The commercial introduction of the continuous wave (*cw*) Ti:sapphire laser in 1988 revolutionized scientific laser applications, and has expanded beyond the laboratory to industrial, medical and many other applications in the "real world".

While in ruby, neodymium, and erbium lasers the upper and lower energy levels are single states, in vibronic lasers the levels are bands that consist of many vibrational sublevels within a single electronic energy state (Fig. 4.11). The name "vibronic" comes from the fact that the considerable band broadening of the absorption and emission transitions results from the strong coupling between electrons and the vibrational modes of the crystal host lattice surrounding the doped ion. As a result, vibronic lasers can emit over a broad range around a central wavelength. By

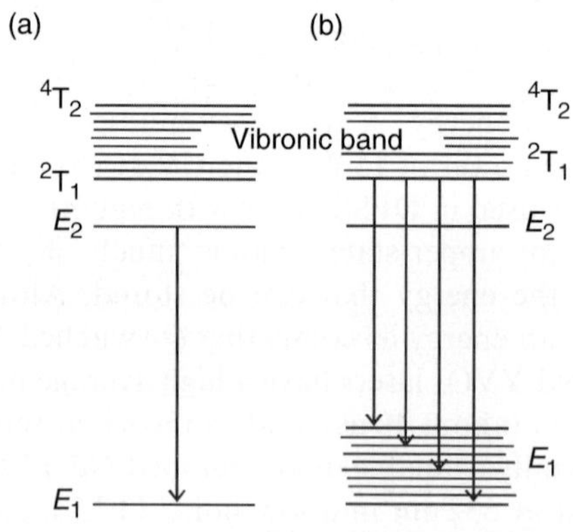

Fig. 4.11 Scheme of electronic levels involved in a laser transition in; (a) non-tunable lasers, (b) tunable lasers.

adjusting the optical cavity length, any laser wavelength within this band can be selected, thereby producing a tunable output.

The main difference between quantum transitions for non-tunable and tunable lasers is presented in Fig. 4.11. In non-tunable lasers the laser transition occurs between single states E_1 and E_2: the level E_2 is not a band. In tunable lasers, the laser transitions occur between the broad vibronic band (T) and the energy levels of the E_1 state, which is also a broad vibronic band consisting of many vibrational sublevels.

The laser transitions occurring in the broad bands lead to wide spectral tunability. Tunable lasers that can produce a broad range of wavelengths are essential for ultrafast pulses. We know from the uncertainty principle that the broader is the band, the shorter is the pulse. For example, a 5-fs pulse requires a bandwidth of nearly 200 nm at an 800 nm central wavelength.

All solid-state tunable lasers belong to the four-level lasers (Fig. 4.12) discussed in the previous section 4.4.1.

The pumping light (1) transfers the ion from the ground vibrational level ($\upsilon = 0$, E_1) of the lower electronic state E_1 to the excited vibrational level of the upper electronic state T ($\upsilon > 0$, T). The excited system relaxes quickly in a radiationless way (2) to the ground vibrational state ($\upsilon = 0$, T). In the third step the emission (laser action) (3) occurs between the upper electronic state ($\upsilon = 0$, T) and the lower electronic state ($\upsilon > 0$, E_1). The lower state ($\upsilon > 0$, E_1) is quickly depopulated by radiationless relaxation to the ground vibrational level of the lower electronic state ($\upsilon = 0$, E_1) (4).

The optical pumping of the solid-state tunable lasers is delivered by:

a) pulsed Q–switched lasers,
b) continuous wave lasers,
c) pulsed flash lamps.

It is well known that in cases where electrons are strongly coupled to vibrations, the fluorescence band maximum does not cover the absorption maximum. For example, in the titanium–sapphire laser the fluorescence maximum occurs at 790 nm whereas the absorption maximum is at about 500 nm. This means that the titanium–sapphire laser can be pumped with the ion argon laser (514 nm) or with the second harmonic

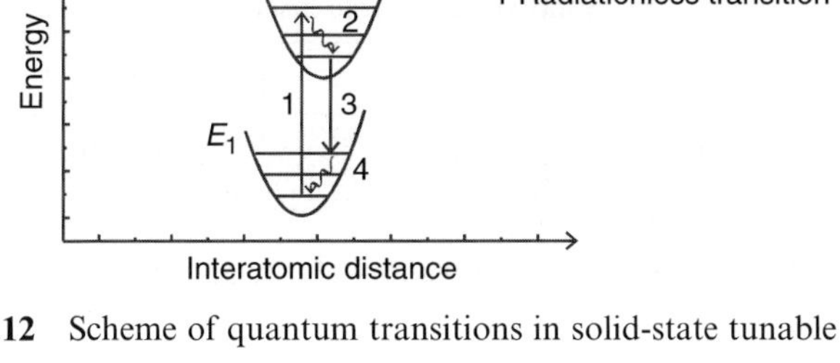

Fig. 4.12 Scheme of quantum transitions in solid-state tunable lasers.

of the Nd:YAG laser ($\frac{1}{2} \cdot 1064\,\text{nm} = 532\,\text{nm}$) or any other neodymium laser, such as Nd:YVO_4. Because the emission bandwidth is exceptionally large (~180 nm), the titanium–sapphire lasers working in the modelocking regime can generate very short femtosecond pulses. The commercially available lasers can generate pulses of 30–80 fs, while research laboratories have already reached much shorter pulses, of the order of several femtosecond—only a few light oscillations. A single cycle of 800 nm light lasts 2.7 fs.

Most of the vibronic lasers are tunable in the near infrared. The titanium–sapphire (Ti^{+3}:Al_2O_3) laser has an extremely broad range of tenability at 670–1070 nm, which is wider than any other tunable medium. To achieve tunability over the entire spectral range one should exchange some optical elements to work in separate spectral ranges—a procedure similar to exchanging of a dye in dye lasers, but definitely much less cumbersome.

Other vibronic crystals include alexandrite in the near-IR, Co:MgF_2, thulium:YAG, and Cr:fosterite emitting deeper in the IR. Among the vibronic solid-state lasers Ce:YLF has the shortest wavelength range (309–325 nm). Table 4.2 shows the tunability range for the most common solid-state vibronic lasers.

The solid-state tunable lasers find many applications in basic spectroscopic investigations and in the physics and chemistry of ultrafast processes. They became workhorses in research laboratories. They are more compact, simple, and convenient than tunable dye lasers and can cover a wide spectral range essential for research. More recently, solid-state tunable lasers are being introduced into medical and industrial applications where their unique characteristics provide benefits in areas for which tunability is essential. They are important in applications such as pollution detection, environmental monitoring, and other areas of remote sensing from aircraft or satellites. Medical applications such as laser lithotripsy with an alexandrite laser have been investigated but not exploited commercially. Recently, several companies have developed commercial alexandrite lasers for the removal of tattoos and hair.

So far, we have discussed the tunability features of the vibronic lasers. Now, we shall stress their ultrafast applications. If we want to extract a specific wavelength from the broadband output of a tunable solid-state laser, external adjustable cavity tuners, such as gratings, prisms and etalons, are used. Conversely, if we are interested in the generation of ultrashort pulses, the full bandwidth of a solid-state laser is employed. Fundamentally, the broader is the band, the shorter id the pulse. We showed in Chapter 3 that the pulse-duration depends upon the number of

Table 4.2

Solid-state tunable lasers

Active medium	Operation	Spectral range [nm]
Alexandrite	continuous	730–810
Ce:YLF	pulsed	309–325
Co:MgF_2	pulsed	1750–2500
Cr:$LiSrAlF_6$	continuous, pulsed	760–920
Ti:sapphire	continuous, pulsed	670–1070

longitudinal modes available for synchronization. The number of modes depends, in turn, on the bandwidth of the laser gain. The broader is the band, the greater is the number of modes, and the shorter is the pulse.

We need to ask which pulses can be termed "ultrafast"–those with a pulse duration shorter than the time required for heat creation and conduction mechanisms? Taking into account the fact that most materials are characterized by an electron–phonon coupling with typical times between picoseconds and nanoseconds, whereas typical heat conduction times are between nanoseconds and microseconds, lasers emitting pulses with durations of femtoseconds and picoseconds are regarded as ultrafast lasers.

Historically, ultrafast *femtosecond lasers* have been based on the *colliding pulse modelocked* dye laser. It has taken about 20 years for ultrafast sources to evolve from the early sub-100 femtosecond colliding-pulse modelocked dye laser to today's compact solid-state diode-pumped lasers, combined with the advent of chirped pulse-amplification that will be discussed in Chapter 6. Such systems can produce light-pulses shorter than a few femtoseconds—only a few light oscillations, of extremely high peak powers, making possible bench-top terawatt lasers [2]. Ti:sapphire is nearly ideal for ultrafast applications. The extremely broad tuning-range indicates the possibility of producing ultrafast pulses. It can produce very short pulses through Kerr-lens modelocking (see Chapter 3). This material has excellent mechanical and thermal properties, and high gain, which can be used in either *cw*- or pulsed lasers. It has a short-lived metastable state (3.2 μs) which makes flash-lamp-pumping inefficient. For *cw* operation, Ti:sapphire is pumped with an ion–argon laser or the second harmonic of a Nd:YVO_4 laser. Pulsed pumping is typically done with the second harmonic of Nd:YAG- or Nd:YLF-lasers.

The development and commercialization of femtosecond lasers has opened up new applications in medicine, particularly in ultrafast surgery. The fundamental feature of tissues' interaction with light is that the heat produced by the laser diffuses from the illuminated zone during the pulse duration, which may lead to undesirable burning. The transfer of heat generated by a laser beam can be overcome with ultrafast lasers. Femtosecond lasers allow far more precise cutting than nanosecond lasers, and produce much less damage to surrounding tissues than do lasers based on photoablation or thermal effects. A femtosecond laser's beam focused on the tissue generates such high electric fields that atoms are torn apart, producing a plasma. The plasma expands rapidly, creating a bubble in the tissue under surgery, and limits tissue damage to the target area. The shorter the pulse, the lower the energy required for photodisruption, and the smaller is the size of the interaction area. For example, a 500 fs pulse with the fluence of 1.6 J/cm^2 produces a fine cut of only 3–13 μm; a 60 ps pulse produces a cavity of 30–120 μm, while a 10 ns pulse creates a 300–1200 μm cavity, requiring much higher energy and a fluence of 185 J/cm^2. So, ultrafast interactions are characterized by negligible heat-affected zones. In Chapter 10 we describe in more detail the main mechanisms responsible for interactions between a tissue and a laser beam, including thermal effects, photoablation, and plasma disruption. The non-thermal ultrafast interactions opened the way for fine-feature femtosecond laser micro-machining. The currently available stents, which are used when a blood clot or a piece of atherosclerotic plaque lodges in an artery and reduces or totally blocks blood flow, have limited usefulness. Femtosecond lasers are promising tools for manufacturing stents, such as cutting the intricate mesh patterns required for making new intracranial stent designs.

The amplified ultrafast lasers can be used in materials processing for non-thermal precise cutting—for example, cutting a high explosive without detonating it, or for photomask repair in the semiconductor industry. Although ultrashort and powerful systems are in the early phase of application, and are just beginning to move out of the laboratory, the rapid progress—with no limit yet in sight—is perhaps as exciting as the advances in computer or telecommunication technologies.

The apparently irresistible desire of the laser community to produce even shorter pulses has led to the generation of attosecond (as) pulses in the soft X-ray spectral range. This is the timescale for electronic motions. Spectroscopic access to the attosecond timescale would allow one to see how an atom reconfigures its own electrons. Much progress has been made to approach this pulse-duration barrier in recent years. At first, trains of subfemtosecond pulses were reported [3]. Then single subfemtosecond pulses were achieved [4]. The most recent results obtained in a collaborative effort in Vienna, Austria provided 650 attosecond pulses that were generated to resolve the excitation dynamics of strongly bound electrons [5]. The pulses of 650 as are reportedly the shortest ones measured to date. While a femtosecond pulse lasts only a few molecular vibration cycles, attosecond pulses are shorter than vibration cycles and are comparable with the periods of electrons revolving in atomic orbitals. This time-domain was, until recently, not accessible by direct experiment. Such a resolution in X-ray pump-probe spectroscopy could soon lead to the direct study of ultrafast processes such as ionization. Focusing a femtosecond laser into a gas (helium, neon, argon) leads to high-harmonic generation (HHG) comprising the spectral region of soft X-rays with the pulse-duration falling into the attosecond range. To measure such short pulses a new attosecond metrology is required because the carrier-envelope phase-relations become extremely important for light pulses interacting with matter [6]. The main problem comes from the fact that the generated pulse has only a fraction of an oscillation period of 750 nm femtosecond light, and typical cross-correlating techniques may not be able to differentiate a single pulse from a train of pulses.

The question arises of whether there is any barrier to producing ultrashort pulses. The duration of any pulse cannot be smaller than the oscillation period of the electromagnetic waves within it, so these cutting-edge pulses are in the extreme ultraviolet (EUV) or X-ray spectral region.

4.4.3. Fiber Lasers

Making a transatlantic phone call twenty years ago was a complicated task. You had to ask an operator for a connection, and when the connection was established you had to speak very loudly to be heard above the static noise level. Now, you are offered a crystal-clear direct-dialed call, with the quality of a local call. That is all due to the "magic" cable that is called a *fiber cable*. Fiber optic links utilizing low-loss fiber cables replaced the standard copper wire previously used to support long distance calls across the Atlantic Ocean.

The optical fiber uses light pulses to transmit information instead of using electronic pulses to transmit information down copper wires. A light-emitting diode (LED) or an injection-laser diode (ILD) can be used for generating the light pulses. There is a transmitter at one end of the fiber cable, which accepts coded electronic pulse information coming from a copper wire and then converts it from the electrical domain to the optical domain in the form of equivalently coded light pulses for further transmission. A light pulse propagating through a fiber has to be amplified. A very important step in the maturation of telecoms is the development of optical amplifiers. Prior to these optical amplifiers, electrical amplifiers were used, significantly slowing the transmission rate because the optical signal had to be converted into an electrical signal, amplified, and then converted back into an optical signal. The optical amplifiers, mainly erbium-doped amplifiers, are capable of transmitting at a rate of one terabyte per second. All these telecommunication technologies are linked with laser technologies, as they need highly effective and stable sources of light. Moreover, the advent of this field brought about a significant development in fiber lasers.

The first fiber laser was proposed in 1960 by Snitzer. It took more than 20 years to evolve a mature fiber-laser technology that triggered the telecom boom in the 1980s. This evolution has been achieved in a series of distinct stages which have overcome the low output power produced by early fiber lasers.

A typical commercially available fiber cable for telecommunications consists of the following elements: the optic core, optic cladding, a buffer material, a strengthening material, and the outer jacket (Fig. 4.13). The optic core is the light-carrying element at the center of the optical fiber. It is commonly made from a combination of silica and germanium. Typical sizes for the core are 8–10 μm and 62.5 μm for a single mode and multimode fibers, respectively. Surrounding the core is the optic cladding made of pure silica. The international standard for the cladding diameter of optical fibers is 125 μm. Surrounding the cladding is a buffer material used to help shield the core and protect it from damage. A strengthening material surrounds the buffer, preventing stretch problems when the fiber cable is being pulled. The outer jacket is added to protect against abrasion, solvents, and other contaminants.

The cladding glass has a refractive index that is lower than the core material, thereby providing total internal reflection of the core signal. Thus, the difference in the refractive indexes creates an extremely reflective surface at the interface. Light pulses entering the fiber core reflect off the core/cladding interface and thus remain captured within the core as they move down the line.

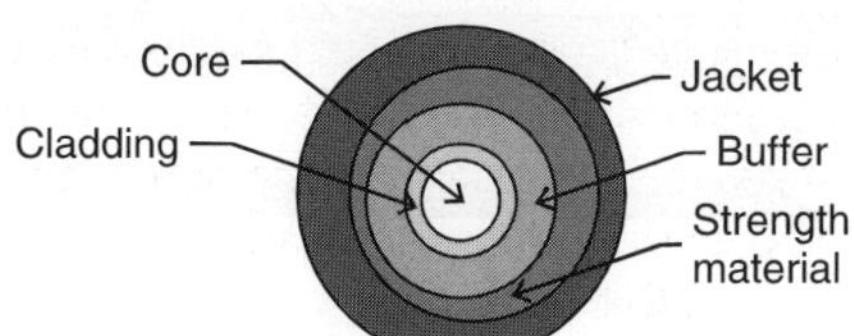

Fig. 4.13 Section of a fiber-optic cable.

We may first recall the concept of total internal reflection from elementary optics. Supposing that the incident monochromatic light wave is planar, the refracted light obeys Snell's law

$$n_{core} \sin \theta_{core} = n_{clad} \sin \theta_{clad} \tag{4.2}$$

where n_{core}, n_{clad}, θ_{core}, θ_{clad} are the refraction indexes for the core and the cladding, respectively. θ_{core}, and θ_{clad} are the incident angle, and the angle of refraction, respectively. Because the incident medium (core) is more dense, $n_{core} > n_{clad}$, and $\theta_{clad} > \theta_{core}$. Thus, as θ_{core} becomes larger, the light transmitted to the cladding layer gradually approaches $\theta_{clad} = 90°$. For incident angles greater than 90° all incoming light is reflected back into the core. This process is known as *total internal reflection.*

A fiber laser consists of a piece of a fiber cable as a resonator cavity sealed with coated ends, that is optically pumped. The cavity length can be tens or even hundreds of meters (in contrast to the much shorter conventional lasers), but the size of the fiber laser does not create a problem, because the flexible fiber cable can be rolled up in a small box measuring only a few centimeters. In contrast to conventional lasers that are based on optical elements which are prone to contamination or misalignment, the fiber laser cavity is nearly perfectly immune to these drawbacks. The fiber cable used in a fiber laser is modified by doping its core with erbium or another rare-earth element (neodymium, erbium, thulium) and adding a cladding layer additional to the typical commercial cable presented in Fig. 4.13. Such a double-cladding consists of an inner cladding (glass) and an outer cladding (polymer). This allows for trapping both the light inside the cavity (core) and the pump-light in the inner layer of cladding, through total internal reflection (Fig. 4.14). In this fiber configuration, multimode diode pump radiation is injected into a fiber through an end-face, then trapped in the inner cladding, propagating along the fiber and producing a population inversion in the core fiber. Another fiber-laser architecture uses multiple points for injection of pump radiation into the cladding.

Rare-earth-doped fibers are well suited to ultrafast applications for several reasons. Their 6-THz amplification bandwidth is broad enough to produce pulses shorter than 100 fs. Now, novel ultrafast fiber lasers are emerging that are capable of delivering average power and pulse durations that may compete with solid-state

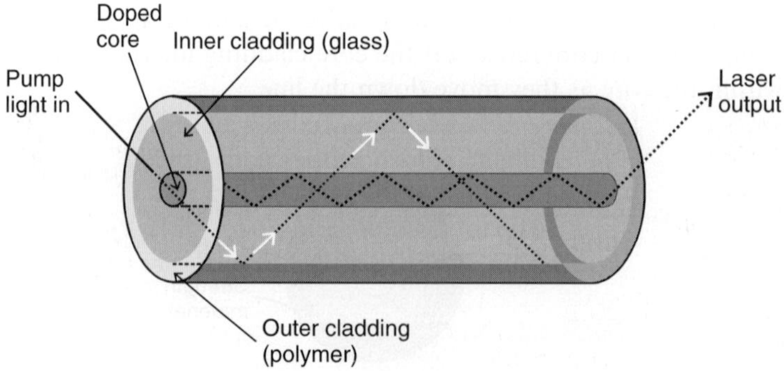

Fig. 4.14 Fiber laser cavity with double cladding.

lasers in certain configurations. Optical communication is a main potential beneficiary of these ultrafast fiber lasers, which can be useful in wavelength division multiplexing (WDM). The development of WDM schemes has created a growing market of higher power devices. In the case of optical amplifiers, the gain must be divided among a large number of channels. Thus, the fiber amplifier power must increase. In recent years, fiber lasers with pulsed output, either modelocked or *Q*-switched, have become available as commercial systems. A typical configuration for a modelocked laser is presented in Fig. 4.15.

In Fig. 4.15 the Yb-doped fiber is pumped with a single diode laser operating at 915 nm with an average power of around 130 mW. A semiconductor saturable-absorber mirror, described in Chapter 3, passively modelocks the laser. The semiconductor mirror uses an AlGaAs/GaAs distributed Bragg reflector, which usually has a bandwidth of 100 nm. A grating pair compensates for dispersion in the fiber. The laser can be tuned over more than 100 nm from 980 to 1100 nm, the wavelength tuning being achieved by adjusting the angular orientation of the high-reflectivity mirror. The duration of the modelocked pulses varies typically from 1.6 to 2.0 ps across the tuning range. A typical output average power is around 3 mW.

Modern optical fibers have extremely low losses, which allows them to travel unattenuated through miles of silica. Unfortunately, light produced through additional effects originating from non-linear phenomena—such as stimulated Raman scattering, four-wave mixing, self phase modulation (SPM), or group velocity dispersion (GVD)—also travel unattenuated through the fiber. The large effective non-linearity of the fiber core can cause severe pulse distortion for fiber lasers working in a pulsed mode, which destroys the ability of the network to achieve high data-rates and long distances. It is critical to counteract or reduce the non-linear effects to make fiber optical communication feasible. When the effects of dispersion, non-linear GVD, and SPM just balance, the result is a pulse, called an *optical soliton*, that travels over very long distances without spreading. These non-linear effects will be described in Chapter 5.

Fiber lasers are being developed in laboratories around the world as alternatives to traditional solid-state lasers. For many laser applications, however, an average power of watts rather than milliwatts is required. Therefore, the Ti:sapphire laser has remained largely unchallenged as a source of broadly tunable, ultrashort pulses, with average power up to 2 W routinely provided from a modelocked oscillator, with

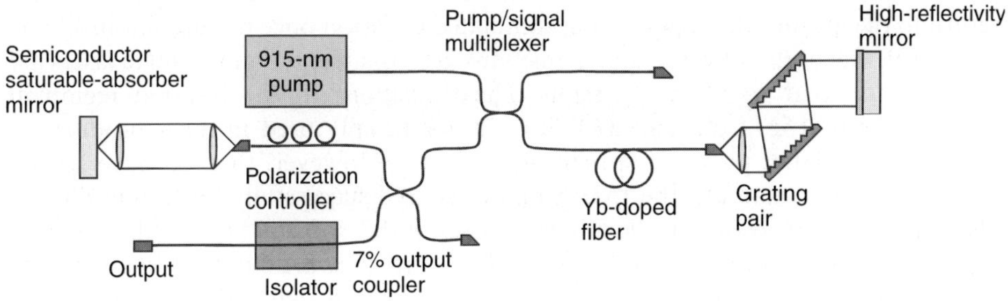

Fig. 4.15 The mode-locked Yb:fiber laser configuration.

pulse durations of the order of 100 fs. The previous generation of high-power fiber systems has never exceeded these limits. However, this situation is changing, owing to improvements in amplification methods. Continuous wave 2 kW fiber lasers have recently been developed for industrial purposes. Powers as large as 13 W were reported as output from a modelocked fiber laser emitting 2–5 ps pulses amplified as a result of an induced chirp from the non-linearity and positive dispersion. Amplification methods will be discussed in Chapter 6.

4.5. GAS LASERS FOR THE VISIBLE RANGE

In *gas lasers* the gain-medium is a gas. Some of the gas lasers, such as the CO_2 laser, the far-infrared molecular laser, and the chemical laser were discussed in sections 4.2.1, 4.2.2 and 4.3. These lasers operate on rotational and vibrational transitions. We shall now discuss gas lasers operating on electronic transitions like solid-state lasers. Gas lasers are facing stiff competition from solid-state lasers, but they are still powerful tools in many scientific, medical and industrial applications. We discuss here the most representative gas lasers including the helium–neon laser and the ion–argon laser.

4.5.1. Helium–Neon Laser

The helium–neon (HeNe) laser can emit several lines in the visible and infrared range, but most commercially available lasers are based on the red line at 632.8 nm, with a power of several milliwatts. HeNe lasers have exceptionally low gain and efficiency, and their *cw* output does not exceed 100 mW. Their applications are limited to low-power tasks.

The active medium of a helium–neon laser is a mixture of helium and neon gases (usually 10:1). The pumping energy is provided by high-voltage dc (direct current) electric discharge at about 10 kV to the bore of a gas-filled tube of length between 10 cm and 2 m, containing gas under a pressure of several torr. The more numerous helium atoms absorb more energy from collisions with electrons produced from the electric discharge than do neon atoms, and they are preferentially excited to higher electronic levels. Then, the helium atoms transfer the excess energy to neon atoms, raising them through resonant collisions to a metastable state, which has an energy nearly identical to that of the excited helium atoms. Then, the excited neon atoms return to the ground state, producing stimulated emission once population-inversion is reached. Generally, several laser transitions are possible, but most HeNe lasers are designed to favor 632.8 nm emission. The transitions in the infrared region at wavelengths of 1523.5 nm and 3392 nm are not usually used in mass production, owing to the higher prices of special optics in IR. However, they do have several specific uses. For example, the 1523.5 nm line is used successfully for testing silicon–glass optical fibers, owing to the minimal losses in this spectral area. Although laser diodes provide an alternative to helium–neon lasers, they are still used in alignment, interferometry, metrology, medical diagnosis such as flow-cytometry, in holography, and in shops to read product codes, because of their good beam coherence and quality. They are also used to determine the position of patients under X-ray irradiation. They serve in industrial alignment to determine straight lines with the laser beam in the construction of buildings, tunnels, and sewer pipes. The compact laser diodes, however, have partially superseded the helium–neon lasers in hand-held barcode readers, but they are still dominant in industrial code readers because of their low divergence.

4.5.2. Ion–Gas Lasers. Argon and Krypton Lasers

In *ion–gas lasers* the gain medium is a *plasma*, an electrically conducting gas consisting of electrons and ions, produced by an electrical discharge. Here we will discuss the most representative ion–gas lasers including argon-ion and krypton-ion lasers. The argon-ion laser emits continuous radiation at powers from several milliwatts to tens of watts in the visible range, and several watts in the UV region. The main lines of the argon-ion laser emission are the green and blue lines corresponding to wavelengths of 514.5 nm and 488 nm, respectively. The strongest line of the krypton laser is at 647.1 nm. Lasers based on ions in plasma are facing stiff competition from solid-state lasers, but there are still many areas where they are used, including confocal microscopy, digital disk mastering, photolithography, and stereolithography.

The laser action of ion–gas lasers working in the visible range (and partly in the UV) occurs between electronic levels, as with other gas lasers from this spectral range. The only difference between the gas lasers and the ion–gas lasers originates from preliminary ionization of the gas by the electrical discharge. Atoms lose one or more electrons, becoming ions that are simultaneously pumped to their excited states. Lasing occurs between the excited- and the ground states of the ions once population inversion is reached. The transitions involved in the argon-ion laser action are shown schematically in Fig. 4.16.

The plasma tubes of the argon-ion laser operate under extreme conditions. The high density of flowing current causes degradation of the materials from which an optical

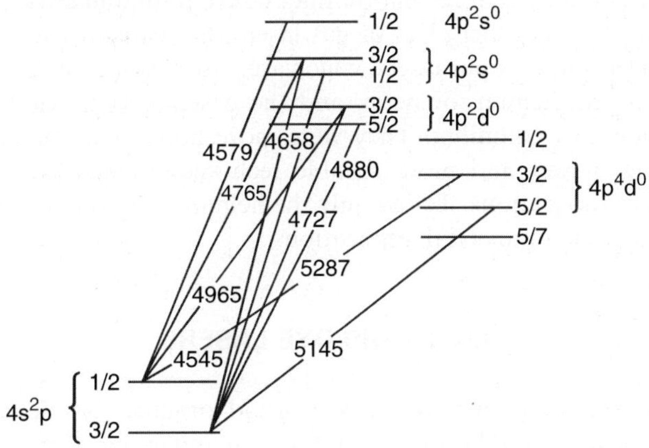

Fig. 4.16 Scheme of energy levels and quantum transitions in argon-ion laser.

cavity and a tube are made. The strong flow of electrons causes ionization and pumping to the excited levels. However, this process generates a large amount of heat, which has to be removed from the system. Moreover, UV radiation generated as a result of laser transitions from doubly ionized atoms causes further degradation of the optical elements such as mirrors and a prism inside the cavity. Many technical advances were required to extend the lifetimes of ion-lasers from hundreds of hours in the early 1980s to the 5000–10000 hours available today. The plasma tube usually has a metal–ceramic construction, and is surrounded by a ceramic envelope that passes heat to a water-cooling system. The helical flow of the cooling water around the tube assures efficient heat removal, and prevents the formation of thermal gradients that might directly distort the resonator. The plasma tube and the outer jacket for the water cooling system are protected by a brass cylinder having high thermal conductivity and excellent rigidity of construction. In typical designs, the plasma discharge occurs between the cathode—consisting of a helical tungsten coil or disk—and the flat copper anode. Copper webs link the disks to the ceramic envelope, facilitating heat transfer from the disk to the envelope where the heat is dissipated into the cooling water. Tungsten disks placed along the length of the laser tube confine the plasma discharge. They are supported by copper disks with holes in them, to ensure proper recirculation and produce an even distribution of pressure along the bore. The plasma tube is placed in a magnetic field perpendicular to the cavity's axis. The magnetic field induces the Lorentz force $\mathbf{F} = e\mathbf{v} \times \mathbf{H}$ directed towards the cavity axis, moving electrons along the axis. As a consequence, the number of electrons in the area of laser action, and the efficiency of laser operation increase. An output coupler and high reflector are attached to the end of the resonator.

The argon-ion lasers are commonly used in scientific laboratories as light sources for Raman spectrometers, and to pump dye and titanium—sapphire lasers. The argon-ion lasers find wide application in data recording on disk matrices (disk mastering) designed to produce replicas of video disks and compact discs. They have found application in forensic analysis, since the visible radiation of the argon-ion laser causes fluorescence of fatty traces found in human fingerprints. Ion lasers are also used in confocal microscopes, which have revolutionized biological imaging—providing much better optical resolution than conventional microscopes.

We should mention another class of gas lasers that can exist in both versions—neutral and ionized-metal vapor lasers, which are based on a combination of a buffer gas such as nitrogen, helium, or neon, and the gaseous phase of a metal such as copper, gold, silver, or cadmium. They were more popular in the past, but strong competition from newer and more reliable technologies has led to a significant reduction in their applications. Today only the helium—cadmium and copper vapor lasers are still used in commercial applications.

4.6. LIQUID DYE LASERS

In *liquid dye lasers* the gain-medium is a liquid organic dye dissolved in liquid solvents. With a broad band between 300–1100 nm and either a pulsed or *cw* output, dye lasers were workhorses in scientific laboratories for many years. The dye lasers were employed commonly to study ultrafast phenomena and were used in all applications where tunability was essential. However, all who worked with them know that the dye-based lasers are generally bulky, inefficient, and difficult to maintain. Organic dyes very quickly deteriorate with use—after tens, or hundreds of hours—in contrast with most other lasing media. Today solid-state, simple, compact and turnkey versions of the fastest laser systems are increasingly used in laboratories, reducing dye-based lasers' applications.

The dyes are characterized by a broad absorption band, resulting in the generation of many longitudinal modes in an optical resonator, that can be favored by tuning continuously across a range of wavelengths. An individual dye can cover a significant spectral range. Moreover, based on knowledge from Chapter 3 concerning modelocking and short pulses generation, we can expect, dyes to be very good candidates for the generation of very short picosecond and femtosecond pulses. Indeed, pulse durations of a few femtoseconds or less have been reported for colliding pulse modelocking (CPM) lasers, which work in the passive modelocking regime with a saturable absorber. The "saturable absorber" passive method was described in Chapter 3.

The dye laser is a four-level laser (Fig. 4.17), and the transitions occur between the electronic levels which are modified strongly by molecular vibrations. The interactions between electrons and molecular vibrations, called *vibronic coupling*, lead to a significant band-broadening of the absorption and emission lines. The vibronic coupling leads, in many cases, to homogeneous broadening which permits all the excited molecules to participate in single frequency light amplification. The homogeneous broadening causes a single-mode's output power from the dye laser to be similar to the power of multimode lasers. Comparison of the scheme of electronic transitions in the dye laser (Fig. 4.17) with that in the vibronic solid-state laser (Fig. 4.12) shows a close similarity. The pumping light (1) transfers the dye from the ground-vibrational level ($v = 0, S_0$) of the lower electronic state S_0, to the excited vibrational level of the upper electronic state S_1 ($v > 0, S_1$). The excited system relaxes quickly in a radiationless way (2) to the ground vibrational state ($v = 0, S_1$). In the third step, the emission (laser action) (3) occurs between the upper electronic state ($v = 0, S_1$) and the lower electronic state ($v > 0, S_0$). The lower state

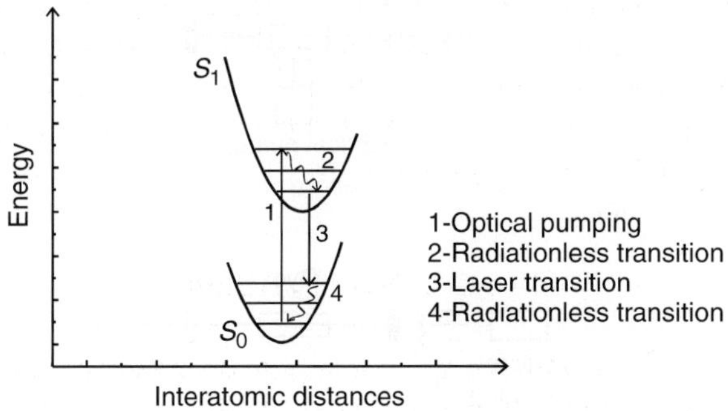

Fig. 4.17 Scheme of energy levels in a dye laser.

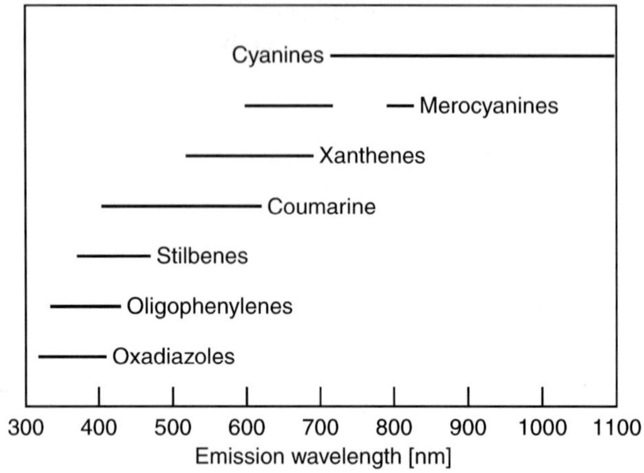

Fig. 4.18 Spectral ranges that can be tuned for different classes of dye molecules [7].

($\upsilon > 0$, S_0) is quickly depopulated by radiationless relaxation to the ground vibrational level of the lower electronic state ($\upsilon = 0$, S_0) (4).

The total losses in a dye laser, beside the typical losses in an optical resonator (mirrors' transmission, optical elements' absorption), consist of losses in an active medium. The losses in a dye result from the intersystem crossing $S_1 \rightarrow T_1$ to the triplet states. These transitions lead to a decrease in the population of the S_1 level, as well as the absorption $T_1 \rightarrow T_m$, because the absorption of the triplet states covers the area of the fluorescence $S_1 \rightarrow S_0$. To avoid the losses caused by the triplet states one should accelerate the triplet states' relaxation, or remove molecules in the triplet states from the active laser area. This can be done by adding triplet-states quenching substances (*quencher*) to the dye solution, or by applying a fast flow of a dye (of the order of 10^{-6} s) through the active area. These times are considerably shorter than the triplet state's lifetime, and the fast flow permits elimination of the losses caused by the triplet states.

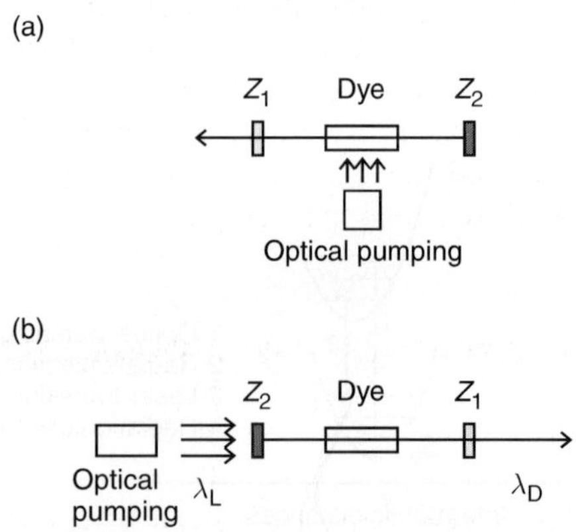

Fig. 4.19 Different geometries of pumping in dye lasers.

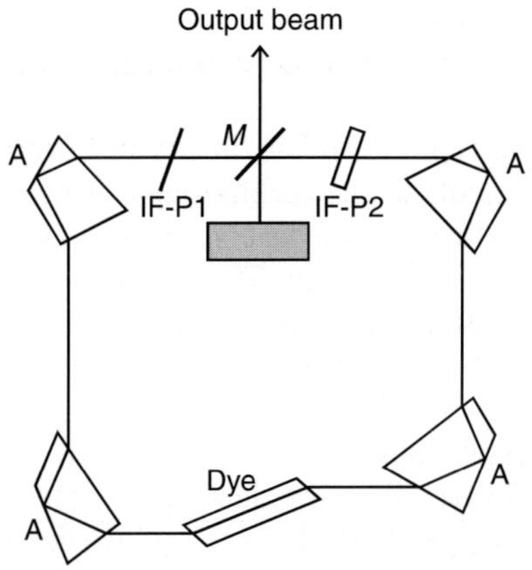

Fig. 4.20 Ring laser, IF-P1 and IF-P2–Fabry-Perot interferometers.

As we have said, an individual dye cannot cover the entire spectrum between 300 nm and 1.1 μm: a range of different types of dyes must be used. Thus, merocyanine and cyanines cover the red region, coumarins and xanthenes cover the greens and yellows, and oxadiazoles, stilbenes and oligophenylenes cover the blue region (Fig. 4.18).

Dyes require high pump powers. The dye molecules are optically pumped by flash lamps, arc lamps, or pulsed and continuous lasers such as the neodymium Nd:YAG (the second harmonic), argon, excimer, and nitrogen types. Two ways of pumping can be distinguished in dye lasers—the transverse (Fig. 4.19a), and the longitudinal (Fig. 4.19b).

The lasers discussed so far have an optical resonator in which the standing wave is generated. A serious drawback of the resonators employed in dye lasers is the spatial phenomenon of "holes burning". This effect makes it difficult to obtain a single-mode operation, and destroys mechanisms based on homogeneous band broadening. To avoid these undesirable effects the ring resonator geometry is applied, in which a traveling wave—instead of the standing wave—is generated (Fig. 4.20). The ring geometry is achieved by internal reflections in the Abbe prisms (A). The laser beam arrives at, and leaves the prisms at the Brewster angle. The output beam is led out either by a beamsplitter *M*, or by total internal reflection.

In order to extract a specific wavelength from the broadband output of a tunable dye laser, external adjustable cavity tuners are used. Three types of external cavity tuners are used: gratings, prisms, and etalons. Conversely, if we are interested in the generation of ultrashort pulses, the full bandwidth of a dye is employed. Fundamentally, the broader the band, the shorter is the pulse.

4.7. GAS LASERS FOR THE ULTRAVIOLET RANGE

4.7.1. Excimer Lasers

Excimer lasers are facing strong competition from solid-state lasers, although they still offer the most efficient access to the ultraviolet spectral region—with high energies, and high peak and average powers in pulsed operation. However, they have some drawbacks, such as poor beam quality (higher-mode structure and high divergence), their size, operating costs, and maintenance requirements. Specialists in this field say that this lingering reputation for finicky performance is now largely undeserved, owing to improvements in the technology during recent years.

Excimer lasers, along with nitrogen lasers, are the most popular gas lasers generating radiation in the ultraviolet range. The active medium is a mixture of a noble gas, halogen gas, and a buffer gas—usually neon. The gas mixture in a typical excimer laser consists of 2–9% of a noble gas, 0.2% of a halogen gas, and 90–98% of a buffer gas which serves as a medium to transport energy. The mixture is confined in a pressure vessel, typically at a pressure of 3500–5000 millibars, and is usually excited by a fast electric discharge lasting a few tens of nanoseconds. Halides of noble gases, such as ArF, KrF, XeF, XeCl which have short lifetimes, and are unstable in the ground state, are created in the gas mixture. The characteristics of typical excimer gases are presented in Table 4.3.

The term "excimer" is an abbreviation of the expression *excited dimer*, and denotes a molecule RH* in the excited state, E_1, which does not exist in the ground state, E_0 (Fig. 4.21). The laser transitions occurring between the excited state, E_1, of the excimer and the ground state, E_0, where the gas reverts to the original separate constituents, lead simultaneously to a reduction in the lower level's population, and an increasing population inversion between the states E_1 and E_0.

The laser gas mixture is usually excited by a fast electric discharge lasting only a few tens of nanoseconds. The gases of the excimer lasers undergo degradation during laser operation. In commercially available lasers the pressure vessel is not sealed off, and is therefore connected to gas bottles that replenish the gases periodically, usually after 10^3–10^6 pulses, once the laser performance deteriorates. One should remember that the excimer lasers—although often used in laboratories, industry and medicine—contain a toxic active medium, and special safety procedures should be

Table 4.3

Typical excimer-laser gases [8]

Gas	Wavelength [nm]	Comments
ArF	193	Requires high operating voltages to transfer energy efficiently to the discharge. Has the shortest absorption depth in organic materials of all excimer wavelengths. The laser beam is absorbed by O_2 molecules in air, so the beam-path requires an N_2 flush for efficient propagation.
KrF	248	The best of all excimer lasers for its combination of power and lasing efficiency: used for machining polymers.
XeCl	308	Very long gas-lifetimes. Often used for marking applications. Lower average powers than KrF generate a lower rate of material-removal overall, although longer absorption lengths in polymers can lead to higher etch rates.
XeF	351	Less attractive wavelength for machining tasks, because of its lower absorption in polymers.

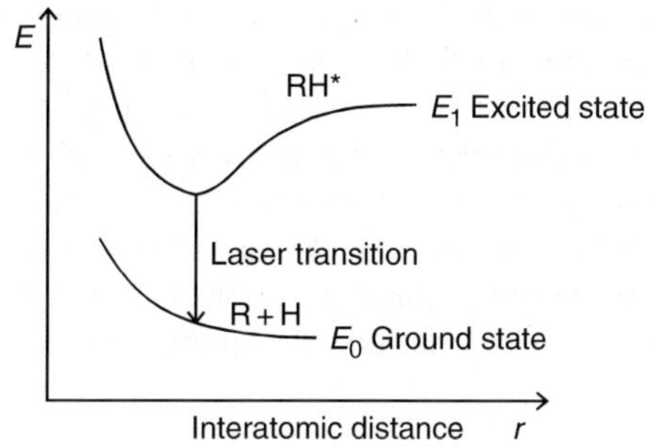

Fig. 4.21 Energy scheme of the ground and the excited electronic states in an excimer laser.

maintained. In the early days of the development of excimer lasers, corrosion of the gas vessel owing to the presence of halogens was a serious problem. In all modern lasers this problem has been removed by careful selection of materials, such as electrodes made of nickel or bromine, and the elimination of organic materials such as lubricants, seals, or insulators. These technical improvements have extended to 10^8 pulses the lifetime between exchanges of laser gas.

The excimer medium has a very high gain, and an output coupler reflectivity of 10–30% is sufficient to achieve an adequate output. The most often used emission lines of excimer lasers are 193 nm (ArF), 248 nm (KrF), 308 nm (XeCl), and 351 nm (XeF). The powers of these lasers are from 1 W to 100 W (for KrF and XeCl) or even more. The typical operating parameters are presented in Table 4.4.

The laser's optics are the second element to require maintenance, and the second factor deciding the laser's lifetime. The optical elements in XeCl lasers are made of quartz or fused silica. In XeF lasers the optical elements used are made of MgF_2, because fluoride etches both quartz and silica, with deterioration of a laser's performance.

Excimer lasers produce high pulse energies, and high average and peak powers. One of the biggest excimer lasers is "Aurora" in the Los Alamos National Laboratory (5 ns, 5 kJ, KrF), used for nuclear fusion control tests. Excimer lasers are used in

Table 4.4

Typical excimer laser output parameters [8]

Parameter	Value
Wavelength [nm]	193, 248, 308, or 351
Pulse energy [mJ]	400–500
Repetition rate [Hz]	100–200
Average power [W]	40–100
Pulse duration [ns]	15–30
Peak power [MW]	10–30
Peak fluence [mJ/cm^2]	100–200
Beam divergence [mrad]	1–3
Beam size [mm]	8–15 × 25–30

scientific laboratories to pump dye lasers, as well as to generate higher harmonics to reach the vacuum-ultraviolet range. They are particularly significant in applications such as metrology and systems alignment. They are important tools in monitoring pollution and ozone concentrations in the upper layers of the atmosphere. Excimer lasers emitting in the deep ultraviolet region currently dominate the lithography industry—serving to draw patterns, marks, and trade codes on ceramics, glass, plastic, and metal. Krypton- and argon-fluoride (KrF and ArF) lasers used in far-UV microlithography facilitates chip design, generating circuit features smaller than 0.18 μm.

4.7.2. Nitrogen Laser

The nitrogen laser emits radiation at 337 nm, and its active medium is gaseous nitrogen. The gas is confined within a pressure vessel, usually at a total pressure between 101 Pa and 2700 Pa. The nitrogen laser generates pulses of duration between 300 ps (at atmospheric pressure) and 10 ns (at 2700 Pa). The repetition rate changes from 1 to 100 Hz. Low-repetition-rate lasers use a configuration of sealed-off devices, and high-repetition-rate lasers employ a flowing version. The pulse energy changes from microjoules to 10 millijoules. For many years, nitrogen lasers were used to pump dye lasers and as sources for many other applications requiring UV radiation. It is well suited to UV laser-induced fluorescence applications in clinical, pharmaceutical, environmental, and process-control work.

4.8. DIODE LASERS

In the lasers discussed so far the laser emission is a result of quantum transitions between electronic, vibrational, or rotational levels. The diode lasers are based on a slightly different principle of operation, although the basic idea of population inversion must be kept. Diode lasers use the phenomena which occur in semiconductors. This is a reason why we need to summarize some basic physics related to semiconductors.

4.8.1. Intrinsic Semiconductors. Doped Semiconductors. *n-p* Junction

Every material has characteristic properties relating to electric current conductivity. Three main classes can be distinguished: a) metals, b) semiconductors, c) insulators. Atoms in a solid are packed into a dense crystalline arrangement, and electron-energy levels are no longer narrow discrete lines. They are significantly broadened, owing to interactions with the condensed environment. However, there are fundamental differences in the energy gaps in metals, *semiconductors*, and *insulators* (Fig. 4.22).

Let us take the example of the metal, sodium (^{11}Na), with the electronic structure: $(1s)^2(2s)^2(2p)^6(3s)^1$. The outer electron level, 3s, is broadened considerably like the

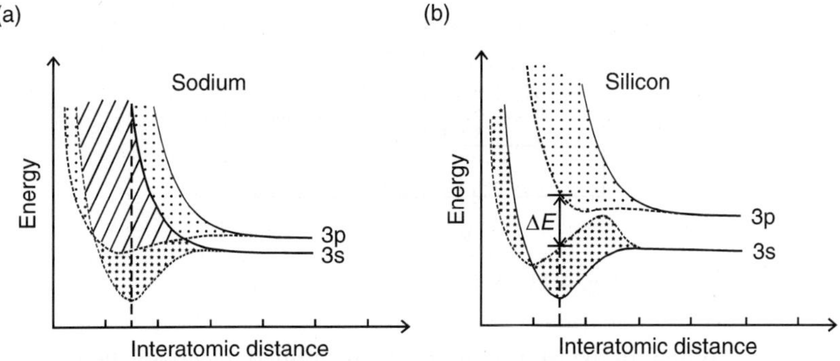

Fig. 4.22 Energy level diagrams for metals and semiconductors.

first upper-lying electron level 3p, which is unoccupied for the isolated atom. This results in overlapping between the neighboring levels (Fig. 4.22a) and the electron can be found either in the 3s or 3p levels. This area of overlapping levels is called, *the conduction band*, because the electrons occupying 3s or 3p orbitals are common for all the atoms and can wander freely through the solid from one atom to another.

In semiconductors, e.g., in silicon ^{14}Si: $(1s)^2(2s)^2(2p)^6(3s)^2(3p)^2$ the outer electron levels are also broadened in a dense crystalline matrix but they do not overlap (Fig. 4.22b). The 3s and 3p orbitals broaden into two distinct bands separated by the energy gap ΔE. We say that the four electrons (originally two from 3s and two from 3p) occupying 3s (or more precisely, an sp^3 orbital) are in *the valence band*, while the level 3p (conduction band) remains empty.

A material, in which the energy gap ΔE becomes very large is called an *insulator*. In *insulators* the energy gap ΔE is of the order of 3 eV or more, which indicates that the collision energy as well as another forms of thermal energy of molecules are insufficient to transport an electron from the valence band to the conduction band. In semiconductors the energy gap ΔE is much smaller (1 eV for silicon), allowing a fraction of the electrons to jump to the conduction band at room temperature. At absolute-zero temperature, semiconductors become insulators.

In a quantum detector, such as a photomultiplier, electrons must escape out of a metal to be detected. An electron trapped in a metal can escape if it gains a minimum energy equal to the work function, W, which is typically of the order of 4–5 eV. Electrons in semiconductors do not need to escape to make the semiconductor a photodetector. To become a photodetector, it is sufficient for the photons irradiating a semiconductor to deliver an energy required to transport electrons from the valence band to the conduction band (Fig. 4.23).

When photons irradiate a semiconductor there is a transfer of electrons from the valence band to the conduction band, leaving a hole (a positive charge) in the valence band. The greater is the light, the more electrons jump the gap, and the material becomes more conductive the. Under these conditions a current will flow that is proportional to the intensity of the incident radiation. Such a photodiode is said to operate as a *photoconductive detector*. Such a photodetector is also called a *photoresistor*, as it is based on the conductivity growth and resistance drop in a

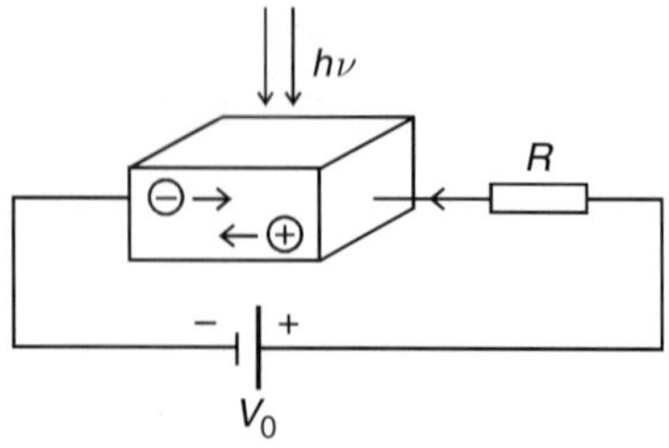

Fig. 4.23 Scheme of a photodetector.

semiconductor. When an external voltage, V_0 is applied, electrons begin to move in one direction (opposite to the applied field), while electron holes travel along the applied field. The electron flux produces a current, I, with intensity proportional to the intensity of the incident light. If a resistor R is placed in series with the circuit (Fig. 4.23), the current will generate a voltage drop, $U = IR$, across the resistor.

Photons irradiating a semiconductor can induce a current flow for considerably lower photon energies than one can get in photomultipliers where the photon energy has to be at least equal to the work-function for electrons from the cathode. The phenomena occurring in semiconductors permit one to construct photodetectors with detectivity covering the near-infrared and infrared regions. For example, the energy gap, ΔE, between the valence band and the conduction band in silicon is 1.11 eV, which corresponds to a wavelength of 1.13 μm (10000 cm^{-1}); in germanium it is 0.66 eV (~6000 cm^{-1}), and in GaAs, 1.42 eV. The exceptionally small gap, $\Delta E = 0.17$ eV (7.3 μm), for InSb (indium antimonide) allows this semiconductor to be applied as an infrared detector.

The semiconductors discussed above are called *intrinsic semiconductors*, in contrast to *extrinsic semiconductors* which will now be discussed. When a silicon matrix, ^{14}Si-($(1s)^2(2s)^2(2p)^6(3s)^2(3p)^2$) is doped with phosphorus ^{15}P-atoms ($(1s)^2(2s)^2(2p)^6(3s)^2(3p)^3$)—whose electrons in the 3s (E_d) state have approximately the same energy as the silicon's electrons in the conduction band 3p, the phosphorus electron (donor) can easily jump to the conduction band of the silicon (acceptor) (Fig. 4.24b). The semiconductors that are doped in such a way gain additional excess electrons, and they are called *semiconductors of the n-type*. On the other hand, if a silicon matrix is doped with aluminum atoms ^{13}Al ($(1s)^2(2s)^2(2p)^6(3s)^2(3p)^1$), which have the electron energy (E_a) close to the energy of the silicon valence band 3s, the

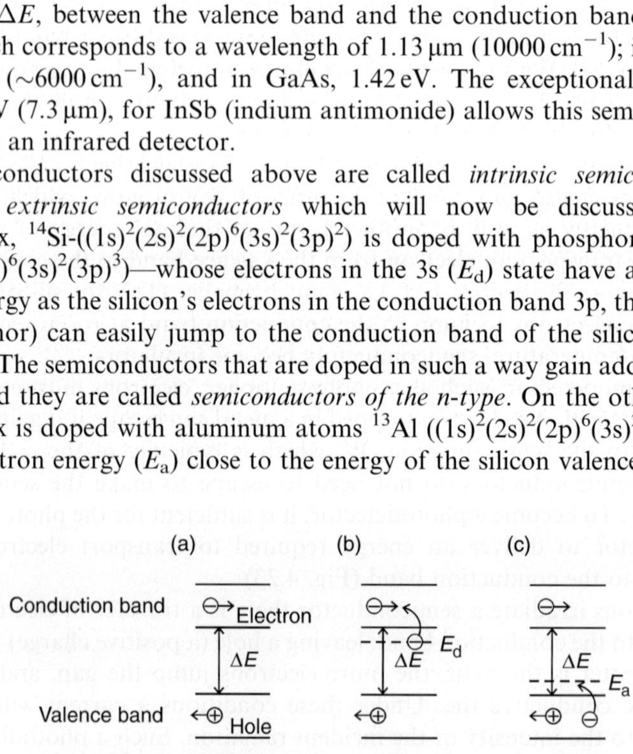

Fig. 4.24 Scheme of intrinsic and extrinsic semiconductors: (a) intrinsic semiconductor, (b) extrinsic semiconductor of n-type, (c) extrinsic semiconductor of p-type.

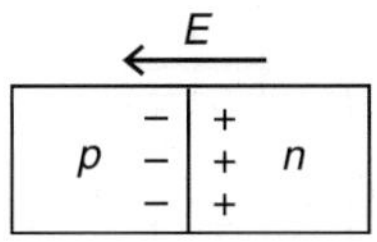

Fig. 4.25 *p-n* junction.

electrons from the silicon jump to the level E_a of aluminum, leaving a hole in the silicon's valence band (Fig. 4.24c). Semiconductors that are doped in such a way are called *semiconductors of p-type*.

If *n*-type and *p*-type semiconductors are brought together to form a *p-n junction*, holes migrate from the *p*-region to the *n*-region, and conduction electrons move in the opposite direction. As a result of this migration, an area in the direct neighborhood of the junction becomes depleted of mobile charge carriers (Fig. 4.25), with an excess of negative charges on one side and an excess of positive charges on the other side of the semiconductor. Consequently, such a charge distribution induces an internal electric field, *E*, directed towards the *p*-region. This internal field prevents further migration of charges across the junction. The *p-n* junction is called the *photodiode*.

When the *p-n* junction is irradiated, new electron-hole pairs are generated, which are moved by the internal field *E*. The holes move towards the area *p*, and electrons towards the area *n*, causing a decrease in the internal voltage potential. When a voltmeter is connected to the junction, it will measure a voltage proportional to the intensity of the radiation incident on the junction (Fig. 4.26a). In an open circuit a photodiode performs as a *photovoltaic detector*. Another situation is shown in Fig. 4.26b, where the *p*- and *n*-regions are connected by a closed external circuit. Under these conditions, a current will flow that is proportional to the intensity of the incident radiation. Such a photodiode works as a *photoconductive detector*.

The detection on the junction can be improved by an external voltage-supply. When a negative terminal of the external voltage source becomes connected to the *n* region, and the positive one to the *p* region, the *p-n* junction is said to be *forward biased*. This means that the charge carriers—both holes from the area *p*, and electrons from the area *n*—flow towards the junction under the external field E_{ex} (Fig. 4.27). The external field E_{ex} is directed just opposite to the internal field *E*. The semiconductor lasers, which will be discussed in the next chapter use the forward-biased junction. On the other hand, when the *negative* terminal of the

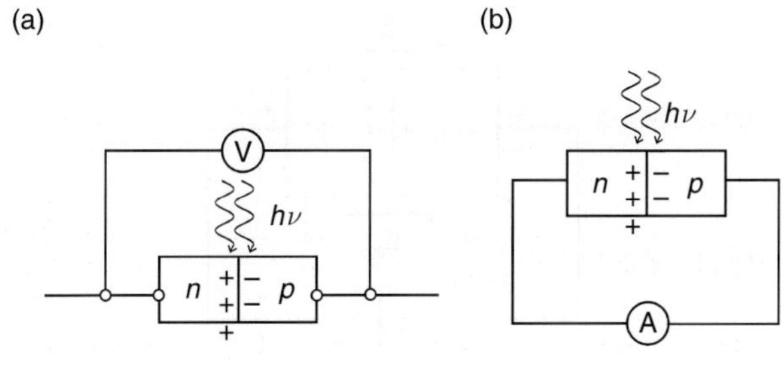

Fig. 4.26 Photodiode types: (a) *photovoltaic detector*, (b) *photoconductive detector*.

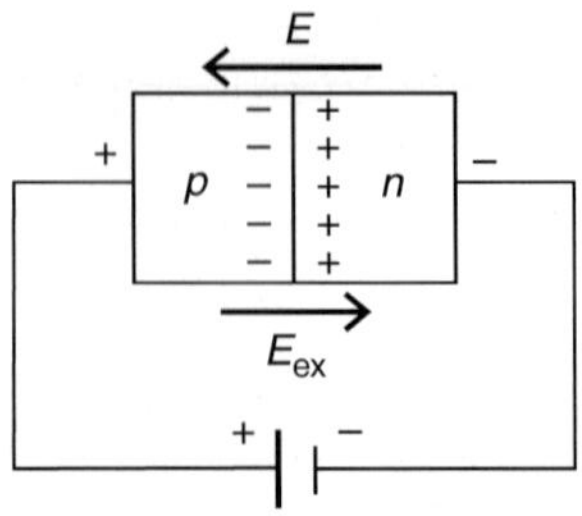

Fig. 4.27 Forward-biased *p-n* junction.

external voltage source is connected to the *p* region, and the positive one to the *n* region, the *p-n* junction is said to be *reverse biased* (Fig. 4.28).

The reverse-biased junction is employed in voltaic and photoconductive photodiodes, as well as in the avalanche photodiodes described in the next section.

4.8.2. Diode Lasers

The simplest light-emitting diode (LED) is the forward biased *p-n* junction discussed in the previous chapter. The external forward biased voltage applied to the *p-n* junction forces holes from the *p*-region and conduction electrons from the *n*-region towards the junction. The movement of the charge carriers in the opposite directions creates an electric current through the device, and recombination of the electrons and the holes at the junction. The energy gained in the recombination can be released as non-selective heat (silicon and germanium semiconductors) or as light (GaAs), with the energy of photons being equal to the gap ΔE between the valence band and the conduction band. The different patterns of relaxation by radiative or radiationless decay depend on the band-gain structure. Semiconductor materials with a *direct band gap* dissipate the recombination energy mostly as light, in contrast to semiconductors with an *indirect band gap* in which the energy is released mostly as heat. The radiative way of releasing energy is observed in materials that belong to the III–V groups of elements, as well as in some materials from the II–VI groups. The distinct paths of relaxation originate from the fact that the conduction-electrons and the holes experience separate dynamics in various materials. In the direct-band gain

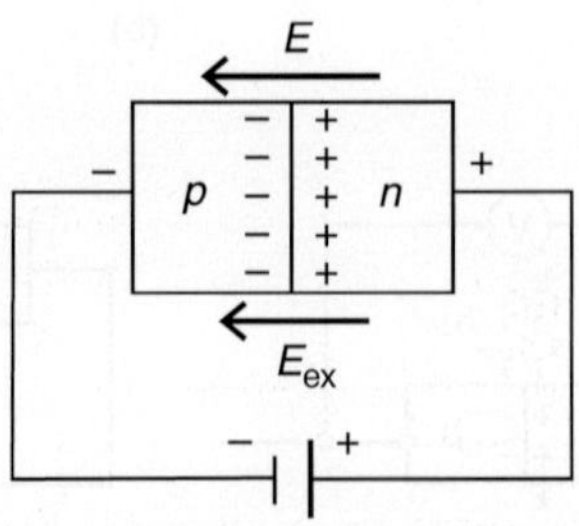

Fig. 4.28 Reverse-biased *p-n* junction.

semiconductor, the momentum of an electron returning from the bottom of the conduction band to the top of the valence band changes very little. This is in contrast to a semiconductor with the indirect band gap, where this electronic transition is accompanied by a significant shift in the momentum.

In principle, a semiconductor laser (also known as a diode laser) can be obtained from a LED in which population-inversion is achieved. The diode laser requires much higher current densities than an LED to reach the laser threshold (Fig. 4.29). The population-inversion between the conduction and valence bands is achieved by applying a strong forward-bias across the device, to inject a large number of electrons and holes into the junction. Operating under the conditions of population inversion means that the number of recombination events considerably exceeds the number of electron-hole pairs generated.

To achieve greater efficiency, some kind of optical feedback is needed—as in the other lasers we have described so far. Usually, two opposing crystal faces cleaved perpendicularly to the junction play the role of an optical cavity, just like mirrors in conventional lasers. Therefore, the light emitted on the *p-n* junction is reflected on the diode edge, and turns back to the junction again: additional electrons and holes are produced with electrons jumping from the valence band to the conduction band. The electrons again recombine with the holes under an applied forward-bias voltage: they return to the valence band, emitting light having energy equal to the difference, ΔE, between the valence band and the conduction band. Moreover, the optical feedback between the cleaved parallel facets produces a more directional, coherent beam of light from the junction plane—in contrast to an LED, which emits in nearly all directions. The simplest laser diode is shown in Fig. 4.30.

The earliest diode lasers, shown in Fig. 4.30, were a simple *p-n* junction and required extremely high current densities to reach a laser threshold. Although the first semiconductor laser was built in 1962, many years elapsed before the technology matured enough to meet the demands of real-world applications. The first diode lasers had to operate only in a pulsed regime under cryogenic cooling. Better performance was attained by inserting a few layers of a semiconductor material, instead of the simple *p-n* junction described above. The first continuous wave operation at room temperature was achieved in 1970.

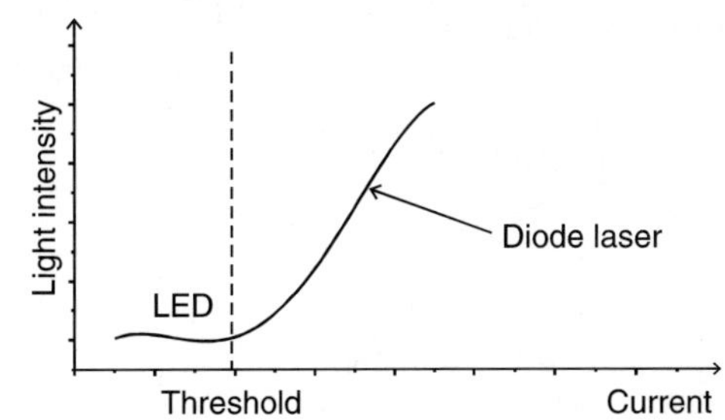

Fig. 4.29 Scheme illustrating the difference between the light-emitting diode (LED) and the diode laser.

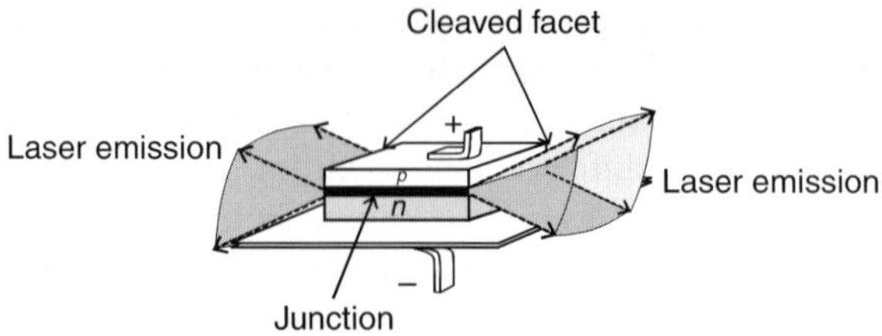

Fig. 4.30 Laser diode scheme.

The *cw* operation becomes possible through the advent of the technology of *double heterostructures*.

A huge variety of diode laser architectures exists, and the differences result from technological constructions to confine a carrier, current, or light inside the cavity. The first laser diode looked something like the device shown in Fig. 4.30. Since the early 1960s, when the first laser diodes were developed, semiconductor lasers have evolved significantly, invoking sophisticated semiconductor fabrication techniques. Several junctions of different materials are used in these designs—not just a single junction as shown in Fig. 4.30. Some types of diode lasers are shown in Fig. 4.31. Besides the classification in Fig. 4.31 we can distinguish lasers in which the interatomic distances of the crystalline layers that are grown on the substrate are perfectly matched to the crystal lattice of the substrate, and others that are based on large deviations (about 1%) from the size of the substrate crystal lattice, producing so-called *strained layers material.*

In order to achieve a performance goal equaling the reliability and stability of other lasers, the diode lasers required some important improvements. First, there was a need to reduce heat while maintaining a population inversion. Secondly, to improve the beam's quality and divergence. Third, to provide higher power levels

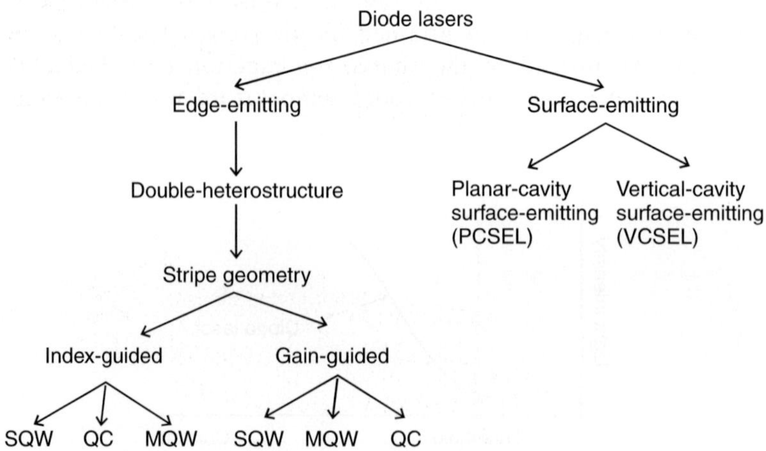

Fig. 4.31 Diode lasers; SQW–single quantum wells, MQW–multiple quantum wells, QC–quantum cascade.

while maintaining a single transverse mode output. Fourth, to control the wavelength of the output.

The first goal can be achieved by increasing the effectiveness of population inversion at a given current or/and by reducing the current-flow into a small region. The double heterostructure architecture in Fig. 4.32a helps to confine charge carriers, and enhances population inversion. A double heterostructure junction consists of a very thin *p*-type layer (approximately 0.2 μm) of a direct-band-gap material having a smaller band-gain than the thicker *p*-type and *n*-type layers above and below it. When a forward-biased voltage is applied to this *p-p-n* junction, holes and electrons are injected from the outer regions into a central active layer, where they become trapped in the potential well created by the wider-band-gain materials. "Carrier confinement" into a small volume increases population-inversion, which means a higher efficiency, lower threshold current, less heat, and more light. This effect can be enhanced by adding "current confinement" to the double heterostructure. One way to introduce the current confinement is to use a stripe electrode instead of a wide

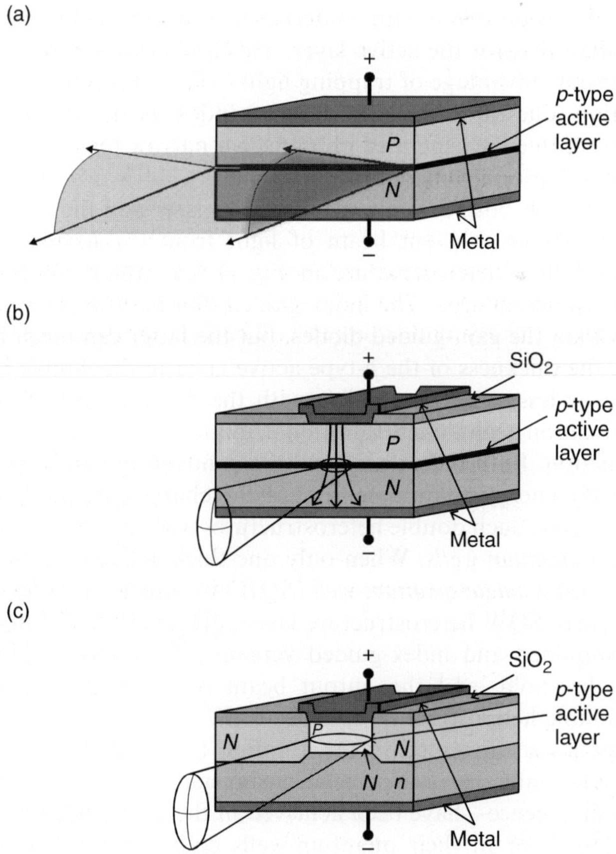

Fig. 4.32 (a) the basic double heterostructure diode; (b) gain-guided, striped geometry diode; (c) index-guided, striped geometry diode [9].

area of the diode's surface. Confining the current into a small region perpendicular to the plane of a junction leads to a higher current density, higher concentration of charge carriers inside a diode, higher gain, and less heat. A number of techniques can be used to confine the current flow to a small area, but the simplest method is presented in Fig. 4.32b. A thin ribbon of positively charged metal is put between two insulator layers such as silicon dioxide (SiO_2) and covers the *p-p-n* junction instead of a monolithic layer of a metal as in Fig. 4.32a.

The stripe-geometry diodes can be divided into two main groups:

1. *gain-guided diodes* (Fig. 4.32b),
2. *index-guided diodes* (Fig. 4.32c).

The effects of "carriers confinement" and "current confinement" can be enhanced further by introducing "photons confinement". The index-guided technique is commonly applied to trap photons inside the optical cavity, and to increase the density of the cavity optical power. The principle of index-guided diodes is illustrated in Fig. 4.32c. A thin active layer of *p*-type small-band-gain material is surrounded on four sides by the wider-band-gain materials characterized by having refractive indexes lower than that for the active layer. This architecture is a gain guiding, but it has the additional advantage of trapping light inside a narrow area through total-internal reflection, due to the higher refractive index inside the cavity. This architecture helps to confine the generated photons in a narrow region, enhancing optical feedback and population inversion.

One may notice that both the gain-guided diode laser and the index-guided diode laser produce a tighter coherent beam of light from the narrow active area, in contrast to the double heterostructure in Fig. 4.32a, which emits a broad beam across the entire junction area. The index-guided diodes generate a narrower, more coherent beam than the gain-guided diodes, but the latter can reach higher powers.

If we reduce the thickness of the *p*-type active layer in the double heterostructure down to 50 nm or less, which is feasible with the development of advanced semiconductor fabrication techniques, additional quantum effects occur in the active layer of the junction. Both the valence and the conduction bands split into discrete sublevels, with the energy and momentum of the charge-carriers depending on the thickness of the layer. Such double heterostructures with an active layer thicker than 50 nm are called *quantum wells*. When only one thick active layer is employed, the diode laser is called a *single-quantum-well* (*SQW*) in contrast to a *multiple-quantum-well* (*MQW*), where SQW heterostructure layers are stacked. The SQW and MQW have both gain-guided and index-guided versions. They have a higher gain, and lower current threshold, and the output beam is much more coherent than in conventional double heterostructure architectures.

From the discussion presented so far, the first two goals formulated above—to reduce heat while maintaining a population inversion, and to improve the beam quality and its divergence—have been achieved in the gain-guided and in the index-guided diode lasers, or in their quantum-wells counterparts. The third goal—to provide higher power levels—is also achieved, because the powers typical of these architectures are comparable with many conventional lasers. However, to reach very high powers, the single diode lasers must be packaged into arrays. A one-dimensional array consists of a collection of active stripes arranged in a line and separated by a short distance of about 10 μm. A two-dimensional array is obtained by stacking one-dimensional arrays on top of each other. Such array designs can reach *cw* powers of 100 W at room temperature, and are used to pump solid-state lasers such as Nd:YAG or Nd:YVO_4. The quality of the beam is poor, because the separate diodes are not usually synchronized in phase, but it is sufficient for efficient pumping. To improve phase-synchronization, phase-locked architectures have been invented (*ROW arrays – resonant optical waveguide array*), in which the short distances between stripes are equal to an odd number of half-wavelengths, facilitating optical coupling from the adjacent diodes.

We now return to the classification presented in Fig. 4.31. So far, we have discussed diode lasers that emit from their edges. They are called *edge-emitting diode lasers*. There is, however, another important class of diode lasers which emit from their surfaces (*surface-emitting diode lasers*), shown in Fig. 4.33. They are divided into two basic classes: a) *planar-cavity surface-emitting diode laser* (*PCSEL*), and, b) *vertical-cavity surface-emitting diode laser* (*VCSEL*). One can see from Fig. 4.33 that *PCSEL* is practically an edge-emitting laser diode, where additional angled facets are added to deflect the light vertically from the surface. VCSELs, in contrast to the *edge-emitting diode lasers*, have the resonator cavity in the plane perpendicular

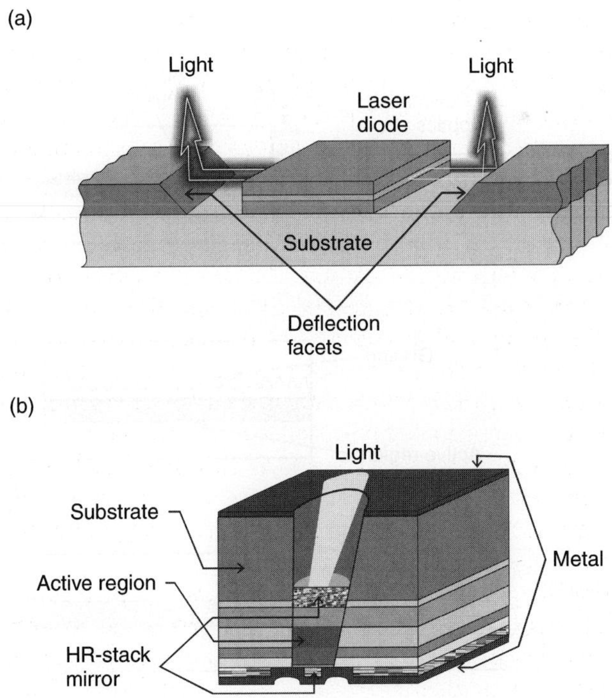

Fig. 4.33 (a) planar-cavity surface-emitting diode laser (PCSEL); (b) vertical-cavity surface-emitting diode laser (VCSEL) [9].

to the junction, with mirrors built of thin-film HR (high-reflector) stacks, above and below the junction layer. The VCSELs' powers are typically of milliwatts, and are generally lower than those of edge emitters, because the beam passes through only a very short gain-region, vertically through the junction layer. However, most of the drawbacks of edge-emitting diode lasers have been reduced in the VCSELs – hopping among the multiple longitudinal modes has been eliminated, and the divergent elliptical beam has been replaced by a round beam of better quality. Therefore, one of the benefits of the VCSEL design is in the control of the output wavelength. The resonator cavity in VCSEL is much shorter than that in edge-emitting diode lasers, reducing significantly the number of longitudinal modes. This eliminates hopping instability, typical for conventional edge-emitting diode lasers, where the resonator length, usually of hundreds of micrometers, is long enough to produce many modes.

Another way of controlling the output wavelength is to use designs that employ diffraction gratings that ensure an effective feedback only for a specific wavelength. The three configurations used most widely to achieve narrow-band light output from laser diodes are: a) external-cavity lasers, b) distributed-feedback (DFB), and c) distributed Bragg-reflection (DBR).

The first semiconductor materials used in diode lasers were arsenic, aluminum and gallium (As, Al, Ga). The first GaAs diode laser emitted at about 900 nm. Binary-

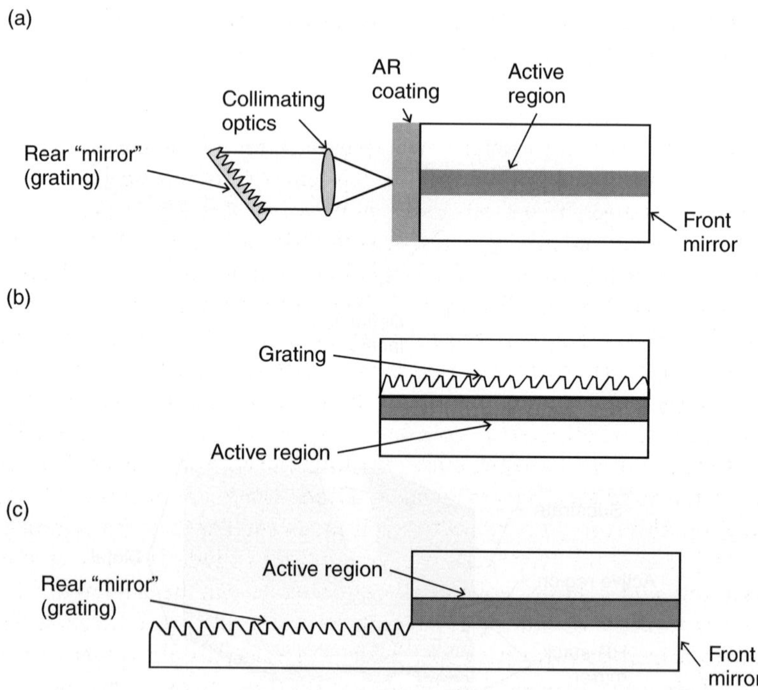

Fig. 4.34 Configurations used for narrow band light output from laser diodes, (a) external cavity; (b) distributed-feedback (DFB); (c) distributed-Bragg-reflection (DBR) [9].

compound semiconductors have a fixed-wavelength output. Ternary and quaternary compounds can cover a wide spectral range depending on the mixing ratio. For example, by changing the aluminum concentration, x, in ($Ga_{1-x}Al_xAs$) from 0 to 0.45 we can change the laser emission from 870 to 620 nm. Semiconductor materials with a larger energy gap, ΔE permit the emission of light at shorter wavelengths. Thus, employing materials from the groups III–V–GaInP/GaInAlP, was a spectacular success. Different indium-gallium-based systems can operate in a broad range from the visible to the IR beyond 4 μm. The family of semiconductors belonging to the groups II–VI (Ze, Se) has a larger gap, ΔE, than have GaAs and GaInP. These materials make it possible to construct lasers emitting in the blue and green regions, but their lives are limited to a few hundred hours. Other candidates in the race to achieve a commercial blue-light laser were semiconductors from the III–V nitride family (Ga, N). The first blue LED and diode laser, based on gallium nitride (GaN) grown on sapphire substrates and emitting at 350 nm, was invented in the 1990s by Nakamura, and became commercially available in 1999. Since then, GaN VCSELvs have been demonstrated in the laboratory, but serious problems with finding means to inject current through the dielectric mirrors has made them commercially unavailable. Adding indium to the active layer of gallium nitride lasers shifts the *cw* output to 420 nm at room temperature, but with a significantly reduced lifetime, far from the million-hour lifetimes common for commercial diodes. The race to shift light emission from blue at around 400 nm, to the more desirable longer wavelengths in the far-blue or green regions is obvious. Although blue light at 400 nm has enormous benefits for optical storage, it may create some problems for the durability of the optics in digital video disks. Moreover, the human eye's sensitivity is better at somewhat longer wavelengths. Unfortunately, the area of mid-blue and green light from diode lasers is still an immature technology.

Recent advances have shown that nitrides are also promising materials, not only in the ultraviolet region, but also in a spectral region far from the blue – in the infrared. The continuous change of the refractive index of $Al_xGa_{1-x}N$ between $n = 2.34$ to $n = 2.03$, by changing the mixing ratio, x, shifts the emission to 1.55 μm and makes group III-nitrides ideal materials for optical communications. They have a better refractive-index match with silica optical fibers than does indium phosphide (InP). Recent studies [10] have shown that III-nitrides are excellent hosts for erbium ions. Table 4.5 and Fig. 4.35 shows the most typical materials used in laser diodes.

We have mentioned that the multiple longitudinal modes generated in edge-emitting diode lasers can lead to unstable hopping among these modes. However, the same effect can actually be beneficial for tunability. In section 4.4.3 we discussed tunability in the solid-state vibronic lasers. Diode lasers are also tunable, although the range of tunability for typical materials is much smaller than the extremely broad range of the Ti:sapphire vibronic laser. Tunability in diode lasers is very useful in communication and remote-sensing applications. It can be achieved by varying either the temperature or the input current. Selection of a specific wavelength can be made with dispersion elements provided by the external-cavity lasers, distributed-feedback (DFB), and distributed-Bragg-reflection (DBR) configurations.

So far, we have discussed only the principles of various heterostructure designs and configurations. Figure 4.36 shows details of the complicated nature of

Table 4.5

Semiconductor materials used in laser diodes [10]

Material	Wavelength [nm]	
GaN	350	
ZnSe	525	
AlGaInP	650–680	
$Ga_{0.5}In_{0.5}P$	670	
$Ga_{1-x}Al_xAs$	620–900	
GaAs	900	
$In_{0.2}Ga_{0.8}As$	980	
$In_{1-x}Ga_xAs_yP_{1-y}$	1100–1650	
$In_{0.73}Ga_{0.27}As_{0.58}P_{0.42}$	1310	
$In_{0.58}Ga_{0.42}As_{0.9}P_{0.1}$	1550	
InGaAsSb	1700–4400	
PbEuSeTe	3300–5800	cryogenic temperature
PbSSe	4200–8000	cryogenic temperature
PbSnTe	6300–29000	

semiconductor layers, including the two-quantum-well active region, that are applied in "the real world" technology.

As for all laser systems, the techniques such as *Q*-switching or modelocking described in Chapter 3 can be applied to semiconductor lasers. The generation of ultrashort optical pulses from semiconductor lasers has been the subject of intense activity since the end of the 1970s. Many observers predict that the true revolution in

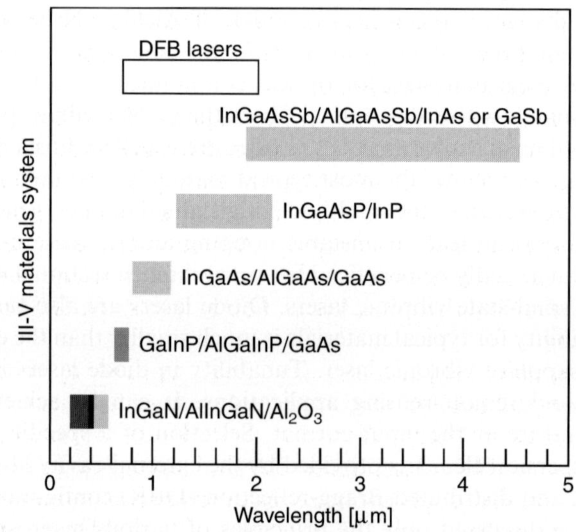

Fig. 4.35 Spectral range for different semiconductor materials [7].

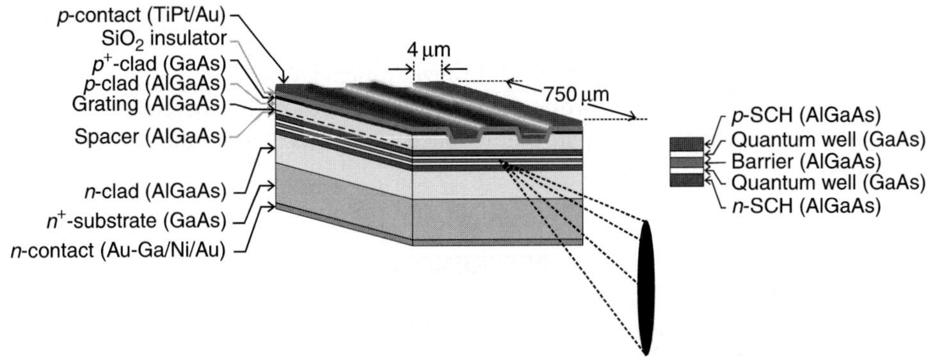

Fig. 4.36 Distributed-feedback ridge-waveguide diode laser. Details of semiconductor layers, including the two-quantum-well active region are shown [7].

optical telecommunication and time-resolved spectroscopy will depend on the pulsed-mode operation of diode lasers. Those who are interested in details of semiconductor lasers in pulsed-mode operation are referred to ref. [11].

Advanced diode-laser designs were developed in the 1980s and 1990s to produce stable, less divergent, reduced bandwidth, and blue and red extended laser diodes. Diode lasers became powerful devices in the development of the "information superhighway". It is nearly impossible to make a long-distance telephone call without them. DFB and DBR diode lasers from the 1.3–1.55 μm range are used commercially in fiber-optic communication. Strained-layer architectures with InGaAs as an ultra-thin active layer, and GaAlAs as confinement layers, all grown on a GaAs substrate, are used for pumping erbium-doped optical amplifiers at 980 nm. The efficient management of optical networks through wavelength-division multiplexed (WDM) fiber optic systems requires broadly tunable diode lasers.

Diode lasers are used by practically everybody. They are used in every CD-player, every CD-ROM, and every laser printer. The GaAlAs diodes, which emit light at about 780 nm, are used in CD-players.

Laser-diode arrays can produce powers as high as 100 W and more. Therefore, these devices compete with the solid-state lasers that were traditionally used in most industrial and medical applications. They are ideally suited for applications that require efficient high power-delivery, such as the pumping of solid-state lasers, providing an alternative to flash-pumping. Although the poor beam quality has excluded them so far from traditional high-power applications, such as cutting or deep-penetration welding, these barriers will be overcome by further improvements in their architecture.

Many visible gas-laser applications have been replaced by diode lasers. Their reduced prices, compact designs, and comparable powers have caused diode lasers to relegate gas helium-neon lasers. Shops use diode lasers in hand-held scanners. In medical diagnostics the HeNe laser line at 630 nm has been replaced with the 670 nm line of the diode laser. Visible gas lasers are regarded as much more suitable than diode lasers for applications that require high beam quality and long coherence distance, such as holography or confocal laser-scanning microscopes, but improvements in diode-laser technology will enable them to replace traditional laser types for more demanding tasks.

The blue diode-lasers will truly revolutionize optical storage. The optical storage-capacity increases in inverse proportion to the square of the wavelength. So the blue diode system may have a storage capacity of 15 GB—20 times more than a CD-ROM.

REFERENCES

1. W. Koechner, *Solid-State Laser Engineering*, 5th Edition, vol. 1, Springer-Verlag (1999)
2. E.J. Lerner, *Ultrafast lasers deliver powerful, precise pulses*, Laser Focus World 34(12) (December, 1998) 77–84
3. N.A. Papadogiannis, B. Witzel, C. Kalpouzos, D. Charalambidis, *Observation of Attosecond Light Localization in Higher Order Harmonic Generation*, Phys. Rev. Lett. 83(21) (1999) 4289–4292
4. M. Hentschel, R. Kienberger, Ch. Spielmann, G.A. Reider, N. Milosevic, T. Brabec, P. Corkum, U. Heinzmann, M. Drescher, F. Krausz, *Attosecond metrology*, Nature 414 (2001) 509–513
5. M. Drescher, M. Hentschel, R. Kienberger, M. Uiberacker, V. Yakovlev, A. Scrinzi, Th. Westerwalbesloh, U. Kleineberg, U. Heinzmann, F. Krausz, *Time-resolved atomic inner-shell spectroscopy*, Nature 419 (2002) 803–807
6. A. Baltuska, Th. Udem, M. Uiberacker, M. Hentschel, E. Goulielmakis, Ch. Gohle, R. Holzwarth, V.S. Yakovlev, A. Scrinzi, T.W. Hänsch, F. Krausz, *Attosecond Control of Electronic Processes by Intense Light Fields*, Nature 421 (2003) 611–615
7. Eric J. Lerner, *Tunable lasers cover the spectrum*, Laser Focus World 34(10) (October, 1998) 143–152
8. J. Higgins, *Excimer lasers machine in the UV*, Laser Focus World 37(6) (June, 2001) 183–188
9. Th.V. Higgins, *The smaller, cheaper, faster world of the laser diode*, Laser Focus World 31(4) (April, 1995) 65–76
10. R.G. Wilson, R.N. Schwartz, C.R. Abernathy, S.J. Pearton, N. Newman, M. Rubin, T. Fu, J.M. Zavada, *1.54*-μm photoluminescence from Er-implanted GaN and AlN, Appl. Phys. Lett. 65(8) (1994) 992–994
11. *Femtosecond Laser Pulses. Principles and Experiments*, C. Rulliere ed., Springer-Verlag, Berlin Heidelberg New York (1998)
12. J. Hecht, *Understanding Lasers*, 2nd ed., IEEE Press, New York (1994)
13. J. Hecht, *The Laser Guidebook*, 2nd ed., Tab books, Blue Ridge Summit, PA (1992)
14. D.C. O'Shea, W.R. Callen, W.T. Rhodes, *Introduction to Lasers and Their Applications*, Addison-Wesley Publishing Co., Reading, MA (1997)
15. *Handbook of Molecular Lasers*, P.K. Cheo, ed., Marcel Dekker Inc., New York, NY (1987)
16. W.J. Witteman, *The CO_2 Laser*, Springer Verlag, Berlin (1987)
17. Stephen J. Matthews, *Still going strong*, Laser Focus World 37(5) (May, 2001) 185–190
18. R.W. Duffner, *Airborne Laser: Bullets of Light* (Plenum, New York, 1997); Jeff Hecht, *Beam Weapons: The Next Arms Race* (Iuniverse com., reprint of 1984 Plenum book)
19. G. Allcock, P.E. Dyer, G. Elliner, H.V. Snelling, *Experimental observations and analysis of CO_2 laser-induced microcracking of glass*, J. Appl. Phys. 78(12) (1995) 7295–7303
20. A.C. Tam, J. Brannon, P. Baumgart, I.K. Pour, *Laser texturing of glass disk substrates*, IEEE Trans. Mag. 33(5) (1997) 3181–3183
21. D. Kuo, S.D. Vierk, O. Rauch, D. Polensky, *Laser zone texturing on glass and glass-ceramic substrates*, IEEE Trans. Mag. 33(1) (1997) 944–949

22. Paul Mortensen, *Remote Sensing*, Laser Focus World 32(7) (July 1996) 22–23
23. K. Bondelie, *Sealed carbon dioxide lasers achieve new power levels*, Laser Focus World 32 (8) (August 1996) 95–100
24. M.R. Emmert-Buck, R.F. Chuaqui, Z. Zhuang, R.A. Weiss, L.A. Liotta, *Laser Capture Microdissection*, Science 274 (1996) 998–1001
25. P. Repond, M.W. Sigrist, Appl. Opt. 35 (July 1996) 4065
26. V.T. Volov, SPIE 2907 (Feb. 1996) 256
27. Eric J. Lerner, *Carbon dioxide lasers deliver flexibility and power*, Laser Focus World 34 (9) (September, 1998) 131–138
28. V.V. Apollonov, S. Yu Kazantsev, V.F. Oreshkin, K.N. Firsov, Elektron (Moscow) 24 (1997) 213; *Feasibility of increasing the output energy of a nonchain HF (DF) laser*, Quantum Electron. 27 (3) (1997) 207–209
29. M. Endo, SPIE 3268 (1998) 106
30. J. Vetrovec, SPIE 3092 (1996) 723
31. J. Hecht, *The Airborne Laser shoots for ballistic missile defense*, Laser Focus World. 40 (1) (January, 2004) 105–109
32. Do-Kyeong Ko, Gwon Lim, Sung-Ho Kim, Byung Heon Cha, Jongmin Lee, *Self-seeding in a dual-cavity-type pulsed Ti:sapphire laser oscillator*, Opt. Lett. 20 (7) (1995) 710
33. R. Battis, J. Kuper, C. Anderson, A. Beger, *Solid-State Lasers: Alexandrite lasers make their mark in industrial applications*, Laser Focus World 36 (5) (May, 2000) 257–262
34. F.J. Duarte, *Tunable Lasers*, Academic Press, New York (1995)
35. Wentao Hu, Hui Ye, Chuangdong Li, Zhonghong Jiang, Fuzheng Zhou, *All-solid-state tunable DCM dye laser pumped by a diode-pumped Nd:YAG laser*, Appl. Opt. 36 (3) (1997) 579–583
36. D. Kopf, A. Prasad, G. Zhang, M. Moser, U. Keller, *Broadly Tunable Femtosecond Cr:LiSAF Laser*, Opt. Lett. 22 (9) (1997) 621–623
37. Ramon U. Martinelli, Raymond J. Menna, Pamela K. York, Dmitri Z. Garbuzov, Hao Lee, Joseph H. Abeles, Nancy A. Morris, John C. Connolly, S. Yegna Narayan, Jacobus S. Vermaak, Gregory H. Olsen, David E. Cooper, Clinton B. Carlisle, Haris Riris, Anthony L. Cook, *Tunable single-frequency III-V semiconductor diode lasers with wavelengths from 0.76 to 2.7* μm, SPIE 2834 (1996) 2–16
38. Donna Z. Berns, *Air-cooled ion lasers see strong biomedical demand*, Laser Focus World 35 (11) (November, 1999) 147–152
39. Lisa Tsufura, Orest Baransky, *Helium-neon lasers have a multicolored future*, Laser Focus World 31 (3) (March, 1995) 83–87
40. P.F. Moulton, *Spectroscopic and laser characteristics of* $Ti{:}Al_2O_3$, J. Opt. Soc. Am. B 3 (1) (1986) 125–132
41. Eric J. Lerner, *Ion-laser enhancements find novel applications*, Laser Focus World 34 (6) (June, 1998) 103–105
42. J. Hecht, *Rare earths create useful long-wavelength lasers*, Laser Focus World 29 (11) (November, 1993) 135–142
43. Th.V. Higgins, *Creating Laser Light*, Laser Focus World 30 (6) (June, 1994) 127–133
44. W. Demtröder, *Laser spectroscopy*, Basic Concepts and Instrumentation, 3rd ed., Springer-Verlag, Berlin (2003)
45. J. Wilson, J.F.B. Hawkes, Optoelectronics; An introduction, 2nd ed., Prentice Hall, New York, NY (1989)
46. *Quantum Well Lasers*, P.S. Zory Jr. ed., Academic Press Inc., San Diego, CA (1993)
47. *Surface Emitting Semiconductor Lasers and Arrays*, G.A. Evans, J.M. Hammer eds., Academic Press Inc., San Diego, CA (1993)
48. B.A. Saleh, M.C. Teich, *Fundamentals of Photonics*, John Wiley & Sons, New York, NY (1991)
49. *Handbook of Solid–State Lasers*, Peter K. Cheo ed., Marcel Dekker Inc., New York, NY (1989)
50. *The Blue Laser Diode : The Complete Story*, Shuji Nakamura, Stephen Pearton, Gerhard Fasol, 2nd updated and extended ed., Springer, Berlin New York (2000)

51. R. Hui, S. Taherion, Y. Wan, J. Li, S.X. Jin, J.Y. Lin, H.X. Jiang, *GaN-based waveguide devices for long-wavelength optical communications*, Appl. Phys. Lett. 82 (9) (2003) 1326–1328
52. Claire Gmachl, Hock M. Ng, S.-N. George Chu, Alfred Y. Cho, *Intersubband absorption at* $\lambda \sim 1.55\,\mu$m in well- and modulation-doped GaN/AlGaN multiple quantum wells with superlattice barriers, Appl. Phys. Lett. 77 (23) (2000) 3722–3724
53. Norio Iizuka, Kei Kaneko, Nobuo Suzuki, Takashi Asano, Susumu Noda, Osamu Wada, *Ultrafast intersubband relaxation (≤150 fs) in AlGaN/GaN multiple quantum wells*, Appl. Phys. Lett. 77 (5) (2000) 648–650
54. Katsumi Kishino, Akihiko Kikuchi, Hidekazu Kanazawa, Tetsuo Tachibana, *Intersubband transition in* $(GaN)_m/(AlN)_n$ *superlattices in the wavelength range from 1.08 to 1.61* μm, Appl. Phys. Lett. 81(7) (2002) 1234–1236
55. J.E. Colon, D.W. Elsaesser, Y.K. Yeo, R.L. Hengehold, and G.S. Pomrenke, *Luminescence Study of the Intra-4f Emissions from GaAs:(Er + O) and AlGaAs:(Er + O)*, Mat. Res. Soc. Symp. Proc. 301 (1993) 169
56. T. Inushima, V.V. Mamutin, V.A. Vekshin, S.V. Ivanov, T. Sakon, M. Motokawa, S. Ohoya, *Physical properties of InN with the band gap energy of 1.1 eV*, J. Cryst. Growth 227–228 (2001) 481–485
57. J. Wu, W. Walukiewicz, K.M. Yu, J.W. Ager III, E.E. Haller, Hai Lu, William J. Schaff, Yoshiki Saito, Yasushi Nanishi, *Unusual properties of the fundamental band gap of InN*, Appl. Phys. Lett. 80 (21) (2002) 3967–3969

– 5 –

Nonlinear Optics

We usually experience intensities of light that are weak or moderate with a conventional irradiance of several W/m^2. Light having such intensities interacts with matter by reflection, refraction, scattering or absorption, with little or no modification of the frequency. Moreover, when beams of different frequencies meet in a material, they do not interact with one another. The material responds proportionately—the higher the intensity of the light, the higher is the absorption or other linear response.

This situation is altered dramatically way when the intensity of light exceeds a certain threshold. A wide variety of phenomena can occur when intense light interacts with matter, resulting in the creation of new beams of different wavelengths. In this way we enter the exciting world of nonlinear optics, a term which describes the interaction of intense light with matter. Most nonlinear effects require light intensities that can be produced by high-power lasers.

Laser spectroscopy, photochemistry, and femtosecond laser chemistry could not have developed so dynamically if they had not employed the phenomena of nonlinear optics, and these phenomena have extended considerably the spectral range of laser sources. Nonlinear interactions in crystals permit the generation of the second- and third-harmonic, as well as the higher harmonics. This indicates, for example, that when intense radiation of frequency ω is focused onto a crystal, two components are generated: radiation at the incident frequencies ω and a second beam having exactly twice that frequency, 2ω. The component having frequency 2ω is called the *second harmonic* (Fig. 5.1).

Besides the generation of the second harmonic, the sum $(\omega_1 + \omega_2)$, and the difference $(\omega_1 - \omega_2)$ frequencies, mixing may occur, as well as many other nonlinear effects such as the third harmonic at 3ω, four-wave mixing, the optical Kerr effect, the Pockels electro-optical effect, stimulated Raman scattering, stimulated Brillouin scattering, phase conjugation, self focusing, self phase modulation, two-photon absorption, two-photon ionization, and two-photon emission. Some of these effects (Raman, Pockels, Kerr effects) were discovered before the laser's invention.

However, the usefulness of nonlinear optics is not limited to providing new light-sources or to extending the spectral range. Some spectroscopic methods which provide valuable information about the dynamics of molecular systems would not exist if nonlinear phenomena have not occurred. We only mention here stimulated

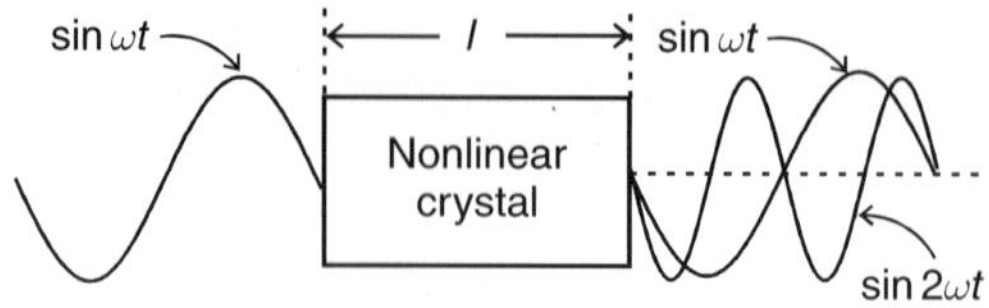

Fig. 5.1 Generation of the second harmonic in a nonlinear crystal.

Raman scattering, the pump-probe methods, or photon echoes. These phenomena will be discussed in Chapter 8.

We shall now describe the basic principles of nonlinear optics. The light emitted from conventional sources, such as a bulb or a flash lamp, is neither monochromatic nor coherent in time and space. The intensity of the electric field emitted from the conventional sources of light is small (10–10^3 V/cm). When radiation of such intensity interacts with matter by reflection, scattering, absorption, or refraction, it does not change significantly its macroscopic and microscopic properties, because it is several orders of magnitude smaller than the intensity of the electric field in matter (about 10^9 V/cm). The laser beam's intensity, especially from the pulsed lasers generating short pulses, can achieve values higher than 10^{12} W/cm^2, corresponding to an electric field of the order of 10^5–10^8 V/cm, which is comparable to the intensity of the electric field in the matter.

A weak electric field, $\boldsymbol{E}$, induces in a material an electric induction, $\boldsymbol{D}$, where

$$\boldsymbol{D} = \boldsymbol{E} + 4\pi\boldsymbol{P}, \tag{5.1}$$

which depends on the medium's *polarization* $\boldsymbol{P}$. For a small intensity of light the polarization induced in the material depends linearly on the electric field intensity, $\boldsymbol{E}$

$$\boldsymbol{P} = \chi^{(1)}\boldsymbol{E}, \tag{5.2}$$

where $\chi^{(1)}$ is called the *electric susceptibility* of the first order. This term represents the linear response used in conventional linear optics. The above expression arises from the fact that the medium polarization, P, is a sum of the orientational polarization, P^{orient} ($P^{orient} = N\frac{\mu^2}{3kT}E$), and the induced polarization, P^{ind} ($P^{ind} = N\alpha E$). When light interacts with a dielectric, the electrons are displaced from their equilibrium positions and begin to oscillate. These oscillations give rise to the induced polarization. Light having small intensity causes only small displacements—the electrons behave like harmonic oscillators and the response is linear, as described by eq. (5.2). When the light entering a dielectric is intense, the displacements become large, and the electrons which are driven far from the equilibrium position no longer oscillate in a simple harmonic way.

When the electric field $\boldsymbol{E}$ of the incident light increases, the polarization of the medium is no longer linearly dependent on $\boldsymbol{E}$ and can be expanded in a power series

$$\boldsymbol{P} = \chi^{(1)}\boldsymbol{E} + \chi^{(2)}\boldsymbol{E}^2 + \chi^{(3)}\boldsymbol{E}^3 + \ldots \tag{5.3}$$

where each term depends on the susceptibility of the n-th order. The susceptibility is a complex quantity, and contains all the information about the optical properties of the dielectric material—its real part, about the index of refraction (refractive index), the imaginary part, about the absorption coefficient.

The susceptibility is a tensor because the electric field applied in the x direction may result in an electron moving in the y or z directions as well. It results from the fact that electrons in the material oscillating in the direction of the light wave experience not only a restoring force, but also forces from the neighboring molecules. Thus, eq. (5.3) should be written in the following form

$$P_i = \chi_{ij}^{(1)} E_j + \chi_{ijk}^{(2)} E_j E_k + \chi_{ijkl}^{(3)} E_j E_k E_l + \dots \tag{5.4}$$

where i, j, k denote the x, y, z Cartesian components. For the repeated indices of the susceptibility χ and the field intensity E summation must be performed over the components of the electric field.

If several weak electric fields act on a medium, in linear approximation the polarization is the sum of polarizations from the individual fields

$$\boldsymbol{P}(\boldsymbol{r}, t) = \boldsymbol{P}_1(\boldsymbol{r}, t) + \boldsymbol{P}_2(\boldsymbol{r}, t) + \boldsymbol{P}_3(\boldsymbol{r}, t) + \dots \tag{5.5}$$

This indicates that the weak electric fields obey the superposition principle according to which electromagnetic waves spread in a medium independently, without interacting with one another. All optical phenomena undergoing this principle are called linear, and belong to the field of linear optics. If electric fields are strong, this assumption is no longer valid. The intense electric fields violate the superposition principle because they interact with each other, and all optical phenomena for which the superposition principle of electric fields is violated are called *nonlinear optic phenomena.* One example of a nonlinear phenomenon is the dependence of the index of refraction n on the light intensity. For small intensities, the index of refraction has a constant value. However, if a liquid medium is illuminated with light of large intensity, the index of refraction depends directly on the electric field of the light E^2, and the liquid becomes birefringent, like a uniaxial crystal. This effect is known as the *Kerr effect*. Another example is an absorption coefficient, which in the linear regime is expressed in the form of the *Lambert-Beer formula*:

$$T = I/I_0 = \exp(-\alpha c l) \tag{5.6}$$

or

$$A = \ln\frac{1}{T} = \alpha c l, \tag{5.7}$$

where T is the transmission expressed as the ratio of the transmitted beam's intensity, I, to the intensity of the incident beam, I_0; α is the absorption coefficient, c is the concentration of the absorbing medium, and l is length of the optical path. In the regime of linear optics the absorption coefficient does not depend on the light's intensity. For larger intensities this approximation is no longer valid.

5.1. SECOND ORDER NONLINEAR PHENOMENA

The nonlinear phenomena of the second order are described by the second term of the equation (5.4)

$$P_i = \chi^{(2)}_{ijk} E_j E_k. \tag{5.8}$$

This term is responsible for the *second harmonic generation* (*SHG*) and *the sum and difference frequency mixing*, as well as *parametric conversion*. The second harmonic generation was demonstrated for the first time in 1961, when light from a ruby laser illuminating quartz produced coherent ultraviolet light. The light of the ruby laser at a wavelength of 694.3 nm was directed onto a nonlinear crystal. By changing the angle between the direction of the laser beam and the direction of the crystal optic axis it was noticed that, at certain angles, the output beam from the crystal has two components: 694.3 nm and 347.15 nm (= 694.3 nm/2) . Therefore, besides the fundamental component at 694.3 nm, which does not change its frequency after passing through the crystal, an additional component appears with twice the fundamental frequency (Fig. 5.2), which is called the second harmonic.

A phenomenological explanation is as follows. Let the laser beam be a plane monochromatic wave at frequency ω_1 and with a wave vector $\boldsymbol{k}_1$

$$\begin{aligned} \boldsymbol{E}(\boldsymbol{r}, t) &= \boldsymbol{E}(\omega_1, \boldsymbol{k}_1) e^{i(\boldsymbol{k}_1 \boldsymbol{r} - \omega_1 t)} + \boldsymbol{E}^*(\omega_1, \boldsymbol{k}_1) e^{-i(\boldsymbol{k}_1 \boldsymbol{r} - \omega_1 t)} \\ &= 2\boldsymbol{E}(\omega_1, \boldsymbol{k}_1) \cos(\boldsymbol{k}_1 \boldsymbol{r} - \omega_1 t) = \boldsymbol{E}_0 \cos(\boldsymbol{k}_1 \boldsymbol{r} - \omega_1 t). \end{aligned} \tag{5.9}$$

The electric field $\boldsymbol{E}$ induces second-order polarization in a nonlinear crystal

$$P_i^{(2)}(\boldsymbol{r}, t) = \chi^{(2)}_{ijk} E_j(\boldsymbol{r}, t) E_k(\boldsymbol{r}, t) = \chi^{(2)}_{ijk} E_0^2 \cos^2(\boldsymbol{k}_1 \boldsymbol{r} - \omega_1 t). \tag{5.10}$$

Using the trigonometric relationship, $\cos^2 x = \frac{1+\cos 2x}{2}$, the expression (5.10) may be recast into the following form

$$P_i^{(2)}(\boldsymbol{r}, t) = \chi^{(2)}_{ijk}(0) \frac{E_0^2}{2} + \frac{1}{2} \chi^{(2)}_{ijk}(2\omega_1, 2\boldsymbol{k}_1) E_0^2 \cos 2(\boldsymbol{k}_1 \boldsymbol{r} - \omega_1 t). \tag{5.11}$$

Thus, the polarization induced by the interaction of two waves, each of frequency ω_1, consists of two terms; a constant time-independent polarization $\frac{1}{2}\chi^{(2)}_{ijk}(0)E_0^2$, and the polarization modulated at the frequency $2\omega_1$. The constant polarization (called the

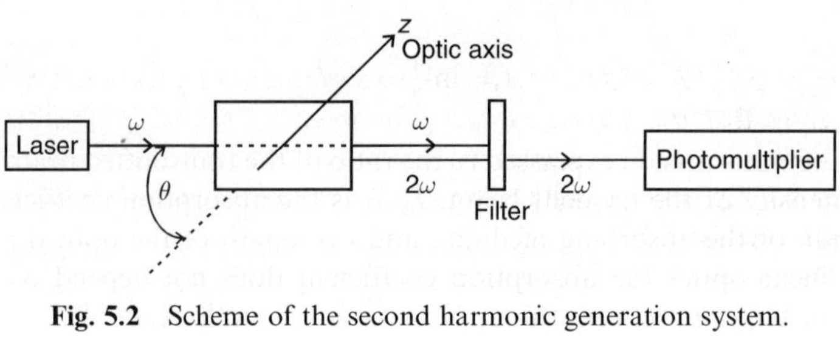

Fig. 5.2 Scheme of the second harmonic generation system.

DC effect) has been employed for laser-beam power measurements. The oscillations of the polarization at $2\omega_1$ generate in the material a new SHG wave at the frequency $2\omega_1$, coming from individual particles in different positions, $\boldsymbol{r}$, in the crystal. These microscopic contributions can interfere positively or destructively, depending on the relative phases. The nonlinear polarization wave travels through the material with a phase velocity that matches the primary refracted wave. This is because the refracted wave induces polarization of the material. However, the SHG wave that is generated by the nonlinear polarization travels more slowly, owing to dispersion. Consequently, the polarization wave and the SHG wave pass in and out of phase with each other as they propagate through the material. The macroscopic SHG wave produces a large intensity only when the phase-velocities are synchronized. In the opposite case, destructive interference takes place, reducing considerably the yield of the SHG process. To maximize the efficiency of the SHG by energy transfer between the nonlinear polarization and the SHG wave, the phase velocities of the two waves must match. This condition is called *phase matching* and plays a crucial role in nonlinear optics.

The phase velocity υ_{ph} is a velocity of propagation of a plane wave through a medium characterized by the index of refraction $n(\omega)$

$$\upsilon_{ph} = \frac{\omega}{k} = \frac{c}{n(\omega)}. \tag{5.12}$$

We need to ask why phase matching is so important in the second harmonic generation? The incident wave characterized by the frequency ω_1 and the wave vector $\boldsymbol{k}_1$ generate a nonlinear polarization wave at the frequency $2\omega_1$, and the wave vector $2\boldsymbol{k}_1$. This polarization induces the SHG wave at the frequency $\omega_2 = 2\omega_1$, and the wave vector $\boldsymbol{k}_2$, which is generally not equal to $2\boldsymbol{k}_1$. It indicates that the phase of the polarization wave ($2\boldsymbol{k}_1$) is generally different from the phase of the generated SHG wave ($\boldsymbol{k}_2$). The phase-matching condition is fulfilled only when

$$\boldsymbol{k}_2(2\omega_1) = \boldsymbol{k}_1(\omega_1) + \boldsymbol{k}_1(\omega_1). \tag{5.13}$$

The phase-matching condition can be written more generally as

$$\boldsymbol{k}_3(\omega_3) = \boldsymbol{k}_1(\omega_1) + \boldsymbol{k}_2(\omega_2) \tag{5.14}$$

or

$$\Delta \boldsymbol{k} = \boldsymbol{k}_3(\omega_3) - \boldsymbol{k}_1(\omega_1) - \boldsymbol{k}_2(\omega_2) = 0. \tag{5.15}$$

Since the wave vector $\boldsymbol{k}$ is related to the photon momentum $\boldsymbol{p}(\boldsymbol{p} = \hbar\boldsymbol{k})$, this condition simply indicates that the energy exchange between the waves is possible only when the law of conservation of momentum for three photons participating in the process of frequency mixing is not violated. When all the beams are collinear, the condition (5.14) may be replaced by

$$n_3\omega_3 = n_1\omega_1 + n_2\omega_2, \tag{5.16}$$

where we employed the relationship (5.12). For the second harmonic, with $\omega_1 = \omega_2$ and $\omega_3 = 2\omega_1$, eq. (5.16) takes the form

$$n(2\omega_1) = n(\omega_1). \tag{5.17}$$

This indicates that the second harmonic may be generated efficiently only when the index of refraction at $2\omega_1$ is equal to the index of refraction at ω_1. In most cases, this is impossible, owing to dispersion in the material, which simply means that the two waves differing in wavelength, have different indices of refraction as well. However, we can use some tricks related to the properties of birefringent crystals.

In the following section 5.2, we will show that one of the most convenient ways to achieve phase matching employs birefringent crystals. Birefringent crystals have two indices of refraction: n^o – the index of refraction of the ordinary beam, and n^e – the index of refraction of the extraordinary beam, which depends on the direction and polarization of the light. We will show that, for a properly chosen angle between the incident beam and the optical axis, the indices of refraction (and the phase velocities) of the induced polarization wave and the second harmonic wave can be made equal

$$n^e(2\omega_1) = n^o(\omega_1), \tag{5.18}$$

which ensures converting the fundamental frequency to its second harmonic with great efficiency.

The phase-matching condition ($\Delta\mathbf{k} = 0$) (eq. 5.15) can be derived in a more formal way by adding up the microscopic polarization contributions described as plane waves with different phase-velocities from different positions of the medium. One can show that, in the ideal case of a plane wave, the second harmonic intensity is expressed in the form

$$I(2\omega) = K(\chi^{(2)}_{eff})^2 l^2 \left(\frac{\sin(\Delta kl/2)}{\Delta kl/2}\right)^2 I_0^2(\omega), \tag{5.19}$$

where: l is the length of a crystal, $I_0(\omega)$ is the intensity of incident radiation at frequency ω, $\chi^{(2)}_{eff}$ is the effective susceptibility of the second order, and K is a constant for a given wavelength and a given material. The expression (5.19) is derived in many textbooks on nonlinear optics [1–2].

One can see that the intensity of the second harmonic is described by the well-known function $(\frac{\sin x}{x})^2$, where $x = \Delta kl/2$ (presented in Fig. 3.4). This function has a maximum at $x = 0$, that is at $\Delta k = 0$, when the phase-matching condition occurs.

The relationship (5.19) shows the possible methods of increasing the second harmonic's intensity. This can be achieved by employing:

a) laser pulses of high incident intensity I_0; the magnitudes of I_0 are limited by a crystal-damage threshold,
b) nonlinear materials of high $\chi^{(2)}$ susceptibility,
c) the phase-matching condition $\Delta\mathbf{k} = 0$,
d) a long optical path, l.

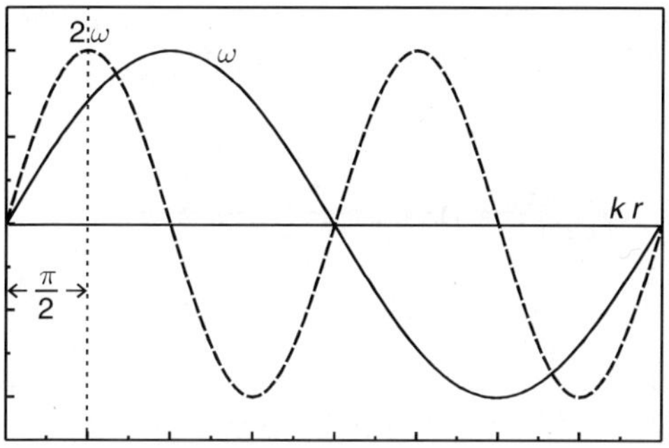

Fig. 5.3 Scheme illustrating the crystal coherence length.

However, the material thickness l is limited by a certain critical value l_{coh}, called the *coherence length*. One can see from Fig. 5.3 that the fundamental wave at frequency ω and the SHG wave traveling along the crystal are increasingly out of phase, leading to destructive interference and reducing the second harmonic's intensity. The distance from the entrance-surface of the crystal to the point where the second harmonic's intensity has its first maximum has been termed the coherence length, l_{coh}. If the crystal length happens to be an odd multiple of the coherence length, no second-harmonic light will be emitted.

The coherence length can be estimated roughly from its definition

$$\Delta k \cdot l_{coh} = \frac{\pi}{2}, \tag{5.20}$$

and

$$l_{coh} = \frac{\pi}{2\Delta k} = \frac{\lambda}{4(n(2\omega) - n(\omega))}. \tag{5.21}$$

However, the plane-wave approximation in eq. (5.20) does not take into account the divergence of the ordinary and extraordinary beams (Fig. 5.4).

The second harmonic generation (SHG) is a special case of the *three-wave interaction*. Two incident waves of frequencies ω_1 and ω_2 interact in a nonlinear crystal and generate a wave at the sum, $\omega_3 = \omega_1 + \omega_2$, or the difference, $\omega_3 = \omega_1 - \omega_2$, frequency. The *sum frequency generation* (*SFG*), $\omega_3 = \omega_1 + \omega_2$, is used for conversion into a shorter-wavelength spectral region, in contrast to the *difference frequency generation*, $\omega_3 = \omega_1 - \omega_2$ (DFG). The second-harmonic generation is a special case of SFG when $\omega_1 = \omega_2$.

5.2. PHASE MATCHING METHODS

It has been shown in the previous section 5.1 that, for the second harmonic generation, the phase-matching condition $\Delta \boldsymbol{k} = 0$ is equivalent to the requirement of $n(\omega) = n(2\omega)$. However, in most cases it is impossible to achieve this condition, owing to dispersion of a material, which simply means that the index of refraction, $n(\omega)$

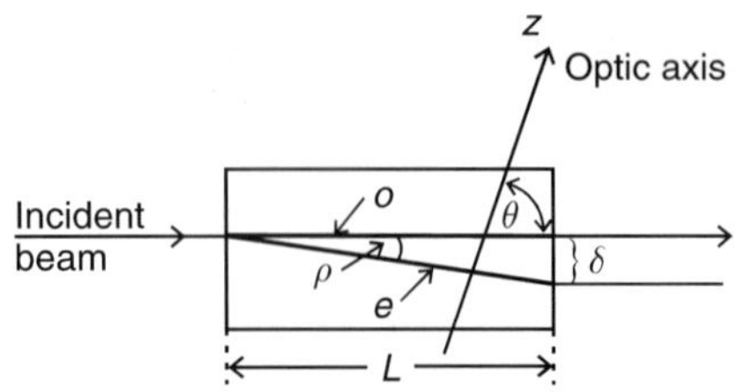

Fig. 5.4 Birefringent uniaxial crystal, o – ordinary beam, e – extraordinary beam, L – crystal length, ρ – birefringence or "walk-off" angle, δ – walk-off distance.

depends on the radiation's frequency. There are a few possibilities for overcoming the dispersion. The common method is to make use of the orientation and polarization dependence of the index of refraction in birefringent crystals.

Birefringence can be observed when light passes through crystals of many materials, including calcite, ice, quartz, mica, or sugar, which are anisotropic and have optical axes. In such crystals the light beam undergoing refraction divides into two rays (Fig. 5.4), in contrast to the isotropic media (liquids, gases, glasses) which show only one ray. The two rays are called the *ordinary* (o) and *extraordinary* (e) rays and have different phase-velocities and different indices of refraction. This phenomenon is called *double refraction* or *birefringence* and was discovered by Bartholinus and Huygens. They found that both rays are linearly polarized, in planes perpendicular to each other.

The ordinary ray obeys Snell's law

$$\frac{\sin\alpha}{\sin\beta} = \frac{n_1}{n_2}, \tag{5.22}$$

where α is the angle of incidence on the crystal surface, β the angle of refraction, n_1 and n_2 are the indexes of refraction of the medium from which the light arrives into the crystal, and of the crystal itself. The extraordinary ray does not obey eq. (5.22). When we rotate the crystal around the incident beam's axis, we notice that the ordinary ray does not change direction, while the extraordinary beam rotates around it (Fig. 5.4). This means that the light of the extraordinary ray has different velocities in different directions of propagation, which depend on the orientation of its polarization with respect to the crystal's optic axis. In contrast, the ordinary beam, polarized in the direction perpendicular to the optical axis, travels with the same velocity, and experiences the same index of refraction regardless of its direction of propagation. After passing through the crystal, the ordinary and extraordinary beams do not propagate along the same line, and the beam is divided onto two orthogonally polarized beams separated by δ, which is called the *walk-off distance* (Fig. 5.4).

One can distinguish *uniaxial* and *biaxial crystals*, and these are further divided into *positive* and *negative crystals*. For the ordinary ray, the surface of the index of refraction n^o is a sphere, because light propagates with the same phase-velocity in each direction. In contrast, the extraordinary ray has preferred directions of propagation. The surface of the index of refraction $n^e(\theta)$ forms an ellipsoid of revolution, with the optical axis being the axis of rotation. If we draw xz- or yz-planes across the

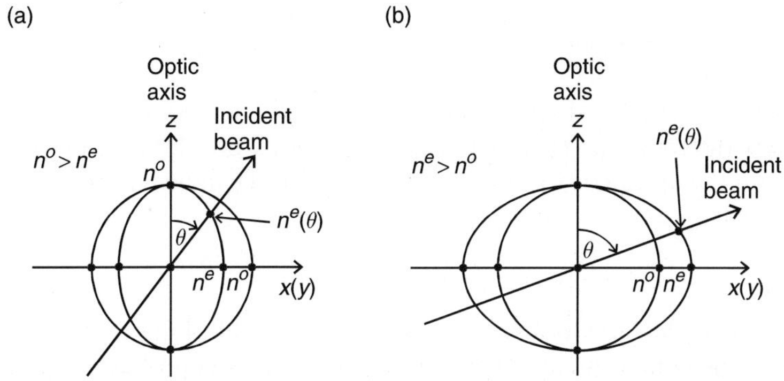

Fig. 5.5 Cross-sections of the index of refraction's surface in the *xz*- or *yz*-planes of a uniaxial birefringent crystal; (a) negative, (b) positive.

index of refraction surfaces, in a uniaxial birefringent crystal we obtain the following cross-sections (Fig. 5.5). One can see that for the ordinary ray the index of refraction, n^o, is independent of the direction of propagation (θ), in contrast to the extraordinary index of refraction, $n^e(\theta)$. The index $n^e(\theta)$ changes between a maximum value equal to the ordinary index, n^o (when the direction of the beam propagation is parallel to the optic axis z) and the minimum value, n^e (when the direction of beam propagation is perpendicular to the optic axis, z). When the beam propagation is at an angle θ with respect to the optic axis the index of refraction $n^e(\theta)$ is given by eq. 5.23 [3]

$$n^e(\theta) = \frac{n^o n^e}{[(n^o)^2 \sin^2\theta + (n^e)^2 \cos^2\theta]^{1/2}}. \tag{5.23}$$

This indicates that the extraordinary ray propagates fastest along the direction perpendicular to the optic axis for the negative crystals (Fig. 5.5a), and along the optic axis for the positive crystals (Fig. 5.5b).

If one draws the cross-sections of the indices of refraction for the fundamental beam and the second harmonic in the negative uniaxial crystal, they will look as shown in Fig. 5.6, where n_1^o, $n_1^e(\theta)$ and n_2^o, $n_2^e(\theta)$ denote the indices of refraction for

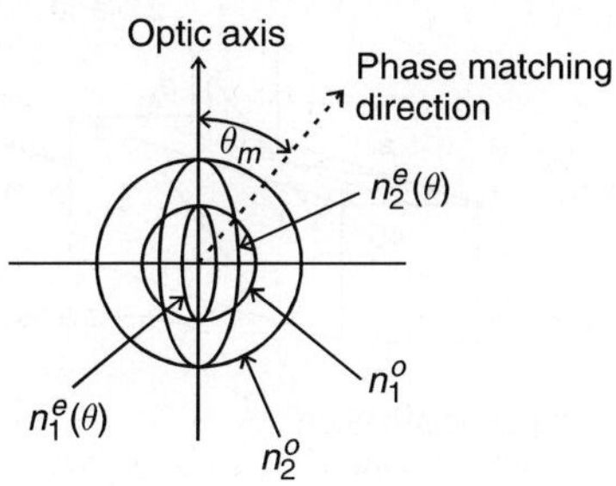

Fig. 5.6 Phase-matching direction in a negative unixial crystal.

the ordinary and the extraordinary rays of the fundamental beam and the second-harmonic beam, respectively.

One may notice that there are certain directions of propagation in the crystal for which the fundamental and second-harmonic beams have the same refractive index, despite their different wavelengths. These directions go along a line forming the angle θ_m with the optic axis for which the surfaces of the indices of refraction, n_1^o and $n_2^e(\theta)$, intersect

$$n_1^o = n_2^e(\theta_m). \tag{5.24}$$

Therefore, when the laser beam propagates in a crystal along these directions the phase-matching condition (5.15) is obeyed and the second-harmonic generation achieves its highest efficiency. This indicates that rotating a nonlinear birefringent crystal makes it possible to find a direction θ_m at which the phase-matching condition $\Delta \boldsymbol{k} = 0$ is valid. The direction θ_m determines the optimal direction of simultaneous propagation of the fundamental wave and the second harmonic wave. The second harmonic intensity observed along this direction is several orders greater than the intensity along other directions. The method, which depends on nonlinear crystal birefringence, and relies on the angular dependence of the extraordinary index by tuning the angle θ, is called the *index matching technique*.

One can distinguish two types of phase matching. If the mixing waves at frequencies ω_1 and ω_2 have the same polarization, and the sum frequency wave $\omega_3 = \omega_1 + \omega_2$ (SFG) is polarized perpendicular to their incident polarization, the I-type of phase matching takes place. In uniaxial negative crystals the I-type phase matching condition is

$$\boldsymbol{k}_1^o + \boldsymbol{k}_2^o = \boldsymbol{k}_3^e(\theta), \tag{5.25}$$

and is called "*ooe*" *phase matching* or "*ooe*" *interaction*, or *type* I^- *phase matching*.

In uniaxial positive crystals the I-type phase matching condition is

$$\boldsymbol{k}_1^e(\theta) + \boldsymbol{k}_2^e(\theta) = \boldsymbol{k}_3^o \tag{5.26}$$

(this is called "*eeo*" *phase matching*, "*eeo*" *interaction*, or *type* I^+ *phase matching*).

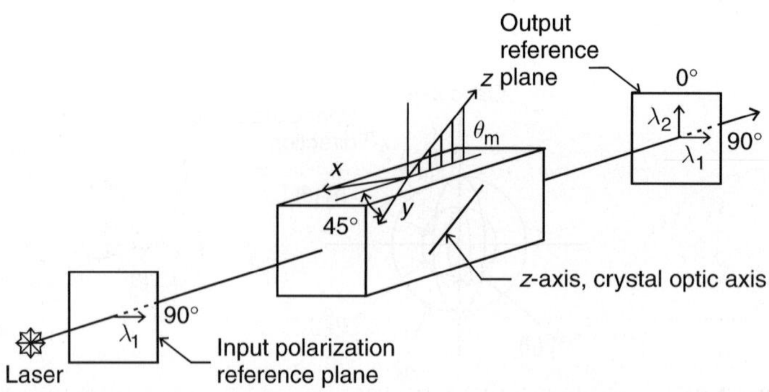

Fig. 5.7 Illustration of I-type angle-tuned phase matching configuration. [4]

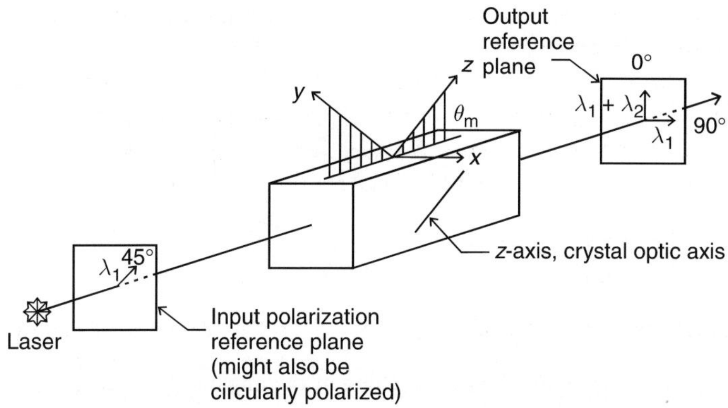

Fig. 5.8 Illustration of II-type angle-tuned phase matching configuration. [4]

If the mixing waves at frequencies ω_1 and ω_2 are of orthogonal polarization, the II-type phase-matching occurs. In uniaxial negative crystals the II-type phase-matching condition occurs when

$$\boldsymbol{k}_1^o + \boldsymbol{k}_2^e(\theta) = \boldsymbol{k}_3^e(\theta), \tag{5.27}$$

and is called "*oee*" *phase matching*, or *type II⁻ phase matching*, or

$$\boldsymbol{k}_1^e(\theta) + \boldsymbol{k}_2^o = \boldsymbol{k}_3^e(\theta), \tag{5.28}$$

and is called "*eoe*" *phase matching*, or *type* II^- *phase matching*.

In uniaxial positive crystals the II type phase matching occurs when,

$$\boldsymbol{k}_1^o + \boldsymbol{k}_2^e(\theta) = \boldsymbol{k}_3^o, \tag{5.29}$$

and is called "*oeo*" *phase matching*, or *type* II^+ *phase matching*, or,

$$\boldsymbol{k}_1^e(\theta) + \boldsymbol{k}_2^o = \boldsymbol{k}_3^o, \tag{5.30}$$

which is called "*eoo*" *phase matching*, or *type* II^+ *phase matching*.

The different configurations of polarization and crystal-angle leading to different phase-matching conditions are presented in Figs. 5.7 and 5.8.

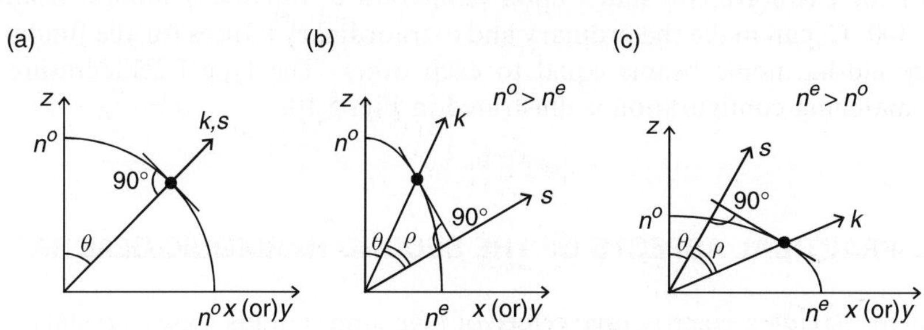

Fig. 5.9 Directions of the wave vector ***k*** and the energy wave (Poynting vector) ***s*** for (a) ordinary beam; and the extraordinary beam in a negative (b), and positive (c), uniaxial crystal.

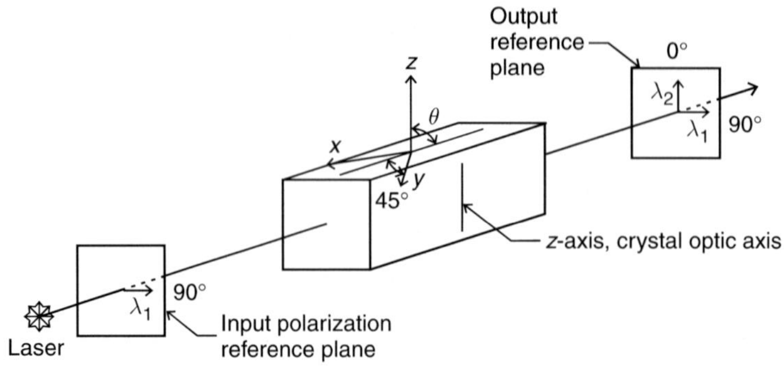

Fig. 5.10 Phase matching at the angle of 90°. [4]

We have shown that the phase-matching ensures a high efficiency in the generation of the second harmonic. However, an additional difficulty with the SHG process is that the energy of the two beams—the fundamental and the second harmonic, that are orthogonally polarized—will propagate in different directions, characterized by the walk-off distance δ in Fig. 5.4. This comes from the fact that in a birefringent crystal the direction of propagation of the wave phase (wave vector $\boldsymbol{k}$) generally does not coincide with that of the wave energy (Poynting vector, $\boldsymbol{s}$). The direction of the Poynting vector, $\boldsymbol{s}$, is defined as normal to the tangent drawn at the point of intersection of the wave vector $\boldsymbol{k}$ with the $n(\theta)$ cross-section (Fig. 5.9). One can see that for the ordinary beam both vectors coincide, which is in contrast to the extraordinary beam.

This disadvantage of the angle-tuned phase-matching technique can be eliminated by employing crystals with the phase-matching angle $\theta_m = 90°$. In this case, the direction of propagation and the optic axis is set to 90° and the angle between the Poynting vectors for the two beams is zero. If this happens, the length of nonlinear interaction in the crystal increases, owing to the beams' overlapping, and enhances the efficiency of SHG. This kind of phase matching is called "*non-critical phase matching*".

The non-critical phase-matching is usually employed in the second method of phase-matching, called a *temperature-tuned phase-matching*. It relies on the dependence of the extraordinary index upon temperature. Relatively modest heating, to about 300 °C, can make the ordinary and extraordinary indices for the fundamental and second-harmonic beams equal to each other. The type I temperature-tuned phase-matching configuration is illustrated in Fig. 5.10.

5.3. PRACTICAL ASPECTS OF THE SECOND HARMONIC GENERATION

In systems having a macroscopic center of inversion (liquids, gases, crystals of some classes) it is impossible to generate the second harmonic. Indeed, if the system has a center of inversion, the polarization $P = f(E)$ is the odd function of the electric field

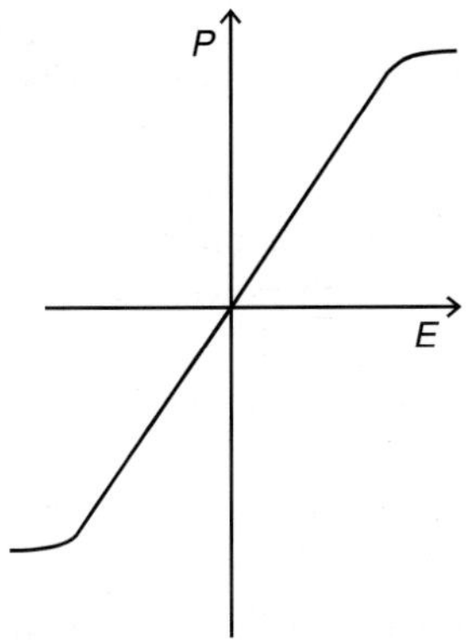

Fig. 5.11 Dependence of polarization, P, on electric field, E, in systems with a macroscopic center of inversion.

(Fig. 5.11). This indicates that the second term, $\chi^{(2)}E^2$, in equation (5.3) has to vanish during one period of the electric-field oscillation. The medium can produce the second harmonic only when the $P = f(E)$ dependence looks like that in Fig. 5.12, which takes place only when there is no macroscopic center of inversion. Therefore, the second harmonic generation is limited to:

a) some classes of crystals (without macroscopic inversion symmetry),
b) surfaces of a medium,
c) gas or liquid systems in which the isotropy has been broken by an electric field or a density gradient.

The lack of an inversion center in a crystal is a necessary but insufficient condition. As was said earlier, the phase-matching $\Delta\boldsymbol{k} = 0$ is required to generate efficiently the second harmonic $I(2\omega)$. The condition $\Delta\boldsymbol{k} = 0$ describes a perfect idealization. In practice it is sufficient when

$$\Delta k \cdot l \leq 2.78, \tag{5.31}$$

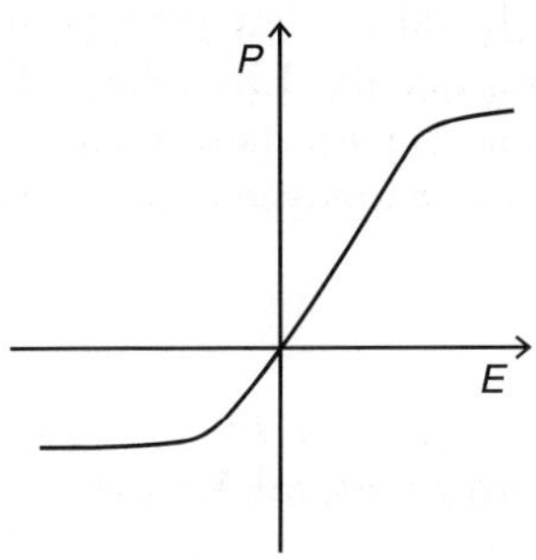

Fig. 5.12 Dependence of polarization, P, on electric field, E, in systems without macroscopic center of inversion.

where the value of 2.78 results from the wave-packet width calculated from $\left(\frac{\sin(\Delta kl/2)}{\Delta kl/2}\right)^2$ in eq. (5.19). Nonlinear processes can be observed by employing *cw* or pulsed lasers. However, the SHG processes occur with higher efficiency for pulsed lasers, such as *Q*-switched and modelocked lasers because these can produce high peak-powers, of the order of megawatts or terrawatts, even if their average powers are modest. The upper limit of high-power lasers is determined by the damage threshold of a crystal.

In the early days of laser development, KDP (potassium dihydrogen phosphate) and its KD*P isomorph (deuterated potassium dihydrogenphosphate) were the most commonly used crystals. KDP has the following refraction indices for the ordinary, n^o, and the extraordinary beams, n^e: $n^o(\omega) = 1.50$; $n^e(\omega) = 1.46$; $n^o(2\omega) = 1.53$; $n^e(2\omega) = 1.48$ at $\lambda = 694.3$ nm (ruby laser line). The phase-matching angle is 50.4°. KDP and ADP (ammonium dihydrogenphosphate) crystals have a typical efficiency of about 30%. They are also utilized in Pockels cells.

The lithium niobate ($LiNbO_3$) crystal is more efficient than KDP. In this crystal, effective phase-matching occurs at the angle 90°. This indicates that the sphere of refractive indices for the ordinary ray of the fundamental beam does not cross the ellipsoid of the refractive indices for the extraordinary beam of the second harmonic, but is tangent to it (Fig. 5.13). The phase-matching angle for $LiNbO_3$ remains 90°, even if the temperature of the crystal is raised to about 160 °C.

The typical SHG conversion efficiency for pico- and sub-picosecond lasers is routinely obtained as about 10%. However, a conversion of 80% is also reported. The typical thicknesses of the crystals are from fractions of millimeters, to centimeters. The most commonly used crystals are: KDP, ADP, $LiNbO_3$, $KTiOPO_4$ (KTP), $LiIO_3$, CsH_2AsO_4 (CDA), $Ba_2NaNb_5O_{15}$, $K_{0.6}Li_{0.4}NbO_3$, lithium triborate (LBO), beta-barium borate (BBO). The $K_{0.6}Li_{0.4}NbO_3$ and $Ba_2NaNb_5O_{15}$ (called banana) have a high efficiency in SHG process. The ability of LBO to achieve temperature-tuned non-critical phase-matching, as well as high damage thresholds and a wide transparency range, are the main reasons for its popularity although it does not possess an exceptionally large susceptibility of the second order $\chi^{(2)}$. Table 5.1 shows some properties and phase-matching parameters for the most frequently used crystals [1].

BBO is a type I crystal in generating SFG; KTP–type II in generating SFG and DFG; $LiNbO_3$–type I (DFG); $AgGaS_2$–type II (DFG). The KDP and KD*P crystals are available both for types I and II, but type II is more efficient despite some requirements on incident-beam quality. LBO, BBO, KTP are used for SHG in the visible and near-visible regions; $AgGaS_2$ (silver thiogallate) is used in the infrared. For detailed properties of crystals employed in nonlinear optics the reader is referred to refs. [5–6].

5.3.1. SHG for Pico- and Femtosecond Pulses

The rapid development of laser technologies made it possible to generate ultrashort pulses of the order of 5 femtoseconds or less. In commercially available lasers the

Table 5.1

Properties of important nonlinear materials and phase matching parameters [1]

Material	Phase-matching type	Effective nonlinear coefficient* $[10^{-12}$ m/V]	Index of refraction $n_0(\omega)$	Damage threshold [GW/cm^2]	Absorption [cm^{-1}]	Phase-matching angle	Walk-off angle
KD*P	II	0.37	1.49	0.5	0.005	53.7°	1.45°
KTP	II	3.18	1.74	0.5	0.010	24.3°	0.26°
LBO	I	1.16	1.56	2.5	0.005	–	–
BBO	I	1.94	1.62	1.5	0.005	22.8°	3.19°
$LiNbO_3$ (5%MgO)	I	4.7	2.23	0.10	0.002	90°(1)	0°

*For 1064 nm to 532 nm second-harmonic generation, (1) at $T = 107\,°C$.

pulses are of the order of tens of femtoseconds. The mechanism of the second-harmonic generation by femtosecond lasers is the same as discussed previously. However, for such short pulses one should take into account an additional effect, which is negligible for longer pulses or in continuous work. An ultrashort pulse corresponds to a large spectral width, $\Delta\omega$. The phase-matching condition should be obeyed, not only for the center of the spectral distribution ω_0 for the fundamental, but also for the whole spectral range of the pulse $\left(\omega_0 - \frac{\Delta\omega}{2};\ \omega_0 + \frac{\Delta\omega}{2}\right)$. If the phase matching is valid only around the center, ω_0, the generated second harmonic will be spectrally narrowed, and the pulse duration will become longer. We shall illustrate this effect with the following example. The approximate phase-matching given by (eq. 5.31) needs $\Delta k(\omega - \omega_0) \leq 2.78/l$. By converting to frequencies in units of cm^{-1}, one can readily derive [7] the condition for the phase-matching bandwidth in the frequency domain is

$$\Delta\nu = |\nu - \nu_0| \leq \frac{0.221}{l}\left[\lambda_1\left(\frac{dn}{d\lambda}\right)_1 - \lambda_2\left(\frac{dn}{d\lambda}\right)_2\right]^{-1} \tag{5.32}$$

where λ_1 and λ_2 are the wavelengths for the fundamental and the second harmonic, respectively. For a KDP crystal of 1 cm length, the width of the phase-matching bandwidth, $\Delta\nu$, is $20\,\mathrm{cm}^{-1}$ at $\lambda_1 = 1.064\,\mu\mathrm{m}$. In contrast, for a lithium niobate, $LiNbO_3$, crystal of the same length the phase-matching bandwidth $\Delta\nu$ is only about $0.5\,\mathrm{cm}^{-1}$. This indicates that KDP is more suitable for short-pulses generation than $LiNbO_3$, despite its having about a 10-times larger susceptibility, $\chi^{(2)}$. Thus, the SHG process with femtosecond and picosecond pulses requires carefully chosen dispersion properties, which are just as important as the nonlinear properties, $\chi^{(2)}$ of crystals.

Another parameter which is important for ultrashort pulses, related to dispersion properties, is the *group delay*. To understand the importance of this effect for ultrashort pulses, we recall the difference between the phase velocity and the group velocity. *The phase velocity* is defined by eq. (5.12), and describes the speed of the front of a plane, monochromatic wave. An ultrashort pulse cannot be represented by a single plane monochromatic wave because it consists of many monochromatic components that correspond to a large distribution of wave vectors, $\Delta\boldsymbol{k}$. The wave function $\psi(z, t)$ of such a state can be described as a wave packet

$$\psi(z,t) = \int_{k_0-\Delta k}^{k_0+\Delta k} A(k)\exp(i(kz - \omega t))dk, \tag{5.33}$$

which describes the packet of plane waves from the range of $k_0 - \Delta k \leq k \leq k_0 + \Delta k$, propagating along the z-axis. Developing $\omega(k)$ in a series;

$$\omega(k) = \omega_0 + \left(\frac{d\omega}{dk}\right)_0 (k - k_0) \tag{5.34}$$

and inserting this into eq. (5.33), one obtains;

$$\psi(z,t) = 2A(k_0)\frac{\sin\left[z - \left(\frac{d\omega}{dk}\right)_0 t\right]\Delta k}{z - \left(\frac{d\omega}{dk}\right)_0 t}\exp(i(k_0 z - \omega_0 t)). \tag{5.35}$$

The term $2A(k_0)\frac{\sin[z-(\frac{d\omega}{dk})_0 t]\Delta k}{z-(\frac{d\omega}{dk})_0 t}$ can be treated as the amplitude, A, of the wave packet, $\psi(z,t)$. The amplitude, A, is a function of the $\frac{\sin z}{z}$ type, presented in Fig. 5.13, with a maximum at $z=0$, where $z=\left[z-\left(\frac{d\omega}{dk}\right)_0 t\right]$.

The quantity Δz can be treated as a special distribution of the wave packet. Calculating the difference between the first minima, z_1 and z_2, it is easy to show that Δz is equal to $\frac{2\pi}{\Delta k}$. For a monochromatic wave ($\Delta k \to 0$), and the wave packet becomes a plane wave with $\Delta z \to \infty$.

The amplitude, $A = 2A(k_0) = \frac{\sin[z-(\frac{d\omega}{dk})_0 t]\Delta k}{z-(\frac{d\omega}{dk})_0 t}$, reaches a maximum at $\left[z-\left(\frac{d\omega}{dk}\right)_0 t\right]=0$, therefore

$$\frac{z}{t} = \left(\frac{d\omega}{dk}\right)_0 \tag{5.36}$$

Eq. (5.35) says that the wave packet travels with the speed z/t called the *group velocity* v_g

$$v_g = \left(\frac{d\omega}{dk}\right)_0. \tag{5.37}$$

We can assume that the second harmonic is generated under the phase-matching condition in a crystal of length l, by pulses of duration, t_p. It indicates that the phase velocities of the fundamental beam and the second harmonic are equal, ($v_{ph}(\omega) = v_{ph}(2\omega)$), for a given component. It does not mean, however, that the group velocities (5.37) are also equal. The group velocities of the fundamental beam and the second harmonic are usually different, owing to the dispersion properties of the crystal.

A consequence of the different group velocities is the different time, t_g, for the passage through a crystal of length, l, of the fundamental beam and the second harmonic. For the fundamental beam, t_g is given by

$$t_g^1 = \frac{l}{v_g^1} = l\left(\frac{dk}{d\omega_1}\right) = \frac{l}{c}\frac{d(n_1\omega_1)}{d\omega_1} = \frac{l}{c}\left[\frac{dn_1}{d\omega_1}\omega_1 + n_1\right], \tag{5.38}$$

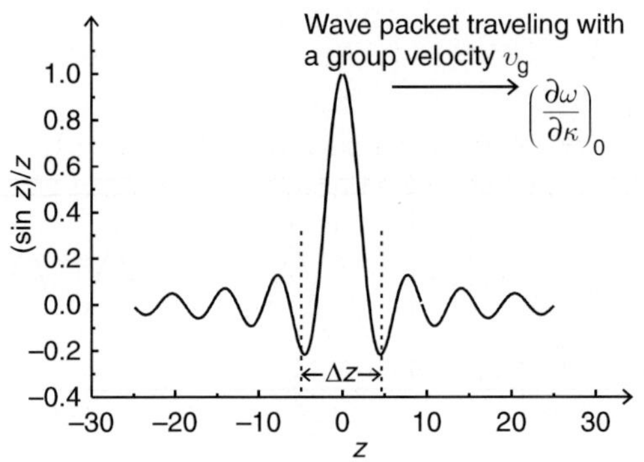

Fig. 5.13 Plot of the function $\frac{\sin z}{z}$.

and for the second harmonic

$$t_g^2 = \frac{l}{v_g^2} = l\left(\frac{dk}{d\omega_2}\right) = \frac{l}{c}\frac{d(n_2\omega_2)}{d\omega_2} = \frac{l}{c}\left[\frac{dn_2}{d\omega_2}\omega_2 + n_2\right]. \tag{5.39}$$

This indicates that the second-harmonic's wave-packet is delayed with respect to the fundamental wave-packet by Δt, where;

$$\Delta t = t_g^2 - t_g^1 = \frac{l}{c}\left[\frac{dn_1}{d\lambda_1}\lambda_1 - \frac{dn_2}{d\lambda_2}\lambda_2\right]. \tag{5.40}$$

The superscripts 1 and 2 in eqs. (5.38–5.40) denote the fundamental and the second harmonic, respectively.

This situation is illustrated in Fig. 5.14. At time $t=0$—that is, at the moment when the fundamental beam (I) arrives at the crystal, the wave packets of the fundamental and the second harmonic overlap in space. After some time, owing to the different group-velocities, the overlapping of the wave packets becomes less and less efficient ($t=t_1$), and finally disappears ($t=t_2$). The shorter the pulse-duration, t_p, the larger is the spectral range, Δk, and the lower the wave-packet dimension, Δz. The greater the *group delay* time, $\Delta t = t_g^2 - t_g^1$, the shorter is the length of interaction between the fundamental and the second harmonic beams.

The interaction length is limited by the condition

$$\Delta t \leq t_p. \tag{5.41}$$

This condition defines the effective crystal length, l. If the crystal is longer than l, leading to the violation of condition (5.41), the rest of the crystal is ineffective because the beams no longer overlap, and the SHG process does not occur. For $\Delta t_g > t_p$ the second-harmonic intensity does not increase with crystal length, but a temporal pulse stretching occurs.

Let us consider the limiting case, where the group delay is much smaller than the pulse duration

$$\Delta t \ll t_p. \tag{5.42}$$

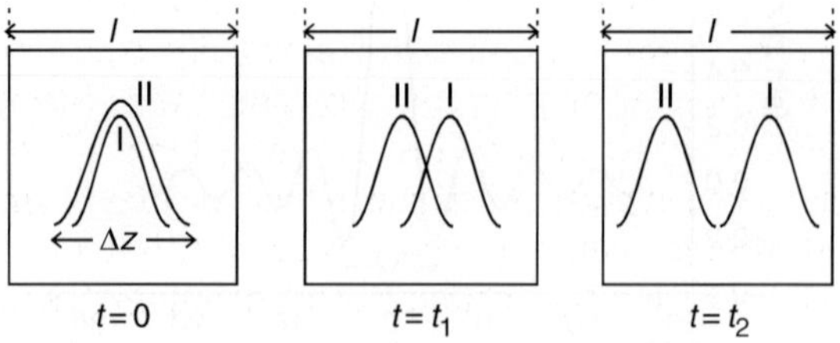

Fig. 5.14 Scheme illustrating the consequence of different group velocities for the fundamental beam (I) and the second harmonic (II).

which approximately means that $\Delta t \rightarrow 0$. Therefore, eq. (5.40) takes the form

$$\frac{dn_1}{d\lambda_1}\lambda_1 = \frac{dn_2}{d\lambda_2}\lambda_2. \tag{5.43}$$

If one takes into account the phase-matching condition ($n_1 = n_2 = n$) and $\lambda_1 = 2\lambda_2$, one obtains

$$2\left(\frac{dn}{d\lambda}\right)_1 = \left(\frac{dn}{d\lambda}\right)_2. \tag{5.44}$$

When Δt increases, eq. (5.40) gives

$$2\left(\frac{dn}{d\lambda}\right)_1 > \left(\frac{dn}{d\lambda}\right)_2. \tag{5.45}$$

This indicates that for crystals characterized by a very small group-delay time, Δt, the second harmonic achieves a maximum of dispersion, Δn. For this maximum dispersion, Δn, one obtains the widest spectral distribution for the second harmonic, $\Delta k_2 = \frac{\Delta n(\omega)\omega_2}{c}$, and the shortest pulse duration, t_p. So, if we want to generate short pulses for the second harmonic, we should choose nonlinear crystals characterized by a small group-delay time, Δt. The group-delay, Δt, values for some typical crystals are as follows: 6 ps/cm for $LiNbO_3$; 0.7 ps/cm for $LiIO_3$; 0.08 ps/cm for KDP.

5.4. PARAMETRIC OSCILLATOR

In sections 5.1 and 5.3 we discussed the conditions for effective generation of the second-order harmonic. The second-order processes, SHG and SFG, result from interaction between two laser beams at frequencies ω_1 and ω_2, to produce a new beam at higher frequency, ω_3

$$\omega_1 + \omega_2 = \omega_3. \tag{5.46}$$

For two incident beams of the same frequency, ω_1, eq. (5.46) describes the frequency-doubling $2\omega_1$ (SHG), a particular case of the sum-frequency mixing (SFG). However, we can look at eq. (5.46) in another way. Formally, when looked at from right to left, the equation represents another phenomenon – the generation of two beams of lower frequencies from one incident beam – some kind of, "frequency splitting" corresponding to a photon energy splitting into two photons of smaller energy. Does this effect predicted from the equation really exist? Yes, it does – and it is known as *parametric conversion*. The incident beam, called the pump beam, is divided into a, "signal" and an "idler" beam.

The generation of the new beams is a nonlinear process of the second-order. The threshold for the parametric conversion is high, and for moderate pump powers the nonlinear crystal must be placed inside a resonator. The scheme of a typical *optical parametric oscillator* (OPO) pumped with a pulsed laser beam with frequency ω_p is shown in Fig. 5.15.

The parametric oscillator has a configuration similar to the laser optical resonator, but the phenomena occurring in it are different. The pumping beam of strong intensity, ω_p, from a laser or a regenerative amplifier interacts with the crystal, which has sufficiently large nonlinear susceptibility. One of the results of such an interaction can be the generation of the second harmonic. The other phenomena including parametric conversion can also occur. The parametric conversion can be described as inelastic scattering of a pumping photon, $\hbar\omega_p$, on a crystal lattice. As a result of interaction, two photons: $\hbar\omega_s$ and $\hbar\omega_i$ are generated. The component with frequency ω_i is called the *idler component*, and the component ω_s, the signal component. The law of conservation of energy requires

$$\omega_p = \omega_i + \omega_s. \tag{5.47}$$

As in the case of SHG, the photons at frequencies ω_i and ω_s being created in the process of parametric generation can participate in the macroscopic wave-generation when the condition of phase-matching occurs;

$$\boldsymbol{k}_p = \boldsymbol{k}_i + \boldsymbol{k}_s \tag{5.48}$$

or,

$$\Delta k = \boldsymbol{k}_p - \boldsymbol{k}_i - \boldsymbol{k}_s = 0; \tag{5.49}$$

This is the condition for conservation of momentum for three photons. The condition (5.47) would give an infinite number of frequency combinations, while the condition (5.48) limits the frequencies ω_i and ω_s to one pair, which simultaneously obeys the conditions (5.47) and (5.49)

$$(\omega_p, \boldsymbol{k}_p) \Leftrightarrow (\omega_i, \boldsymbol{k}_i); (\omega_s, \boldsymbol{k}_s). \tag{5.50}$$

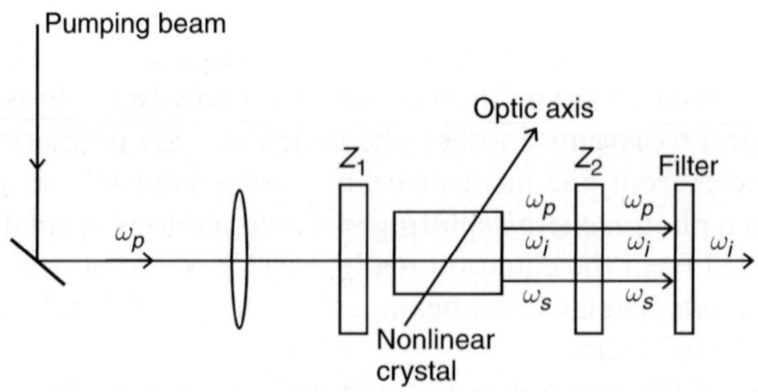

Fig. 5.15 Optical parametric oscillator (OPO).

The highest efficiency of the parametric conversion can be achieved when the beams of $\boldsymbol{k}_p$, $\boldsymbol{k}_i$, as well as $\boldsymbol{k}_s$, are colinear. For this case, the phase-matching condition (5.49) can be written as

$$n_p\omega_p = n_s\omega_s + n_i\omega_i, \tag{5.51}$$

where n_p, n_s and n_i are the indices of refraction at the pump, signal, and idler beams, respectively. In order to achieve the phase matching one can use the method described in the previous paragraph–rotation of a nonlinear crystal. When the conditions (5.47) or (5.51) are fulfilled, and the amplification is higher than the overall losses, light with the frequency of the signal wave, ω_s, or the idler wave, ω_i, is generated in the oscillator. When the parametric oscillator can emit radiation at both frequencies ω_s and ω_i, it is called a *double resonance oscillator*. One should note that both the signal, ω_s, and the idler, ω_i, components have frequencies lower than the frequency, ω_p, of the pumping beam, which makes the parametric oscillators useful sources of radiation in the infrared range. Additionally, the parametric oscillator generates radiation over a wide spectral range. Tuning is usually achieved by rotating a crystal in the optical resonator cavity, or by varying the temperature.

Besides the optical parametric oscillators (OPO), the same nonlinear phenomena occur in the *optical parametric generator* (OPG) and the *optical parametric amplifier* (OPA). The parametric generator differs from the parametric oscillator because the nonlinear crystal is not placed in an optical resonator. In the optical parametric amplifier (OPA) two laser beams, with frequencies ω_p (pump beam) and ω_s, are incident on a nonlinear crystal, where the parametric conversion occurs. As a result of nonlinear interaction in the crystal, the idler component $\omega_i = \omega_p - \omega_s$ is generated, which is much more intense than that achieved in OPG.

Figure 5.16 shows a typical example of the optical parametric amplifier (OPA) for a $LiNbO_3$ crystal. Notice that for the parametric conversion, as for SHG, various types of phase-matching may occur. The $LiNbO_3$ crystal represents the phase-matching of type I ($e=o+o$). Figure 5.17 shows the configuration where all the nonlinear devices discussed so far; SHG, OPO, OPG, and OPA are combined to provide radiation from the IR range that is used in the pump-probe experiments.

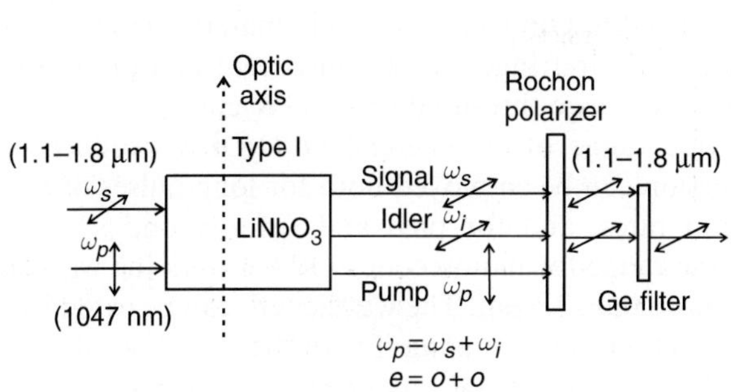

Fig. 5.16 Optical parametric amplifier (OPA) for a $LiNbO_3$ crystal.

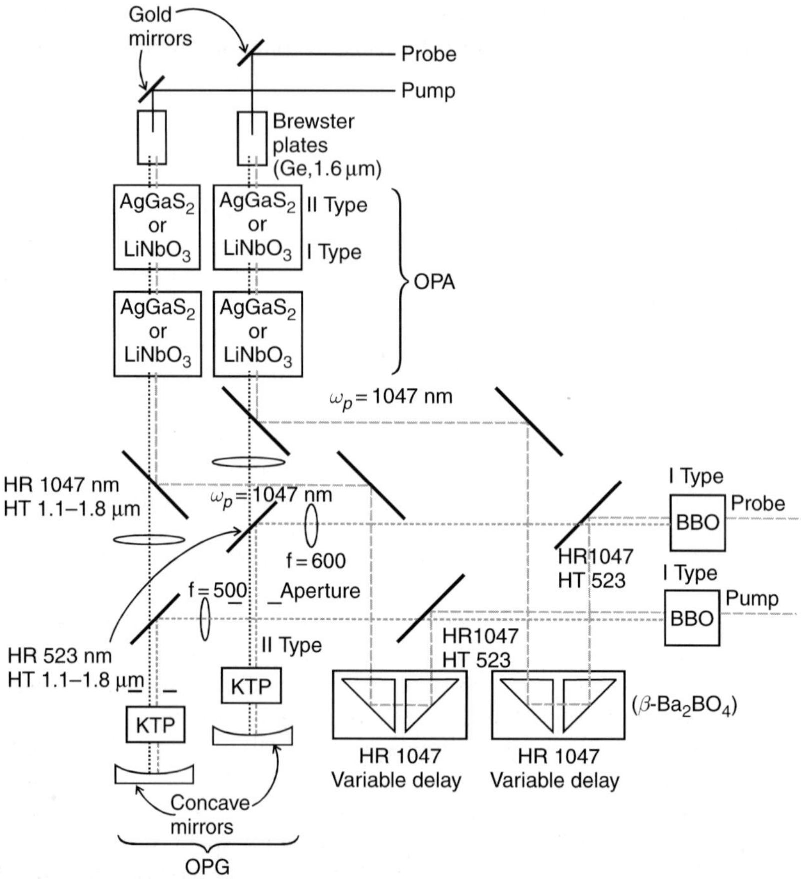

Fig. 5.17 The configuration in which some of the nonlinear devices discussed in this chapter—SHG, OPG and OPA—are combined to provide radiation from the IR range that is used in the pump-probe experiments. [8]

The nonlinear optical parametric conversion offers the most promise as a widely tunable laser source. This is the most powerful source for reaching the infrared region. Development of beta-barium borate (BBO) crystals has pushed the progress in commercially available OPO systems. This material is characterized by a high nonlinear coefficient, is birefringent and transparent in a broad range from 200 nm to 3.5 μm, and has a damage threshold of 1.5 GW/cm^2.

Currently, a wide range of commercial OPO/OPA products is available. The parametric conversion can be employed both for long pulses of a few nanoseconds, and low repetition rates, and for pico and femtosecond systems having a high repetition. A tuning range for nanoseconds OPA covers the wavelength range from about 300 nm to more than 3.5 μm. The wavelength can be extended to the UV range by doubling the OPO output or mixing it with the pump beam. These OPA systems employ BBO, KTP, or KTA crystals, and can generate energies ranging from a few mJ to more than 100 mJ.

Typical femtosecond OPA systems use BBO crystals and offer tunability from 1.1 μm to 1.6 μm (signal) and 1.6 μm to 3.0 μm (idler). The output pulses are near to transform-limited, with pulse-width less than 130 fs, and typical energies about 100 μJ (Fig. 5.18).

The wavelength coverage can be extended further, from <300 nm to >10 μm, via harmonic-generation, sum-frequency mixing, or difference-frequency mixing (Fig. 5.19).

One can get the range between 480 and 580 nm for the sum-frequency mixing, 3–10 μm for the difference-frequency mixing, 600–1200 nm for the II harmonics of the signal and idler components and 300–600 nm for the IV harmonic of the signal and idler components.

Since the OPO system covers the range of infrared frequencies, they are used to study femtosecond dynamics of vibrationally excited states in the pump-probe experiments. Many such experiments use two tunable OPOs that produce a pump pulses to excite the vibrational state under study and the second pulse to probe the dynamics occurring on the pico- and femtosecond time scale. The vibrational relaxation processes and experimental methods to study them will be described in Chapter 8.

The combination of OPA systems with *chirped pulse amplification* (CPA) technology may allow the average power from OPA to be increased by more than an order of magnitude. This method, called OPCPA, was first proposed by researchers from the Rutherford Appleton Laboratory (Oxford, England). In this method, the Ti:sapphire regenerative amplifier is replaced by an OPA. An OPA employing two BBO crystals in series can produce the same energy-gain with a single pass of the seed pulse. Such an OPCPA system could maintain a pulse duration of an OPA system of tens of femtoseconds, while increasing the pulse energies well above a Joule, leading to focused intensities of 10^{24} W/cm^2 or higher. Such intensities are enormously high. Indeed, only 10^{15} W/cm^2 is required to ionize nearly all the material, and at 10^{18} W/cm^2 electrons can be accelerated to nearly the speed of light. As a result of such a high intensity, the light's pressure exceeds the thermal pressure, and generates an

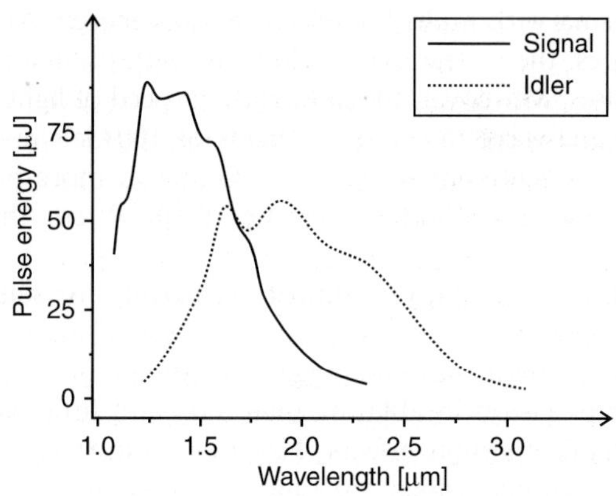

Fig. 5.18 Typical output for commercially available femtosecond OPA system.

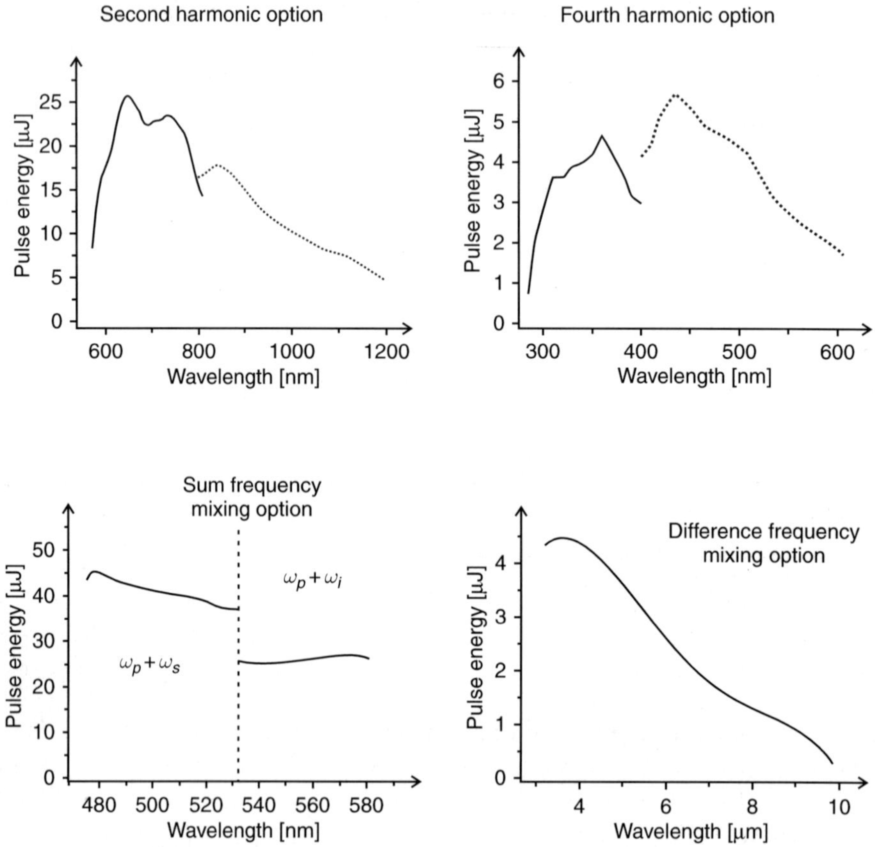

Fig. 5.19 The typical wavelength coverage for femtosecond OPA systems from <300 nm to >10 µm via harmonic generation, sum-frequency mixing or difference-frequency mixing.

extremely dense plasma with multi-kilo electron-volts energy. At such high densities, energies, and pressures, the interaction of electrons with plasma ions generates extremely short X-ray pulses. Moreover, the difference in speed of light electrons and heavy ions produces a plasma-wave that induces fields of 100 GeV/m—much higher than available in today's conventional accelerators. Going to more powerful pulses, self-focusing of the pulse occurs, leading to intensities of 10^{20} W/cm^2 that can ignite a pellet of fusion fuel [9].

While OPO systems are employed almost exclusively for scientific applications, they have begun to generate interest for various areas of materials' processing and surgical applications in medicine. The biggest advantage of ultrafast pulses in these applications is their precision in ablating material, with little damage to the target material, and negligible thermal effects. However, industrial applications usually need higher average powers, which are lower than fractions of a watt for typical OPO femtosecond systems.

5.5. THE THIRD ORDER NONLINEAR PROCESSES

The *Raman scattering* processes can be described by the equation

$$P_i = \chi_{ij}^{(1)} E_j + \chi_{ijk}^{(2)} E_j E_k + \chi_{ijkl}^{(3)} E_j E_k E_l. \tag{5.52}$$

The first term is connected with linear polarization and describes spontaneous, linear Raman scattering, which is studied by the conventional methods of Raman spectroscopy. The second- and third terms describe spontaneous, nonlinear Raman scattering (called *hyper-Raman scattering*). The third is also responsible for the, *stimulated Raman scattering.*

Spontaneous, linear Raman scattering is well-documented in many books on molecular spectroscopy, and is outside the scope of this book. Spontaneous Raman scattering is illustrated in Fig. 5.20.

The levels denoted as E_0 and E_1 represent electronic energy levels, while the levels numbered with a quantum number, υ, represent the vibrational levels. If the sample is illuminated with photons of energy $\hbar\omega_L$, smaller than the resonance energy $\Delta E = E_1 - E_0$, not all of the photons that interact with the sample are absorbed, but cause the potential energy of the interacting molecules to be raised to a virtual state, $\hbar\omega_L$ above the ground state. Almost immediately, most molecules return to the ground state through the emission of photons having the same energy as the incident photons. This elastic scattering is called the *Rayleigh scattering.* A small fraction of the molecules drops back to the first excited vibrational state ($\upsilon = 1$) instead to the ground state. Since the energy of the incident and the scattered photons are different, the scattering is inelastic and the process is known as *Stokes Raman scattering*, with the scattered radiation observed at a lower energy, $\hbar(\omega_L - \omega_{vib})$. Molecules that are already in the excited vibrational state ($\upsilon = 1$) will undergo an analogous effect when illuminated with a laser light. When the excited molecules drops back to the ground-vibrational state ($\upsilon = 0$), the scattered radiation will be observed at higher energy, $\hbar(\omega_L + \omega_{vib})$. This scattering is known as *anti-Stokes Raman scattering.* The frequency ω_{vib} denotes the frequency of a given vibrational mode of the molecule.

To describe Raman scattering a full quantum-mechanical theory is required, but some intuitive description can be also obtained from a classical picture. The electric field, $E_0 \cos(\boldsymbol{k} \cdot \boldsymbol{r} - \omega_L t)$ drives the electron displacements that induce the

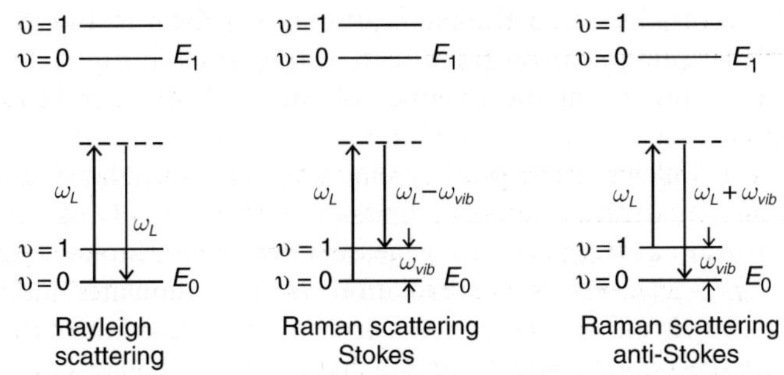

Fig. 5.20 Scheme of the Rayleigh- and linear spontaneous Raman scattering.

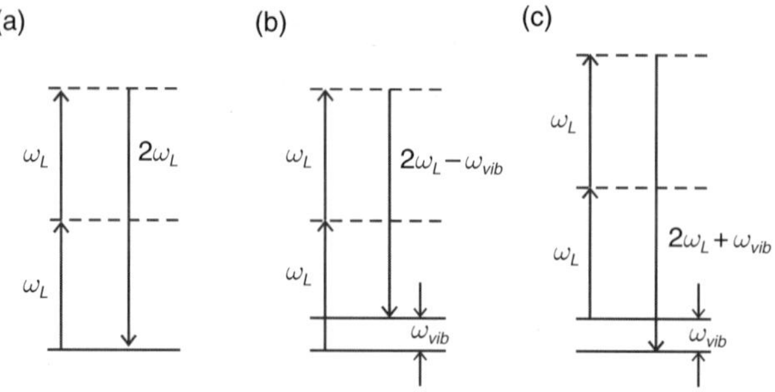

Fig. 5.21 Hyper-Rayleigh and hyper-Raman scattering: (a) hyper-Rayleigh scattering; (b) hyper-Raman scattering, Stokes; (c) hyper-Raman scattering, anti-Stokes.

polarization $\boldsymbol{P}$ in a medium modulated in time that, in turn, generates a wave at the same frequency, ω_L (Rayleigh scattering). When the dipole oscillations are modulated additionally by the molecular vibrations at frequency ω_{vib} the waves are generated at $(\omega_L - \omega_{vib})$ (*Stokes Raman scattering*) or $(\omega_L + \omega_{vib})$ (*Stokes Raman scattering*).

When more intense light is employed, nonlinear Raman processes occur in the sample, such as the *hyper-Raman or stimulated Raman scattering*. When the intensity of the incident radiation increases, the probability also increases that two or three photons will participate in the Raman and Rayleigh scattering. Such phenomena are called the *hyper-Raman* or *hyper-Rayleigh scattering*. Figure 5.21 illustrates hyper-Rayleigh and hyper-Raman scattering related to the second term of eq. (5.52), with the involvement of two photons.

5.5.1. Stimulated Raman Scattering

The *hyper-Raman scattering* belongs to the group of spontaneous nonlinear scattering phenomena. In this section we will concentrate on *stimulated Raman scattering* (*SRS*), which is a third-order nonlinear optical phenomenon, $\chi^{(3)}$. Although described by the third term in equation. (5.52), like the *three-photon hyper-Raman scattering*, the mechanism of *stimulated Raman scattering* is completely different.

The mechanism of stimulated Raman scattering is shown in Fig. 5.22, and does not differ from the spontaneous Raman scattering scheme in Fig. 5.20.

The difference comes from the intensity of the incident laser beam. When the incident light exceeds a certain threshold in the material, it can enhance or stimulate the rate of Raman Stokes (or anti-Stokes) emission. This stimulated Raman scattering shows a similarity to the stimulated emission in the optically pumped laser. The intense incident-beam at frequency ω_L generates the intense Stokes radiation at the frequency $\omega_s = \omega_L - \omega_{vib}$. The Stokes radiation, in turn, stimulates additional Stokes scattering from the virtual scattering state. In contrast to the linear Raman scattering, the stimulated Raman scattering can achieve 50% or higher conversion of the pump wavelength to the Stokes line. There is a full analogy to the phenomena of

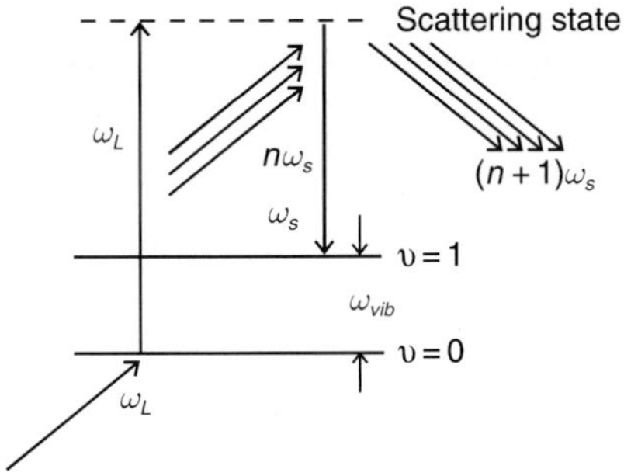

Fig. 5.22 Scheme of stimulated Raman scattering.

spontaneous and stimulated emission discussed in Chapter 1. The only difference is that we previously considered the stationary quantum energy states between which the absorption, and spontaneous and stimulated emission occur, in contrast to the scattering transitions between the virtual scattering state and the ground state. For the intensities above the material threshold, the incident and the scattered radiation at ω_L and ω_s act simultaneously on a system of molecules leading to the creation of phase-coherent vibrations at the frequency $\omega_{vib} = \omega_L - \omega_s$.

The stimulated anti-Stokes scattering is generated in a similar way to that described above. In the spontaneous Raman effect, the intensity of the anti-Stokes scattering is weak, because the population of vibrationally excited levels is negligible (from the Boltzmann distribution). The illumination of a system with an intense beam at frequency ω_L causes a significant deviation from the Boltzmann distribution, and a strong signal from the stimulated anti-Stokes scattering is produced.

The stimulated Raman scattering is a special case of *four-wave interaction*:

$$\omega_4 = \omega_1 + \omega_2 \pm \omega_3, \tag{5.53}$$

$$\Delta \boldsymbol{k} = \boldsymbol{k}_1 + \boldsymbol{k}_2 \pm \boldsymbol{k}_3 - \boldsymbol{k}_4 = 0. \tag{5.54}$$

Equations (5.53) and (5.54) describe a non-degenerate case with four photons of different frequencies. The stimulated Raman scattering represents a partially degenerate case: ($\omega_1 = \omega_3$, $\omega_2 = \omega_4$). Indeed, the four-wave interaction for the stimulated Stokes scattering can be written in the form

$$\omega_4 = \omega_s = \omega_L - \omega_{vib} = \omega_L - (\omega_L - \omega_s) = \omega_L + \omega_s - \omega_L, \tag{5.55}$$

and therefore

$$\omega_4 = \omega_s;\ \omega_1 = \omega_L;\ \omega_2 = \omega_s;\ \omega_3 = \omega_L. \tag{5.56}$$

The intensity of the stimulated Stokes scattering, I_S^{SRS} is expressed as

$$I_S^{SRS} = const \cdot ((\chi^{(3)}))^2 I_L^2 I_S l^2 \left(\frac{\sin \Delta kl/2}{\Delta kl/2}\right)^2, \tag{5.57}$$

where the factor, *const.*, is a constant for a given wavelength and a given nonlinear material, $\chi^{(3)}$ denotes the effective susceptibility of the third order.
The stimulated anti-Stokes scattering, in turn, can be described as fully non-degenerated four-wave interaction

$$\omega_{AS} = \omega_L + \omega_{vib} = \omega_L + (\omega_L - \omega_s), \tag{5.58}$$

therefore

$$\omega_4 = \omega_{AS};\ \omega_1 = \omega_L;\ \omega_2 = \omega_L;\ \omega_3 = \omega_S. \tag{5.59}$$

The intensity of the stimulated anti-Stokes scattering I_{AS}^{SRS} is expressed as

$$I_{AS}^{SRS} = const \cdot ((\chi^{(3)}))^2 I_L^2 I_S l^2 \left(\frac{\sin \Delta kl/2}{\Delta kl/2}\right)^2. \tag{5.60}$$

Although formally the expressions (5.57) and (5.60) do not differ in any way, one should note that the phase-matching condition, Δk, is different for the stimulated Stokes and anti-Stokes scattering. One can see from expression (5.55) that for the stimulated Stokes scattering the phase-matching condition is given by

$$\Delta \boldsymbol{k} = \boldsymbol{k}_L + \boldsymbol{k}_S - \boldsymbol{k}_L - \boldsymbol{k}_S = 0, \tag{5.61}$$

which indicates that the phase-matching between the pumping beam, ω_L, the Stokes beam ω_s and molecular vibrations is established automatically.

In the case of the anti-Stokes scattering the phase-matching obtained from eq. (5.58) is given by

$$\Delta \boldsymbol{k} = \boldsymbol{k}_L + \boldsymbol{k}_L - \boldsymbol{k}_S - \boldsymbol{k}_{AS} = 2\boldsymbol{k}_L - \boldsymbol{k}_S - \boldsymbol{k}_{AS}. \tag{5.62}$$

This indicates that the phase matching condition is achieved only for certain directions, for which the relationship applies

$$\boldsymbol{k}_{AS} = 2\boldsymbol{k}_L - \boldsymbol{k}_S. \tag{5.63}$$

The relationship (5.63) says that the direction of wave-propagation of the stimulated anti-Stokes scattering is limited to a conical surface with its axis parallel to the direction of the incident laser beam (Fig. 5.23).

Why does the phase-matching condition, $\Delta \boldsymbol{k} = 0$, occur automatically for the stimulated Stokes scattering, and not occur for the stimulated anti-Stokes scattering? This results from the fact that the phase of vibrating molecules is defined by the more intense Stokes scattering.

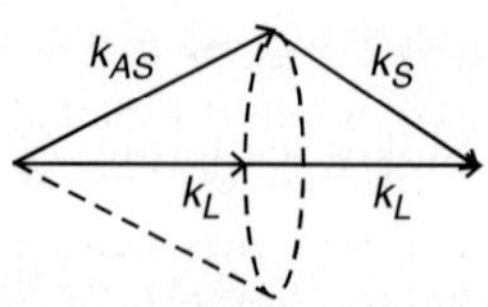

Fig. 5.23 Directions of the stimulated anti-Stokes Raman scattering.

The differences between the linear, spontaneous Raman scattering and the non-linear, stimulated Raman scattering can be summarized as follows:

1. The intensity of the spontaneous Raman scattering (Stokes and anti-Stokes) is several powers of ten lower than the pumping beam intensity, while the intensity of the stimulated Raman scattering (Stokes and anti-Stokes) is comparable with the pumping beam intensity.
2. The stimulated Raman scattering is observed exclusively above the threshold intensities. The threshold depends on the medium gain $g(\omega_s)$ and the length of the pumping area. The gain $g(\omega_s)$ is expressed by

$$g(\omega_S) \propto \left(\frac{\partial\sigma}{\partial\Omega}\right)\frac{\Gamma}{(\omega_L - \omega_S - \omega_{vib})^2 + \Gamma^2}, \tag{5.64}$$

   where $\frac{\partial\sigma}{\partial\Omega}$ is the scattering cross-section, and Γ is the width of the vibrational line.
3. Owing to the small scattering cross-sections $\frac{\partial\sigma}{\partial\Omega}$, most substances emit radiation only for the vibrational frequencies corresponding to the strongest Stokes lines, $\omega_s = \omega_L - \omega_{vib}$. While still much reduced in strength from the first Stokes transition, multiples at $\omega = \omega_L - n\omega_{vib}$ $(n = 1, 2, 3, \ldots)$ can be readily apparent in the stimulated Raman scattering, in contrast to the spontaneous Raman scattering. The frequencies ω are red-shifted with respect to those of the spontaneous lines corresponding to the overtones. This is caused by the anharmonicity of molecular oscillators satisfying the inequality $\omega_{0\to n} < n\omega_{vib}$.
4. The spectral widths of the spontaneous- and stimulated Raman lines depend on the spectral line-width of the pumping laser. For narrow widths of the excitation lasers the stimulated Raman lines are much narrower than the spontaneous lines.

5.5.2. Coherent Anti-Stokes Raman Scattering (CARS)

We showed in the previous chapter 5.5.1 that the intense pumping light at the frequency ω_L can induce intense stimulated Raman scattering: Stokes' at $\omega_S = \omega_L - \omega_{vib}$ and anti-Stokes' at $\omega_{AS} = \omega_L + \omega_{vib}$. The interacting beams ω_L and ω_S induce in a medium the third-order polarization, $P^{(3)} \propto \chi^{(3)}_{ijkl} E_j E_k E_l$ with the components of the frequency at $\omega_{vib} = \omega_L - \omega_S$, $\omega_S = \omega_L - \omega_{vib}$ as well as, $\omega_{AS} = \omega_L + \omega_{vib}$. The polarization components cause the generation of new waves at frequencies ω_S and ω_{AS}, known as the stimulated-Stokes and anti-Stokes Raman scattering. For the stimulated Stokes scattering the phase-matching condition ($\Delta \boldsymbol{k} = 0$) occurs for all directions, and the scattered light can be emitted in every direction. The stimulated anti-Stokes scattering is observed in the directions of $\boldsymbol{k}_{AS}$ determined by the phase-matching condition; $2\boldsymbol{k}_L - \boldsymbol{k}_S = \boldsymbol{k}_{AS}$.

Despite the high intensity of the Stokes and anti-Stokes scattering there have been limited applications of the stimulated Raman scattering (SRS) in molecular spectroscopy. This results from the fact that the stimulated Raman scattering is observed only for several of the strongest Raman lines. To avoid these deficiencies, coherent techniques known as *CARS* (*Coherent Anti-Stokes Raman Spectroscopy*) and *CSRS* (*Coherent Stokes Raman Spectroscopy*) have been developed. CARS and CSRS are techniques based on SRS that use two laser beams to excite a sample at the frequencies ω_L and ω_S (instead of one at the frequency ω_L in SRS). The phenomena are similar to those described in the previous section, but the CARS and CSRS signals are much stronger. Moreover, by tuning the frequency ω_S (or ω_L) to resonance with a molecular vibration $\omega_L - \omega_S = \omega_{vib}$ it is possible to excite most of the vibrations in contrast to SRS for which only the strongest vibrational lines could be observed. The stimulated anti-Stokes scattering, generated with only one beam at the frequency ω_L, is observed on the cone-surface with the cone-axis parallel to the wave vector, $\boldsymbol{k}_L$ in contrast to the CARS method where the stimulated anti-Stokes scattering is observed only in one direction determined by the plane of the vectors $\boldsymbol{k}_L$ and $\boldsymbol{k}_S$ (Fig. 5.24).

The CARS method is particularly useful because:

a) The intensity of the CARS signal can exceed 10^4–10^5 times the intensity of the signal that can be recorded by spontaneous Raman spectroscopy.
b) The anti-Stokes signal frequency $\omega_{AS} = 2\omega_L - \omega_S$ is higher than the frequency of the beams ω_L and ω_S. This means that the CARS signal does not cover the region of fluorescence, in contrast to the frequency of the CSRS signal.
c) The CARS beam propagates within a small spatial divergence, in a well-defined direction, and can be detected even without a monochromator.
d) The high spatial resolution enables examination of different regions within a sample, such as the local composition and temperatures in plasmas. CARS microscopy has also been used to image bacteria and the mitochondria in living cells.
e) CARS has a high spectral resolution. It is easy to obtain resolutions of 0.3–$0.03\,\text{cm}^{-1}$ for pulsed lasers, and even of $0.001\,\text{cm}^{-1}$ for *cw* single-mode lasers.

One of the possible CARS configurations is shown in Fig. 5.25. The beam from laser 1 (e.g., an argon laser) at frequency ω_L is divided by a beam-splitter (*BS*) into two beams. One of these arrives at a sample directly, and the second one pumps the laser 2 (e.g., a dye-laser tunable over a wide range), which emits the line at frequency

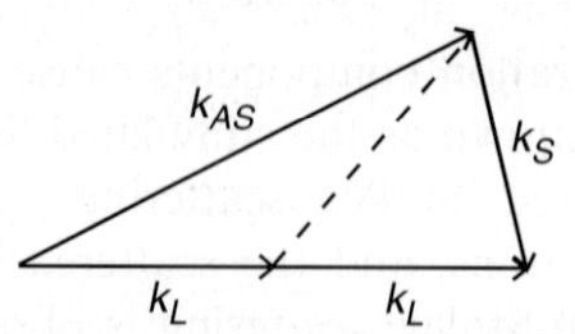

Fig. 5.24 Propagation direction in CARS technique.

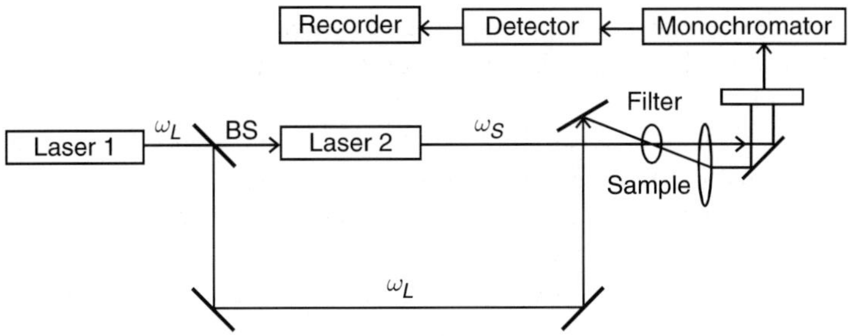

Fig. 5.25 Scheme of CARS configuration.

ω_S. Both beams ω_L and ω_S are focused on the sample. If these frequencies ω_L and ω_S match the resonance condition $\omega_L - \omega_S = \omega_{vib}$, the strong stimulated Stokes and anti-Stokes scattering signals are generated in the sample. By using a filter one can select the anti-Stokes signal at ω_{AS}. With laser beams ω_L and ω_S tunable over a wide spectral range it is possible to study various vibrations, $\omega_{vib} = \omega_L - \omega_S$.

5.5.3. The Other Techniques of Nonlinear Stimulated Raman Scattering

Besides the coherent stimulated anti-Stokes (CARS) and Stokes (CSRS) Raman scattering there are many other techniques for the coherent stimulated Raman effect—for example, SRGS (*Stimulated Raman Gain Spectroscopy*) and IRS (*Inverse Raman Scattering*). The nonlinear four-wave interaction of the incident beams at frequencies ω_L or ω_S with the electric field originating from the coherent vibrations of molecules at the frequency ω_{vib} can cause a gain, not only at the frequencies $2\omega_S - \omega_L$ (CSRS) and $2\omega_L - \omega_S$ (CARS), but also the Stokes gain $\omega_S(I(\omega_S) + \Delta S(\omega_S))$ or the intensity loss of the incident beam $\omega_L(I(\omega_L) - \Delta S(\omega_L))$. The former method related to the Stokes gain is known as the SRGS (Stimulated Raman Gain Stokes) method, and the latter is known as IRS (Inverse Raman Scattering). A schematic diagram of the most commonly used Raman techniques is given in Fig. 5.26.

The stimulated Raman scattering discussed in this chapter is emerging as a crucial technology for high-speed, long-distance fiber-optic transmission [10–12]. The stimulated Raman scattering amplifies the input signal in an optical fiber in the same way as the stimulated emission by erbium atoms amplifies signals in EDFA

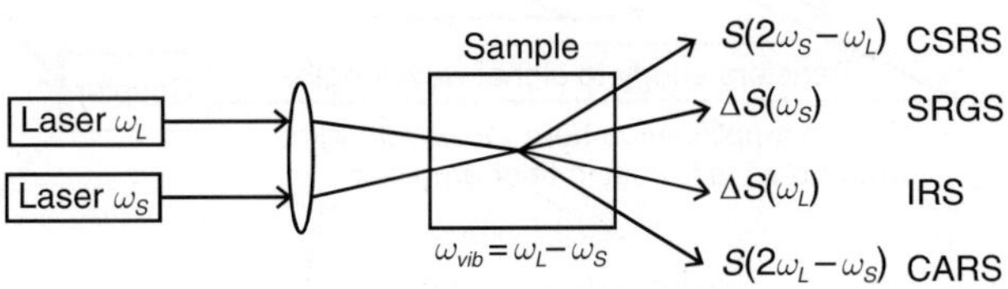

Fig. 5.26 Schematic diagram of the most commonly used techniques of nonlinear coherent stimulated Raman scattering.

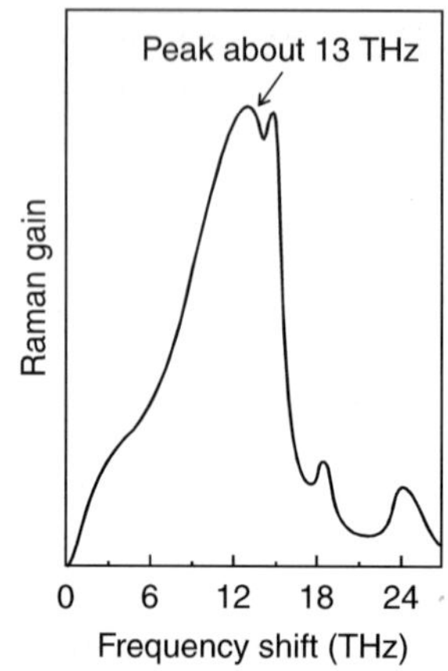

Fig. 5.27 Cross-section for stimulated Raman scattering in silica, showing the shift peaking at a frequency 13 THz from the stimulating wavelength, about 100 nm in the 1550 nm window. [11]

(erbium-doped fiber amplifiers). The crucial difference comes from the fact that the wavelength of the Raman gain is shifted to higher wavelengths (Stokes component) with respect to the pump-pulse ($\omega_S = \omega_p - \omega_{vib}$; where ω_{vib} is the vibrational frequency of molecules in the glass fiber itself, or a dopant). For silica, with the pump near 1550 nm, the amplified wavelength is about 100 nm longer. Through the transfer of energy by optical-phonon interaction, a Raman gain is available at any frequency in a silica fiber, with a peak near 13 THz, shifted from the pump frequency (Fig. 5.27). The maximum gain-transfer occurs if the signal and pump beams have the same polarization.

In the usual configuration the Raman amplification occurs in the final length of fiber before the receiver or EDFA (Fig. 5.28). The signal and pump beams travel in opposite directions, with the pump coupled into the fiber at the receiver end. A coupler directs the pump-light into the transmitting fiber, while diverting the signals which arrive through the fiber to the receiver or EDFA.

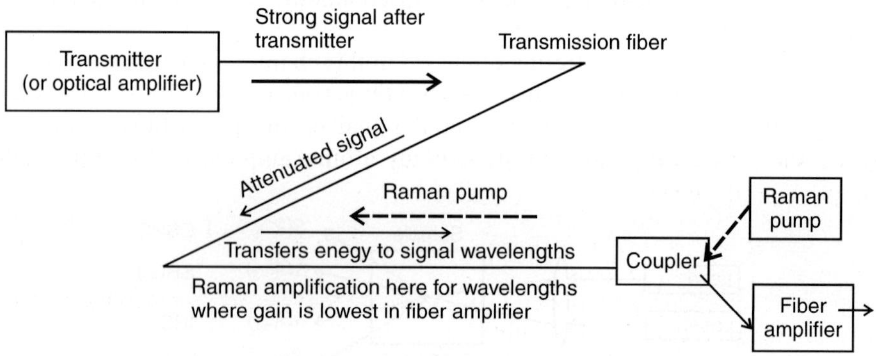

Fig. 5.28 Distributed Raman amplification amplifies the signals passing through the fiber in the transmission cable, by transferring energy from a strong pump beam to the less powerful signal wavelengths. [10]

5.6. NONLINEAR DISPERSION PHENOMENA AFFECTING PICOSECOND AND FEMTOSECOND PULSE DURATION – GROUP VELOCITY DISPERSION (GVD) AND SELF PHASE MODULATION (SPM)

In section 5.3 we have shown that the mechanisms of the second harmonic generation for continuous-wave and pulsed lasers are similar. However, the dispersion properties of an active medium become increasingly more important for short pulses. For femtosecond lasers, and particularly for pulses shorter than 100 fs, the dispersion properties of the active medium and the optical elements in the resonator begin to affect the pulse duration significantly. Moreover, the output beam emitted from the laser usually travels through additional optical elements (mirrors, prisms, crystals, beam-splitters, filters, etc.) before it reaches a detector. Here, we will show that the optical elements have a tremendous influence on the pulse duration.

Two main mechanisms result in modification of the pulse: group velocity dispersion (GVD) and self phase modulation (SPM). SPM arises from the third-order susceptibility, and is a nonlinear third-order effect. The influence of GVD and SPM on the pulse duration is described below.

We first consider the influence of selective transmission (i.e., a filter) or selective reflection (i.e., dichroic mirror) on the temporal shape of a pulse. First, assume that the intensity profile of the laser pulse can be described by a Gaussian function

$$E(t) = \frac{E_0}{\tau} e^{-\frac{t^2}{2\tau^2}}, \tag{5.65}$$

with the pulse duration t_{imp} defined by

$$t_{imp} = 2\tau(\ln 2)^{1/2}. \tag{5.66}$$

As the pulse propagates through the dispersive medium (filter) or is reflected by a dispersive surface (mirror), different spectral components $E(\omega)$ of the pulse are subject to different modifications of the amplitude $A(\omega)$ and the phase $\Phi(\omega)$. The pulse duration of such a modified pulse can be calculated from the reverse Fourier transform

$$E'(t) = \int_{-\infty}^{\infty} E(\omega)A(\omega)\mathrm{e}^{-\mathrm{i}\Phi(\omega)}\mathrm{e}^{\mathrm{i}\omega t}\,d\omega. \tag{5.67}$$

First, assume that only the amplitude $A(\omega)$ is modified when the pulse travels through the optical element

$$A(\omega) = \exp\{-(\omega - \omega_0)^2/2\omega_F^2\}, \tag{5.68}$$

where ω_F is a spectral width of the filter and ω_0 is the frequency corresponding to the maximum spectral density of the pulse.

By substituting eq. (5.68) into (5.67) we obtain the modified pulse temporal shape

$$E'(t) = \frac{E_0}{\tau'} e^{-t^2/2\tau'^2}, \tag{5.69}$$

where

$$\tau' = \tau\left(1 + \frac{1}{\omega_{\mathrm{F}}^2\tau^2}\right)^{1/2} \tag{5.70}$$

is the modified pulse duration. It can be seen that the pulse traveling through the filter becomes longer. The smaller the spectral width of the filter, ω_{F}, the longer is the pulse duration. Only when $\tau \gg 1/\omega_F$, is the pulse duration unaltered.

So far, we have not considered the change in phase which occurs when the pulse passes through the optical element. The expression for the phase $\Phi(\omega)$ of the wave passing through the optical path-length, L, through the medium characterized by the refractive index $n(\omega)$, is given by

$$\Phi(\omega) = \frac{\omega n(\omega)}{c} L. \tag{5.71}$$

We now express the phase $\Phi(\omega)$ as a series-expansion around the central frequency, ω_0

$$\Phi(\omega) = \Phi_0 + \frac{d\Phi}{d\omega}(\omega - \omega_0) + \frac{1}{2}\frac{d^2\Phi}{d\omega^2}(\omega - \omega_0)^2 + \dots \tag{5.72}$$

After substituting (5.72) into (5.67) it becomes evident that the first term Φ_0 does not influence the temporal shape of the pulse, but only contributes to the phase shift. The second term also has no influence on the temporal pulse shape; it only describes the time-delay of the pulse propagating through the medium. Indeed, by comparing the first derivative of eq. (5.71) with eq. (5.38), we obtain

$$\frac{d\Phi}{d\omega} = \frac{n}{c}\left(1 + \frac{\omega}{n}\frac{dn}{d\omega}\right)L = \frac{dk}{d\omega}L = \frac{L}{v_g} = t_g, \tag{5.73}$$

where v_g is the group velocity. The term $\frac{d\Phi}{d\omega}$ has a clear physical meaning; it is the time, t_{g}, required for the pulse spectral component traveling with the group velocity, v_{g}, to pass through the medium of length L. The third term of the expression (5.72), describing the second-order dispersion $\frac{1}{2}\frac{d^2\Phi}{d\omega^2}(\omega - \omega_0)^2$, affects the pulse's temporal shape. Assuming $\delta = \frac{d^2\Phi}{d\omega^2} = \mathrm{const}$ and substituting into (5.67) one obtains

$$E'(t) = \frac{E_0}{\sqrt{\tau^2 + i\delta}} e^{-t^2/2\tau'^2} e^{i\frac{\delta}{\tau^2}\frac{t^2}{2\tau'^2}}, \tag{5.74}$$

and the modified pulse-duration is given by

$$\tau' = \tau\sqrt{1 + \frac{\tau_c^4}{\tau^4}}, \tag{5.75}$$

where $\tau_c = \sqrt{|\delta|}$.

The derivation of the expressions (5.69)–(5.70) and (5.74)–(5.75) can be found in ref. [13].

It is clear from (5.75) that the second-order dispersion term causes lengthening of the pulse duration, i.e., the shorter the input pulse, τ, the longer is the output pulse, τ'. The second-order dispersion plays an important role mainly for short femto-second pulses. For pulses longer than hundreds of femtoseconds this effect becomes negligible. Using (5.71) we can express the second-order dispersion term $\frac{d^2\Phi}{d\omega^2}$ as

$$\frac{d^2\Phi}{d\omega^2} = \left(\frac{2}{c}\frac{dn}{d\omega} + \frac{\omega}{c}\frac{d^2n}{d\omega^2}\right)L = \frac{d^2k}{d\omega^2}L. \quad (5.76)$$

If $\frac{d^2n}{d\omega^2}$ does not vanish, the group velocities corresponding to various frequencies are different, and the medium is said to be characterized by *group velocity dispersion* (GVD).

Some values of $\frac{d^2\Phi}{d\omega^2}$ at 800 nm are: 580 fs^2/cm for Ti:sapphire crystal; 360 fs^2/cm for fused silica, and 1500 fs^2/cm for SF10 glass.

We now summarize the influence of the GVD on the shape and the duration of a laser pulse. The group velocity dispersion is the result of the various Fourier components of a pulse traveling at different phase velocities. The shorter the pulse, the greater is the number of components at different wavelengths, and the greater the influence of GVD on the shape and the pulse duration. This is intuitively obvious if we keep in mind the fact that the shorter the pulse, the broader is the spectral range involved. Since the refractive index, $n(\omega)$, of any material depends in a non-linear way on frequency, each frequency within the pulse experiences a slightly different group velocity, v_g, as it propagates. The broader the spectral range, the greater is the GVD effect for the longest and the shortest wavelengths within a pulse.

Figure 5.29 shows a typical plot of the refractive index, $n(\lambda)$, versus the wavelength λ. For any given wavelength, the refractive index $n(\lambda)$ determines the phase-velocity according to the relationship (5.12). The slope of the curve, $\frac{dn(\lambda)}{d\lambda}$, determines the group velocity according to

$$v_g = \frac{c}{\left(n(\lambda) - \lambda\frac{dn}{d\lambda}\right)} \quad (5.77)$$

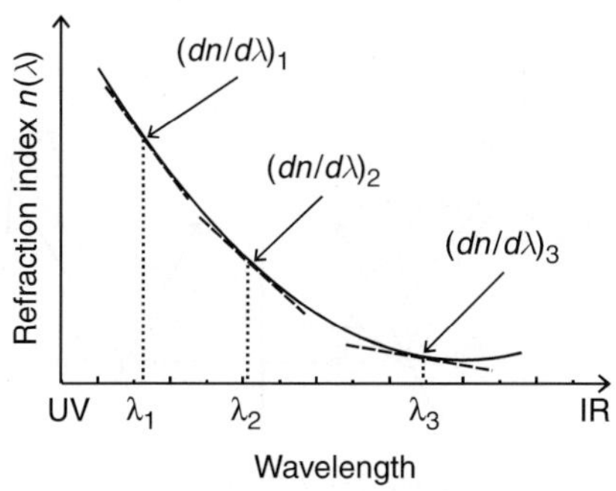

Fig. 5.29 A typical plot of the refractive index $n(\lambda)$ versus wavelength λ.

for the wave-packet at the wavelength λ, which can easily be derived from the definition of the group velocity (5.37).

The second derivative of the curve, $\frac{d^2n}{d\lambda^2}$, determines the value of the GVD for the material. When $\frac{d^2n}{d\lambda^2}$ is zero, which indicates that the relationship in Fig. 5.29 is linear, the GVD is zero, and all the spectral components travel with the same group velocity. For a material with a positive GVD the longer-wavelength components travel faster than the shorter wavelength components. For a negative GVD the relationship is reversed. It should be noted that some textbooks tend to use a different nomenclature, with a reversed GVD sign. The positive GVD is called a normal GVD, and the negative GVD is called an anomalous GVD. The greater the group-velocity dispersion, the greater is the modification in the pulse shape and the pulse duration. The pulse is said to be positively chirped when the longer wavelengths travel faster in the medium than the shorter wavelengths.

In order to achieve short, stable, and reproducible output pulses from a laser, it is necessary to compensate the effects coming from the GVD. To achieve a GVD equal to zero, the group delay time t_g must be frequency independent, i.e., $t_g = \frac{d\Phi}{d\omega} = const$, as we have shown in eqs. (5.73) and (5.76). However, for most optical elements it is impossible to avoid the GVD effect. For example, Ti:sapphire crystals in femtosecond lasers exhibit a positive GVD, inducing a positive chirp on the pulse, and broadening the temporal spectrum of the pulse as it propagates through the cavity of the resonator. In order to achieve the ideally short pulse with a non-deformed shape, the positive GVD effects occurring in the laser system need to be compensated with negative GVD effects having the same absolute magnitude.

The two most common methods for compensating the GVD effect are to use prism pairs for the femtosecond pulses, and a Gires-Tournois interferometer for picosecond pulses.

Figure 5.30 illustrates the principle of the first method. A positively chirped pulse from the cavity of the resonator travels to the prism P_1, where the different spectral components of the pulse are separated. Then, the broadened pulse enters the prisms P_2 and P_3 at the Brewster angle (to minimize losses). Since the glass of the prisms exhibits a positive GVD, the red wavelengths exhibit larger group velocities than the

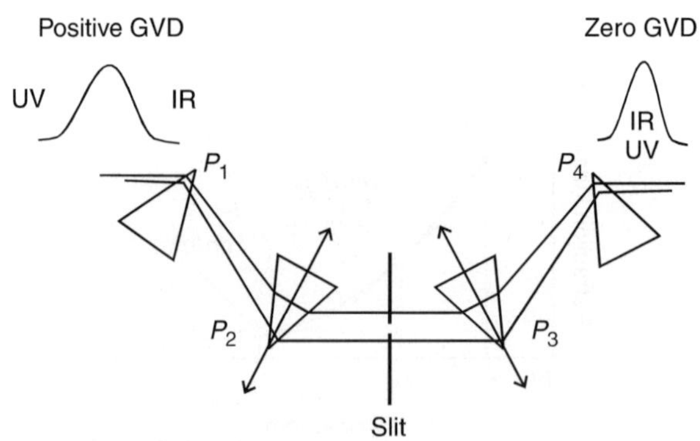

Fig. 5.30 Four prisms used to compensate the positive GVD.

blue wavelengths. However, the red wavelengths traverse a longer path in the prisms P_2 and P_3 than do the blue wavelengths. By translating prisms P_2 and P_3 perpendicular to their bases, it is possible to select a length for the optical path such that the group-velocity dispersion will be compensated. As a consequence, all spectral components of the pulse reach the prism P_4 at the same time, and the pulse is free of chirping. As a consequence, the output pulse is shorter than the input pulse and has an ideal shape, free of GVD effects. The prism P_4 eliminates the spatial distribution of the spectral components. Tuning of femtosecond lasers is achieved by moving the slit perpendicular to the beam and choosing the required wavelength. In the configuration shown in Fig. 5.30 a shorter wavelength is achieved by translating the slit downwards or upwards for a longer wavelength.

In picosecond lasers, GVD compensation is usually performed with a Gires-Tournois interferometer. This consists of a pair of parallel surfaces separated by a distance d. One surface is a partial reflector (reflectivity $r \ll 100\%$), while the other has 100% reflectivity. A typical spacing is of the order of a few tens of micrometers, and the reflectivity, r, is of the order of several percent. The round-trip time, t_i, through the interferometer is given by

$$t_i = \frac{2d}{c}. \tag{5.78}$$

For the Gires-Tournois interferometer, the group-delay-time $t_g = \frac{d\Phi}{d\omega}$ is frequency-dependent and may be expressed as

$$t_g(\omega) = \left[\frac{t_i(1+r(\omega))}{1-r(\omega)}\right] \frac{1}{\left[1+\left(\frac{4r(\omega)}{(1-r(\omega))^2}\sin^2\left(\frac{\omega t_i}{2}\right)\right)\right]}. \tag{5.79}$$

Calculating $\frac{d^2\Phi}{d\omega^2}$ from (5.79) by using $t_g = \frac{d\Phi}{d\omega}$ and $t_i = 2d/c$, we find that $\frac{d^2\Phi}{d\omega^2}$ is proportional to d^2. Figure 5.31 shows a typical group-delay-time, t_g–, dependence as a function of wavelength for a Gires-Tournois interferometer.

The GVD is proportional to the slope of the curve in Fig. 5.31 ($dt_g/d\lambda$) and is periodic, with both positive and negative GVD values in different regions of the spectrum. By increasing the spacing between the interferometer surfaces, the GVD

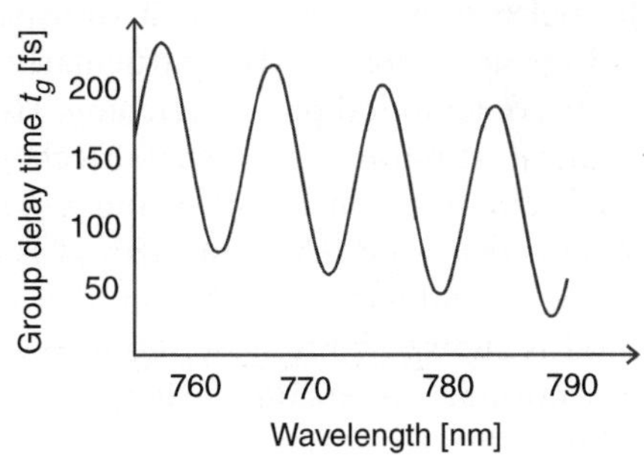

Fig. 5.31 Group delay time, t_g, as a function of wavelength for a Gires-Tournois interferometer.

effect increases. The larger the GVD, the narrower is the spectral bandwidth, and the longer the pulse's duration. For a Gires-Tournois interferometer with a given set of d and r parameters, the pulse duration can increase by as much as a factor of two. It is worth emphasizing that for any given Gires-Tournois interferometer placed in the laser resonator it is possible to make only fine adjustments of the spacing d, and hence the temporal duration of the pulse. In many practical designs the change in the spacing between the plates is achieved by using a piezo-electric transducer. By changing the voltage applied to the transducer, the distance between the plates is varied. By using several Gires-Tournois interferometers, each with different d and r parameters, it is possible to tune the pulses' durations from 1 ps to 80 ps.

There is another effect, self phase modulation (SPM), which causes temporal broadening of an ultrashort pulse. This results from the fact that the non-linear index of refraction $n(\omega)$ depends on the light intensity, I,

$$n(\omega) = n_0(\omega) + n_2(\omega)I(t). \tag{5.80}$$

$$I(t) = e^{-\Gamma t^2} \tag{5.81}$$

When an ultrashort pulse propagates through a nonlinear material the local index of refraction increases. The temporal center of the pulse, which is of course more intense, experiences a larger index of refraction than the leading and trailing edges. Therefore, the central part of the pulse slows, the leading edge speeds ahead, while the trailing edge catches up. In Ti:sapphire crystals, which exhibit a positive (normal) GVD, the longer wavelength pulse components travel faster than the shorter ones. Each of these longer wavelength components is driven by additional dispersion, with red-shifted components resulting from the $n_2(\omega)I$ term in eq. (5.80). A similar effect occurs at the trailing edge, producing additional spectral components shifted towards the blue. This effect is called the self-phase modulation (SPM) of the pulse. Thus, the nonlinear term $n_2(\omega)I$ provides an additional spectral broadening, which imposes an additional chirp on the pulse, which leads to an increase in its duration. The SPM effect on the pulse duration seems to run counter to accepted ways of thinking: the broader the pulse in the frequency domain, the shorter it is in the time domain. To understand this apparent discrepancy with the Fourier transformation we need to stress that the new frequencies created by SPM are not synchronized. Although they are produced under the original pulse envelope, they are no longer transform-limited, as the pulse propagates across a transparent material and the Fourier transformation does not apply. SMP, particularly when combined with positive GVD, leads to effective temporal pulse stretching. However, the same effect can be used to compress ultrafast pulses. Two general techniques are used to compress ultrafast pulses via a combination of SPM and GVD; a) the *grating-fiber method*, b) the *soliton-effect compressor*. The first method is used in the visible and the near infrared ranges. The second method is used in fiber networks. The principle of the grating-fiber method is shown in Fig. 5.32. First, recall that for most wavelengths, materials show "normal dispersion", where the index of refraction n increases as λ decreases (Fig. 5.29). However, for wavelengths close to the maximum of the absorption band, the dependence of the index of refraction on the wavelength, λ, reverses direction. At the maximum of absorption, the GVD is zero. The material

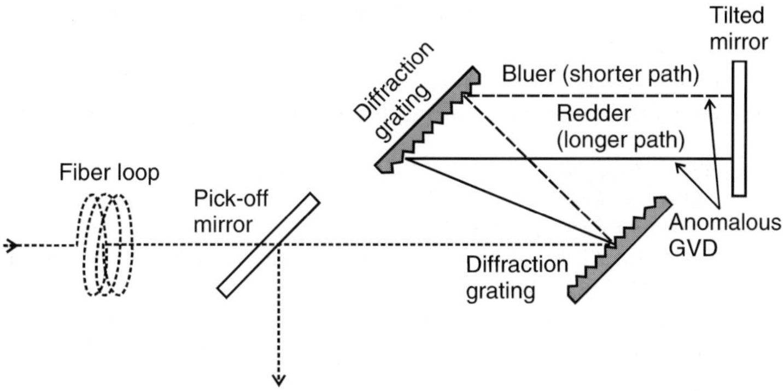

Fig. 5.32 Typical setup for the grating-fiber compression.

shows, "anomalous dispersion" close to the absorption maximum, and the pulse exhibits a negative GVD, with its bluer components traveling faster than the redder components. The grating-fiber method is used for visible and near-IR pulses that are far from the absorption band of the fiber material. So, the input pulse falls in the normal dispersion region, with a positive GVD. During the propagation in the fiber the pulse is affected both by SPM and GVD. The resulting pulse has its spectral width significantly increased, with a significant positive chirp, which is almost linear. Then, the positively chirped pulse passes through an external pair of gratings, which are designed to produce negative GVD. The negative GVD designed pair of grating offset the positive chirp from the positive GVD and SPM by making the optical path length of the redder components substantially greater than the path for the bluer components. Therefore, the trailing bluer edge of the pulse catches up with the leading redder edge, and all components begin to travel at the same velocity. As a result, the output pulse exhibits zero GVD, and is compressed to less than its initial value.

The soliton-effect compressor is used for radiation close to the absorption band of the fiber material. Ordinary silica fiber has the absorption (and zero GVD) at 1.31 μm. Therefore, for radiation close to the maximum, the fiber exhibits anomalous dispersion, and the pulse propagating across the fiber experiences a negative GVD. The soliton-effect compressor consists simply of a piece of fiber having a length chosen properly to balance the positive chirp introduced by SPM with the negative chirp introduced by the fiber. These two effects can offset each other in a properly chosen fiber length, and the result is a compressed pulse that exhibits zero GVD. This pulse is called an optical *soliton*. Such a pulse is stable in shape, power, and duration over extremely long distances, and can be transmitted in telecommunication optical fibers without distortion.

These effects are present in solid-state lasers based on a crystal medium, as well as in fiber lasers with a silica glass as a laser medium. The silica glass has a symmetry center of inversion that rules out the second-order nonlinear effects, but the third-order effects like SPM still exist. The SPM effect in fiber-optic systems is smaller than in solid-state lasers, because of the significantly smaller intensities applied.

REFERENCES

1. W. Koechner, *Solid-State Laser Engineering*, 5th Edition, vol. 1, Springer-Verlag (1999)
2. C. Rullière, *Femtosecond Laser Pulses*, Principles and Experiments, Springer-Verlag (1998)
3. M. Born, E. Wolf, *Principles of Optics*, Macmillan, New York (1964)
4. Lasermetric, Electro-Optics Div., Englewood, New Jersey, Data Sheet 8701, Optical Harmonic Generating Crystals (February 1987)
5. V.G. Dmitriev, G.G. Gurzadyan, D.N. Nikogosyan, *Handbook of Nonlinear Optical Crystals*, Springer-Verlag (1991)
6. A. Laubereau, *Optical Nonlinearities with Ultrashort Pulses*, in: *Ultrashort Laser Pulses and Applications*, W. Kaiser ed., Topics and Applied Physics, vol. 60, Springer-Verlag, Berlin Heidelberg (1988), p. 35
7. R.C. Miller, Phys. Lett. 26A (1968) 177
8. H. Greaner, personal communication
9. E.J. Lerner, *Ultrafast Lasers Deliver Powerful, Precise Pulses*, Laser Focus World 34 (12). (December, 1998) 77–84
10. J. Hecht, *Raman amplifiers boost system margins at high speed*, Laser Focus World 37 (6) (June, 2001) 135–140
11. J. Hecht, *Mitigating nonlinear effects is essential to long-haul transmission systems*, Laser Focus World 38 (5) (May, 2002) 155–162
12. M. Islam, M. Nietubyc, *Not a pretty picture*, WDM solutions 3 (3) (March 2001)
13. H.A. Haus, *Waves and Fields in Opto-electronics*, Prentice Hall (1984)
14. G.A. Mourou, C.P.J. Barty, M.D. Perry, Phys. Today 51 (1) (1998) 22
15. G. Mourou, *The Ultrahigh-Peak-Power Laser: Present and Future*, Appl. Phys. B 65 (1997) 205–211
16. X. Liu, D. Du, G. Mourou, *Laser ablation and micromachining with ultrashort laser pulses*, IEEE Journal of Quantum Electronics 33 (10) (1997) 1706–1716
17. E. Margalith, *The OPO goes from concept to commercial tool*, Laser Focus World 36 (9) (September, 2000) 95–100
18. J.A. Giordmaine, R.C. Miller, *Tunable Coherent Parametric Oscillation in* $LiNbO_3$ *at Optical Frequencies*, Phys. Rev. Lett. 14 (1965) 973–976
19. S.E. Harris, Proc. IEEE 57 (12) (1969) 2096
20. C. Chen et al., Sci. Sin Ser. B 28 (1985) 235
21. Y.X. Fan et al., Proc. Conf. on Lasers and Electro-Optics, postdeadline papers, 11 (1986)
22. Y.X. Fan et al., Proc. Conf. on Lasers and Electro-Optics, 527 (1988)
23. L.K. Cheng, W.R. Bosenberg, C.L. Tang, *Broadly tunable optical parametric oscillation in* β-BaB_2O_4, Appl. Phys. Lett. 53 (3) (1988) 175–177
24. W.R. Bosenberg, L.K. Cheng, C.L. Tang, Ultraviolet optical parametric oscillation in β-BaB_2O_4, Appl. Phys. Lett. 54 (1) (1989) 13–15
25. R. Burnham et al., IEEE Photonics Tech. Papers 1 (1) (1989) 27
26. Yunping Wang, Zuyan Xu, Daoqun Deng, Wanhua Zheng, Baichang Wu, Chuangtian Chen, *Visible optical parametric oscillation in* LiB_3O_5, Appl. Phys. Lett. 59 (5) (1991) 531–533
27. G.P. Agrawal, *Nonlinear Fiber Optics*, 3rd ed., Academic Press (2000)
28. J.A. Buck, *Nonlinear effects in optical fibers*, Chapter 3 in *Handbook of Optics*, Michael Bass ed., vol. 4 Fiber Optics and Nonlinear Optics, McGraw-Hill, NY (2001)

– 6 –

Pulse Amplification

6.1. INTRODUCTION

Output pulse energies from femtosecond lasers typically do not exceed a few nanojoules, and peak powers of megawatts. For many applications, higher energies or peak powers are required. Researchers are seeking methods to shorten pulses, and to increase peak powers and peak intensities on targets. Given the limits on further trimming of the pulse duration, further increases in peak power and peak intensities can only be obtained by increasing output energy. Amplification of energy from the femtosecond lasers makes terawatt peak powers possible. Amplification, combined with the initial pulse stretching and compression as a final stage, can convert terawatt systems into petawatt lasers with sub-picosecond pulses.

So far the highest energies, peak powers and irradiance can be achieved in Nd:glass amplifiers, not those based on Ti:sapphire. The most powerful lasers in the world (in 2003) are "Vulcan" in the Rutherford Appleton Laboratory, UK, delivering 2.5 kJ in two 150 nm beams, 1 pW, 10^{21} W/cm^2 and the Nova system at Lawrence Livermore National Laboratory, CA, USA, delivering 1.3 kJ pulse at 800 ps that can be compressed to 430 fs to achieve 1.3 pW and 10^{21} W/cm^2.

When a laser pulse passes through an optically active material in which the population inversion is maintained by a pumping source, it gains energy from the stimulated emission generated by itself in the medium. As a result the output pulse is amplified (Fig. 6.1).

6.2. THEORETICAL BACKGROUND

Let us consider a mechanism of amplification in the three-level system (Fig. 6.2):

Let us assume that the relaxation time τ_{32} for the $3 \rightarrow 2$ transition is short in comparison with the lifetime of the E_2 state τ_{21}, which is a good approximation in solid-state lasers. This denotes that the number of molecules N_3 occupying the level E_3 is negligible compared with the number of molecules in N_1 and N_2 ($N_3 \approx 0$, and $N_1 + N_2 = N_0$) because the molecules are pumped almost immediately into the metastable level E_2 with only a momentary stay in E_3.

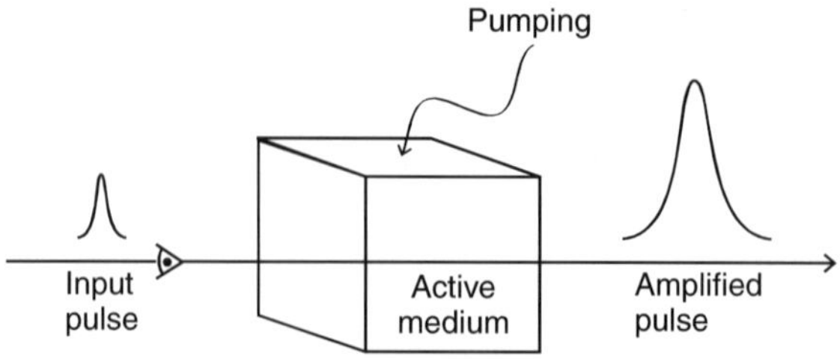

Fig. 6.1 Illustration of amplification.

Based on this assumption, the change in the population is

$$\frac{dN_2}{dt} = W(N_1 - N_2) - \frac{N_2}{\tau_{21}} + W_p N_1, \tag{6.1}$$

$$\frac{dN_1}{dt} = -\frac{dN_2}{dt} \tag{6.2}$$

where

$$W = B_{21}\rho. \tag{6.3}$$

Neglecting spontaneous emission and pumping, W_p, during the pulse duration, which is usually much shorter, eq. (6.1) can be written in the form

$$\frac{dn}{dt} = -\gamma W n, \tag{6.4}$$

where

$$n = N_1 - N_2 \tag{6.5}$$

denotes the population inversion, $\gamma = 1 + \frac{g_2}{g_1}$ (for a three-level system, where g_2, g_1 are the degeneration numbers). One can show [1] that

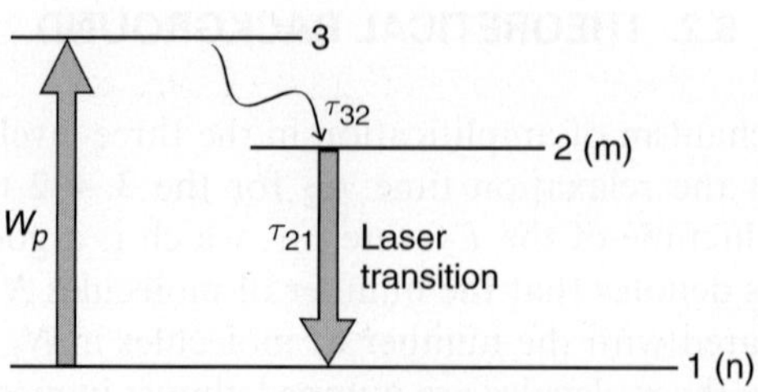

Fig. 6.2 Three-level system.

$$W = \frac{\sigma I}{h\nu} \tag{6.6}$$

where $\sigma\,[\mathrm{cm}^2]$ is the stimulated emission cross section, I is the radiation intensity $\left[\frac{\mathrm{J}}{\mathrm{s\ cm}^2}\right]$, ν is the frequency of the $2 \rightarrow 1$ transition. Substituting (6.6) into (6.4) and expressing the intensity in terms of the photon density ϕ [photons/cm^3]

$$\frac{I}{h\nu c} = \phi, \tag{6.7}$$

one gets

$$\frac{dn}{dt} = -\gamma c\sigma\, n\phi. \tag{6.8}$$

The rate at which the photon density changes in a small volume at x of an active medium is equal to

$$\frac{\partial \phi}{\partial t} = Wn - \frac{\partial \phi}{\partial x} c \tag{6.9}$$

where the first term describes the number of photons generated by the stimulated emission and the second term the flux of photons which flows out from that region. Using once again (6.6) and (6.7) in (6.9) one gets

$$\frac{\partial \phi}{\partial t} = cn\sigma\phi - \frac{\partial \phi}{\partial x} c. \tag{6.10}$$

Using the differential equations (6.8) and (6.10) one can solve them for various types of the input pulse shapes [2–3]. For a square pulse of duration t_p and the initial photon density ϕ_0 one obtains:

$$\frac{\phi(x,t)}{\phi_0} = \left\{1 - [1 - \exp(-\sigma n x)] \exp\left[-\gamma\sigma\phi_0\left(c\left(t - \frac{x}{c}\right)\right)\right]\right\}^{-1}. \tag{6.11}$$

After passing through an active medium of length $x = L$ the amplification is given by

$$G = \frac{\int_{-\infty}^{\infty} \phi\,(L,t) dt}{\phi_0\, t_p}. \tag{6.12}$$

After substituting (6.11) into (6.12) and integrating one gets

$$G = \frac{1}{c\gamma\sigma\phi_0 t_p} \ln\{1 + [\exp(\gamma\sigma\phi_0\tau_0 c) - 1]e^{n\sigma L}\}. \tag{6.13}$$

The equation (6.13) can be cast in a different form using the input energy E_{in}

$$E_{in} = c\phi_0\, t_p\, h\nu, \tag{6.14}$$

and the saturation fluence that we defined in chapter 1 (eq. 1.39)

$$E_s = \frac{h\nu}{\gamma\sigma} = \frac{E_{st}}{\gamma g_0}, \tag{6.15}$$

where $E_{st} = h\nu n$ is the energy stored per volume, $g_0 = n\sigma$ is the small signal gain coefficient (see chapter 1, eq. 1.23). Using the expressions for E_{in} and E_s in (6.13) one obtains the energy gain G

$$G = \frac{E_s}{E_{in}} \ln\left\{1 + \left[\exp\left(\frac{E_{in}}{E_s}\right) - 1\right] G_0\right\}, \tag{6.16}$$

where $G_0 = \exp(g_0 L)$ is the small-signal single-pass gain.
Consider limit cases:

1. The input signal is low, $E_{in}/E_s \ll 1$. Eq. (6.16) can be approximated to the exponential dependence on the active medium length L

$$G \cong G_0 = \exp(g_0 L). \tag{6.17}$$

2. The high input signal, $E_{in}/E_s \gg 1$. Then, eq. (6.16) becomes

$$G \cong 1 + \frac{E_s}{E_{in}} g_0 L \tag{6.18}$$

with the linear dependence of the energy gain.

6.3. DESIGN FEATURES OF AMPLIFIERS

For many scientific and practical applications we need powerful pulses [4–6]. The pulses emitted from the oscillator can be amplified in devices called amplifiers. No single amplifier configuration has emerged because one configuration can be more suitable than other for a given application. Usually, the gain G which can be achieved and the energy which can be extracted from the amplifier are the main parameters in the design of amplifiers.

The four most important configurations for laser light amplification are:

1. multistage power amplifier (Fig. 6.3),
2. multipass amplifier (MPA) (Fig. 6.4),
3. regenerative amplifier (RGA) (Fig. 6.5),
4. chirped pulse amplifier (CPA) (Fig. 6.6).

The multistage power amplifier consists of a laser oscillator and a pumped active medium, the amplifier. The amplifier is driven by the oscillator, which generates an initial pulse of moderate power and energy. The pulse from the oscillator passes through the active medium of the amplifier and its power grows–in extreme cases up to 100 times. The pulse can be amplified further by adding the next amplifier in the configuration presented in Fig. 6.3. These configurations are used for longer pulses.

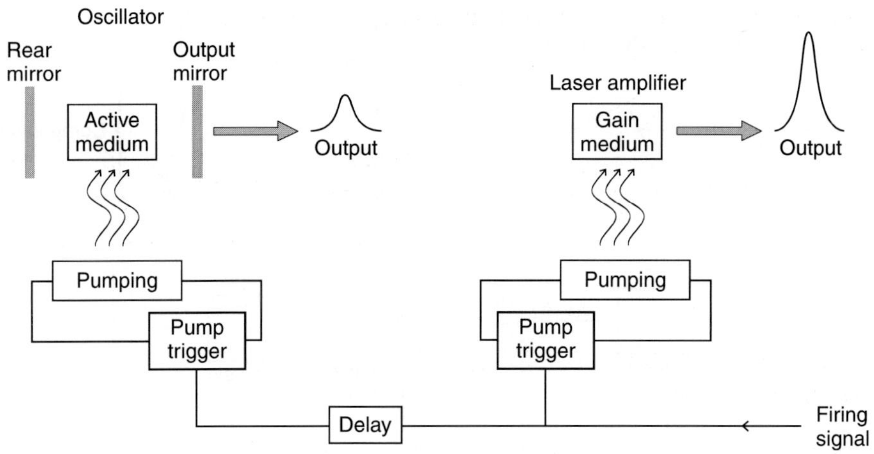

Fig. 6.3 Oscillator-amplifier configuration.

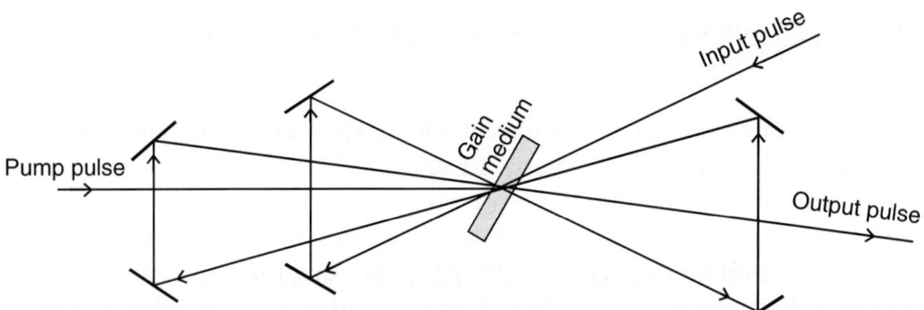

Fig. 6.4 Multipass amplifier.

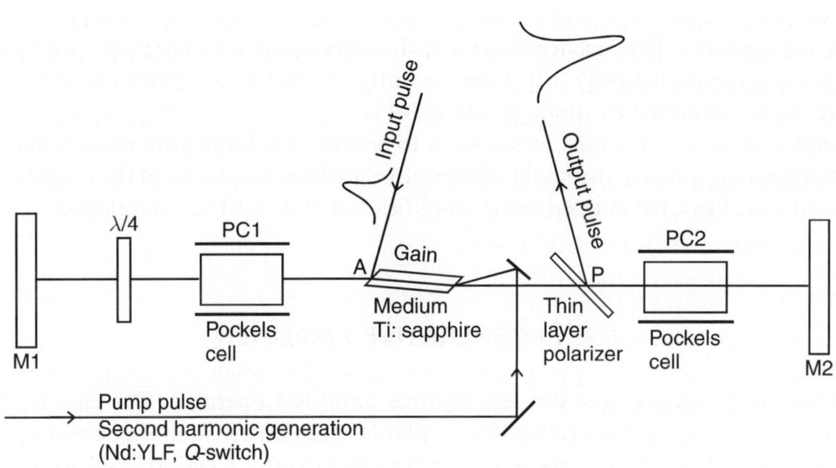

Fig. 6.5 Regenerative amplifier.

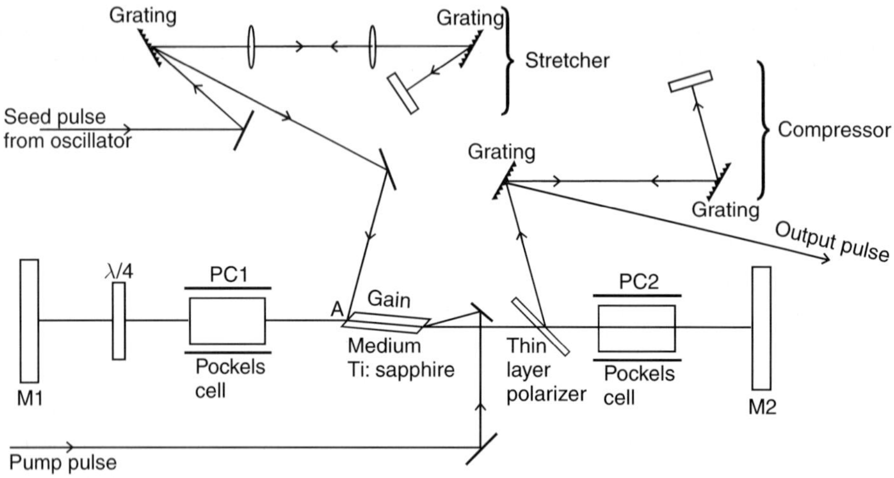

Fig. 6.6 Chirped pulse amplifier.

The double-pass amplifiers are used for typical picosecond systems and can deliver 125 mJ pulses of less than 60 fs at 20 Hz with a solid-state saturable-absorber-based oscillator.

Multipass amplifiers (Fig. 6.4) are used when extremely short pulses are required (shorter than 35 fs at 1 kHz and energy of 1.5 mJ). This configuration is simple and less sensitive to thermal lensing, therefore providing excellent beam quality, and it is relatively easy to calculate and compensate dispersion in the system.

Regenerative amplifiers (Fig. 6.5) are generally employed for ultrafast systems with relatively low output energies from femtosecond oscillators (a few nJ) to achieve the output energy from the amplifier of around 1 mJ at 1 kHz. Higher energy can be obtained from combined RGA/MPA methods that can deliver energy greater than 3.5 mJ per pulse and higher at repetition rates of 1 kHz. Regenerative amplifiers are also used to produce powerful picosecond pulses from a train of low-energy pulses of modelocked lasers. The regenerative amplifiers have originally been pumped by flash lamps or by *cw* arc lamps [7–12]. More recently, the Nd:YAG, Nd:YLF or Nd:glass amplifiers are pumped by diode arrays [13–17].

Amplification of ultrashort pulses leads to enormously large peak powers that may be far above the damage threshold of the active medium employed in the amplifier. To avoid this problem the chirped-pulse amplification (Fig. 6.6) was developed.

6.4. REGENERATIVE AMPLIFIER

Now we shall explain how the regenerative amplifier operates. One can see from Fig. 6.5 that the typical commercially available regenerative amplifier consists of an active medium (Ti:sapphire crystal), two Pockels cells (PC1 and PC2), $\lambda/4$ plate and a thin-film polarizer placed between two mirrors, M1 and M2.

The gain medium is pumped with the second harmonic of the solid-state laser Nd:YLF, Q-switched with pulses durations of 250 ns and an average energy of 8 W at the Brewster's angle.

The regenerative amplifier selects an individual pulse from a train of modelocked pulses (called a seed pulse). The seed pulse is trapped in an amplifier where it is amplified at each pass in the crystal. The pulse passes many times (usually about 10–20 times) through the gain medium to get more energy. Once its energy increases by as much as 10^6, the amplified pulse is removed from the amplifier as the output pulse that can be used for further applications.

The number of passages in the regenerative amplifier depends on the round-trip time between the M1 and M2 mirrors in the amplifier. If the round-trip time is 10 ns and the pumping pulse duration is 250 ns, a typical time evolution of the pulse energy in a regenerative amplifier is given in Fig. 6.7.

The amplified pulse shape follows the shape of the pumping pulse originating from the depletion of the excited-state in the gain medium. It is obvious from Fig. 6.7 that the pulse should be switched out from the amplifier after around 20 passages where there is maximum gain and the energy of the amplified pulse is the highest.

Now we have to understand how the input pulse is trapped in the amplifier. We will show that it is trapped on the polarization basis. Let us recall that:

1. $\lambda/4$ plate changes the polarization from linear to circular,
2. $\lambda/2$ plate rotates the linear polarization by 2α, where α is the angle between the polarization vector and the optic axis,
3. The thin layer polarizer acts as a reflective polarizer (see Chapter 1) that at the Brewster's angle reflects the polarization perpendicular to the plane of reflection, and transmits the polarization parallel and perpendicular to the plane of reflection,
4. The Pockels cell acts like a $\lambda/4$ or $\lambda/2$ plate depending on the external voltage applied.

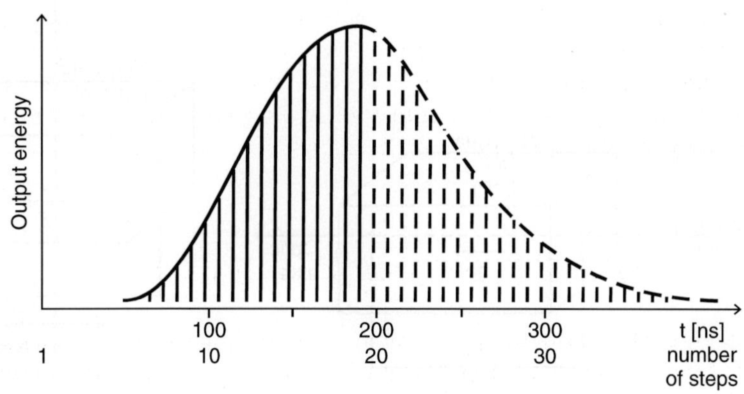

Fig. 6.7 Time evolution of the amplified pulse energy versus time and the number of passages in the regenerative amplifier.

6.4.1. The Pockels Cell

The Pockels cell plays an important role as an electro-optic Q-switch [18]. Now we shall explain its operation. The Pockels cell consists of a nonlinear optical material with the voltage applied to the material. The electric field can be applied along the direction of the optical beam (Fig. 6.8a) or perpendicular to it (Fig. 6.8b).

The crystal becomes birefringent under the influence of the applied electric field. The crystals used for parallel configuration (Fig. 6.8a) are uniaxial in the absence of an electric field with the optic axis along the z direction. The ellipsoid of the refraction index projects as a circle on a plane perpendicular to the optic axis (Fig. 6.9a).

So, the laser beams having polarizations along the x or y directions propagate with the same velocities along the z axis because the crystal is not birefringent in the direction of the optic axis. For the situation presented in Fig. 6.8a, the laser beam linearly polarized along y passes through the KD*P crystal unchanged when no voltage is applied. When an electric field is applied parallel to the crystal optic axis z, the ellipsoid of the refraction index projects as an ellipse, not a circle, on the plane perpendicular to the optic axis with the axes x' and y' rotating by 45° with respect to the x and y crystallographic axes (Fig. 6.9b). The angle of 45° is independent of the magnitude of the electric field. Therefore, when a voltage is applied, the KD*P crystal becomes birefringent along the z axis, and divides the laser beam into two components (along x' and y') that travel through the crystal at different velocities. The polarization of the output beam is dependent on the phase difference between the two orthogonally polarized components–ordinary and extraordinary rays, which depend on the applied voltage

$$\delta = \frac{2\pi}{\lambda} l \Delta n, \tag{6.19}$$

where Δn is the difference in the indices of refraction for the ordinary and extraordinary beams, and l is the crystal length.

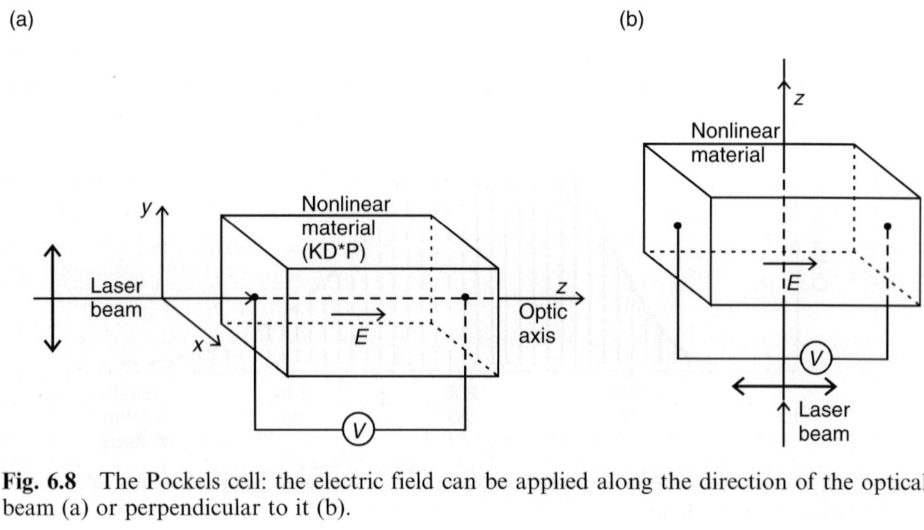

Fig. 6.8 The Pockels cell: the electric field can be applied along the direction of the optical beam (a) or perpendicular to it (b).

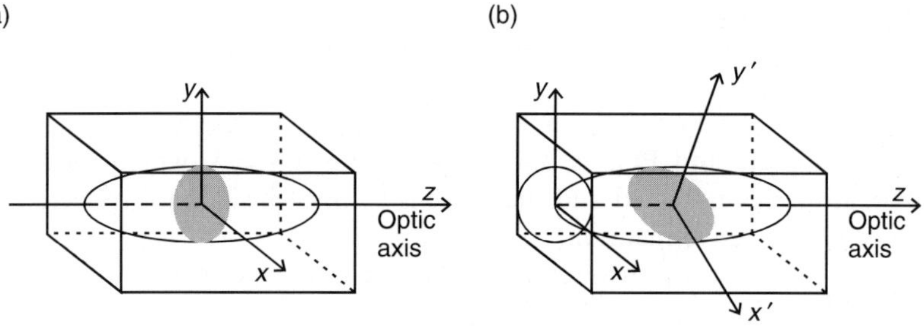

Fig. 6.9 Change of the refractive index in a KD*P crystal, x, y, z – the crystallographic axes (a) without an electric field, (b) when an electric field is applied ($E \neq 0$), x', y', z – the electrically induced axes.

It has been shown [19] that Δn can be expressed as

$$\Delta n = n_0^3 r_{63} E_z, \tag{6.20}$$

where r_{63} is the element of the electro-optic tensor of the third rank that is the response to the applied field E in the z direction (E_z), and n_0 is the index of refraction for the ordinary ray. Employing the relation between the voltage V and the applied electric field E_z, $V_z = E_z l$ and inserting (6.20) in eq. (6.19), one gets

$$\delta = \frac{2\pi}{\lambda} n_0^3 r_{63} V_z. \tag{6.21}$$

When the applied voltage V_z is adjusted to generate the phase difference $\delta = \pi/4$ or $\delta = \pi/2$ the Pockels cell operates as a quarter-wave or half-wave plate.

The Pockels cell belongs to the fastest all-optical switching devices, and is highly reliable. Typical commercially available Pockels cells employ KD*P crystals with $\lambda/4$ voltage between 3.5 and 4 kV at 0.69 μm and 5 to 6 kV at 1.06 μm [1]. As a particular example, Pockels cells are used to select and retain high-peak-power pulses in regenerative amplifiers.

Let us now return to Fig. 6.5 to explain the mechanisms of trapping the seed pulse from a train of modelocked pulses inside the amplifier. The operation of a regenerative amplifier presented in Fig. 6.5 can be divided into the following steps:

1. A seed pulse arrives in the amplifier at the Brewster's angle by reflection from the surface of the active medium (Ti^{+3}:Al_2O_3). The polarization of the seed pulse is perpendicular to the plane of drawing. The polarization of the pulse passing through the Pockels cell, PC1, which is initially switched off, is unchanged but double passing through $\lambda/4$ (before and after reflection from M1) changes the polarization by 90°. So, the beam reflected from M1 is not reflected from the crystal surface at point A but is refracted and passes the crystal leading to its amplification by the stimulated emission induced in the crystal (Ti^{+3}:Al_2O_3) in which the population inversion is generated by the pumping source. The pulse passes through the thin-layer polarizer and the Pockels cell (PC2), inactive at the moment. The pulse is reflected from the mirror M2.

2. When PC1 (and PC2) is still switched off, the beam twice passes the $\lambda/4$ plate and the pulse is removed from the amplifier by reflection from the crystal surface at the point A.
3. When PC1 is switched on (as a $\lambda/4$) before the seed pulse arrives, the total effect of the $\lambda/4$ plate and PC1 is $\lambda/4+\lambda/4+\lambda/4+\lambda/4=\lambda$, which means that the polarization is unaffected and the pulse is trapped inside, not reflected from the surface of the active medium.
4. Each pass of the pulse through the active medium results in amplification of the pulse. When the pulse has been amplified to the desired level ($\sim 10^6$ times), the voltage $\lambda/4$ is applied to the Pockels cell, PC2. The pulse going to the mirror M2 and returning through PC2 ($\lambda/4+\lambda/4=\lambda/2$) changes its polarization and is removed from the amplifier at the point P by reflection from the thin layer polarizer.

6.5. CHIRPED PULSE AMPLIFICATION (CPA)

The configuration of the typical chirped pulse amplifier is presented in Fig. 6.6. One can see that it consists of the regenerative amplifier described in section 6.4, the stretcher and the compressor. The stretcher placed before the regenerative amplifier expands the pulse duration by many orders of magnitude (from femtoseconds to hundreds of picoseconds) and thereby reduces peak power.

Employing the stretcher solves the problem of high intensities in the amplifiers for ultrashort pulses that are above the damage threshold of the active medium of the amplifier. For example, the Ti:sapphire crystal has a high saturation fluence and a large gain-bandwidth needed to generate relatively high energies for sub-picosecond pulses. Self-focusing in the crystal, described in Chapter 3, is a desirable effect when employed in modelocking, but it has also its disadvantages. Kerr lens focusing of the beam can lead to catastrophic beam collapse that may destroy the crystal, which makes it necessary to limit the intensity present in amplifiers to a reasonable magnitude of less than $10\,\mathrm{GW/cm^2}$. This obstacle can be overcome by the technique called *chirped pulse amplification* (*CPA*).

We have discussed in Chapter 5 the factors that lead to stretching the temporal pulse duration. The group velocity dispersion (GVD) is the most important factor affecting the temporal pulse broadening. Due to the GVD, each frequency component that comprises the spectrum of the pulse travels through the medium at a different group velocity. For so-called positive GVD materials, the red components travel faster than the blue components. As we have shown, a pair of prisms (Fig. 5.28) can be used to compensate for the excess GVD that results in the pulse compression. The same can be achieved with the diffraction gratings [20].

The pair of parallel gratings shows a simplified pulse compressor, which demonstrates the concept of compression (Fig. 6.10). One can see that the optical path through the grating pair is longer for the longer wavelengths than for the shorter ones. The mirror M reflects the beam back into the grating pair, recovering the spatial distribution but still increasing the difference in the optical paths for the red and the blue components. Thus, the pair of parallel gratings provides the negative

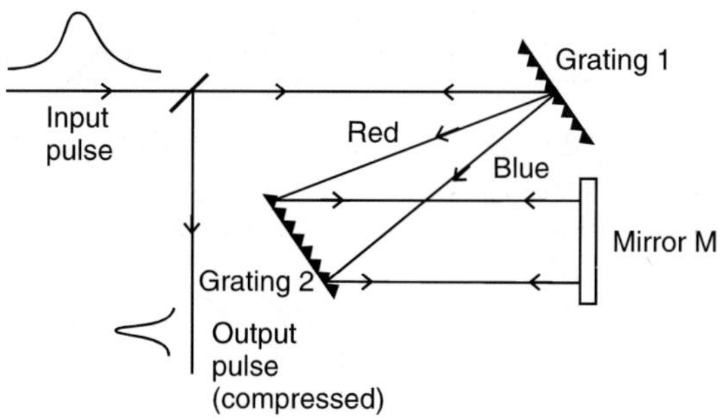

Fig. 6.10 Illustration of the principle of the pulse compression.

group-velocity. If the input pulse that is positively chirped travels through the gratings, the output pulse becomes shorter due to partial cancellation of the positive GVD effect by the negative GVD of the gratings in the configuration presented in Fig. 6.10. Pulse stretching is essentially the reverse of the pulse compression. The gratings can be configured in such a way that the bluer components have to travel a longer path through the stretcher than the redder components. The result is that the stretcher generates a *positive GVD effect* that results in the redder components travelling even faster than with the input pulse and exiting the stretcher with the temporal pulse significantly longer.

One of many possible configurations of the pair of gratings that generate the positive GVD is presented in Fig. 6.11. One can see that the main difference between the compressor (Fig. 6.10) and the stretcher (Fig. 6.11) is that a telescope is added between the gratings to invert the sign of the dispersion from the negative in Fig. 6.10 to the positive in Fig. 6.11.

In practical applications some other configurations are used. For example, instead of using a pair of gratings, a single grating combined with the curved mirror can be used as presented in Fig. 6.12. One can see that this multi-pass configuration reduces the complexity and ensures the four-pass grating that is necessary for the spatial reconstruction of the stretched beam.

The 1990s witnessed a dramatic improvement in the amplification of ultrashort pulses. Amplification of the femtosecond pulses using the chirped pulse amplification (CPA) technique [21] is now commercially available. Amplification of ultrashort

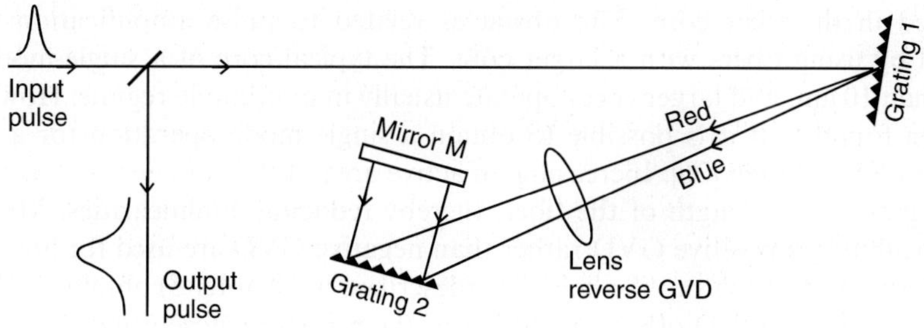

Fig. 6.11 Illustration of the principle of pulse stretching.

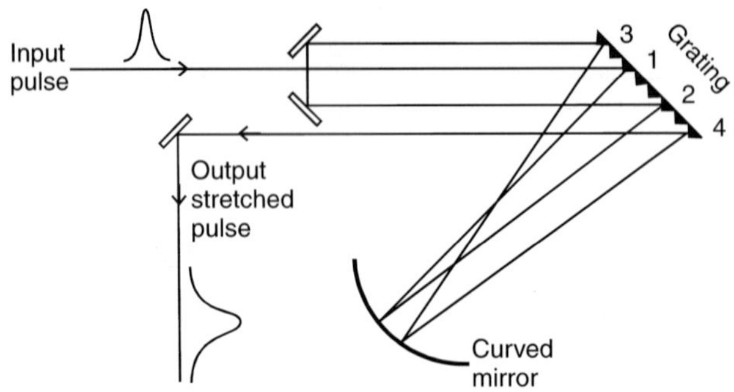

Fig. 6.12 The pulse stretched with a single grating – curved mirror combination.

pulses to mJ, corresponding to multiterawatt peak powers, becomes a routine task. The CPA technique is used both in the multipass configuration [22] and regenerative amplification [23–25]. Quite often these configurations are combined together to reach terawatt powers. Usually the regenerative amplifier is used as a first stage of amplification followed by multi-pass configuration used for the output power-boost stage.

Generally, the multipass amplifiers are used for very short pulses (<50 fs). They need much lower amounts of optical material for amplification thereby reducing nonlinear distortion of the pulse that usually comes at the expense of a rather complicated alignment procedure and a higher level of ASE (amplified spontaneous emission). The regenerative amplifiers are usually used for pulses longer then 50 fs. They have many advantages, including much smaller ASE and a simple operation and maintenance. No realignment is necessary to adjust the number of passages through the amplifier since the beam follows the some path in the cavity.

So far we have discussed the conventional configurations based on bulk-optic solid-state amplifiers. The development of optical communication has created a growing market for higher-power *optical amplifiers* based entirely on fiber technology. The development in double-cladding fibers technology and multi-mode diode pumping has increased the output of *cw* fiber lasers to a level comparable with solid-state lasers. However, achieving a few watts of average power from fiber-based ultrashort laser systems, routinely produced by the solid-state femtosecond systems, is not so straightforward. Limitations are due to the large effective nonlinearities (SPM and GVD) described in Chapter 5 that can destroy short femtosecond pulses produced in the fiber core. The obstacles related to pulse amplification can be reduced by using fibers with a larger core. The typical core of a single-mode fiber is less than 10 μm, and larger cores operate usually in multimode regime. However, it has been found that it is possible to obtain a single mode operation for a careful large-core fiber design [26]. Increasing an active area of the core enables an efficient pumping at shorter length of the fiber, thereby reducing nonlinearities. Moreover, fibers exhibiting a positive GVD rather than negative GVD are used for high-power amplifications. The soliton-like behavior (described in Chapter 5) obtained when the SPM and negative GVD offset each other at the properly chosen fiber lengths does not apply at high powers. A typical optical amplifier is presented in Fig. 6.13 [27].

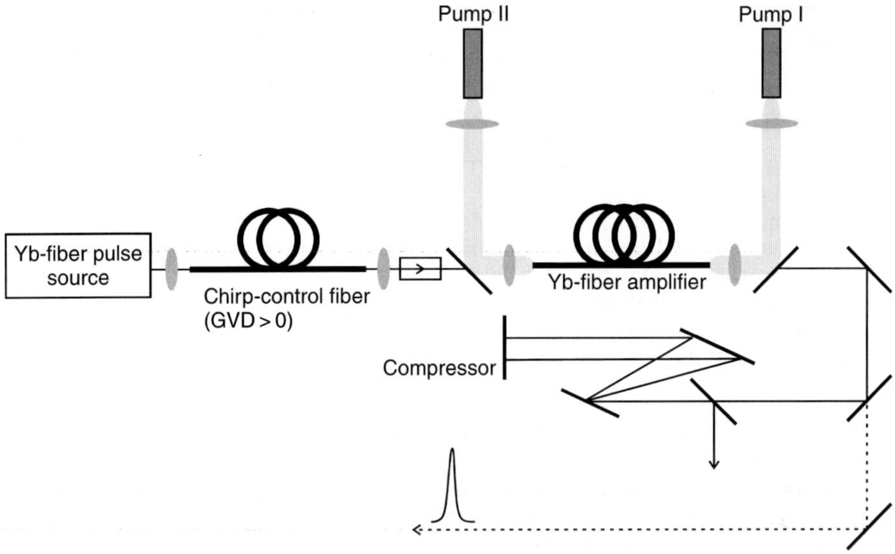

Fig. 6.13 Fiber laser amplification system.

The system consists of a fiber-based ultrashort laser (Yb-doped fiber laser) that emits a seed pulse for further amplification. The seed pulse (1050 nm, 2 ps, 300 mW, 50 MHz) passes through a piece of a single-mode fiber exhibiting a positive GVD to stretch the seed pulse. Then, the stretched linearly-positive chirped seed pulse is amplified in a fiber amplifier. The fiber amplifier (Yb-doped fiber, 4.3 m length, 25 μm-diameter core) is pumped from both ends with a two fiber-bundle-coupled laser-diode operating at 976 nm and 14 W. The seed pulse is amplified to 13 W and the width increases from 2 ps to 5 ps. Then, the pulse is recompressed down to 100 fs, passing through a conventional diffraction grating compressor (negative GVD designed) achieving 5 W output at 1050 nm.

REFERENCES

1. W. Koechner, *Solid-State Laser Engineering*, 5th Edition, vol. 1, Springer-Verlag (1999)
2. L.M. Frantz, J.S. Nodvik, J. Appl. Phys. 34 (1963) 2346
3. R. Bellman, G. Birnbaum, W.G. Wagner, J. Appl. Phys. 34 (1963) 780
4. M.V. Ortiz, *Amplified Ultrafast lasers move out of the laboratory*, Laser Focus World 39 (7) (July 2003) 69–74
5. Andrew T. Jay, Jason S. Robins, *Interventional Neuroradiology, Stroke and Carotid Artery Stenting*, First Union Securities, Inc. and Estimates Company Reports (Aug. 10, 2000)
6. S.K. Gayen, M. Alrubaiee, H.E. Savage, S.P. Schantz, R.R. Alfano, *Parotid Gland Tissues investigated by Picosecond Time-Gated and Spectroscopic Imaging Techniques*, J. Select. Topics Quantum Electron 7 (6) (Nov./Dec. 2001) 906
7. W.H. Lowdermilk, J.E. Murray, J. Appl. Phys. 51 (1980) 2436
8. J.E. Murray, W.H. Lowdermilk, J. Appl. Phys. 51 (1980) 3548
9. I.N. Duling, P. Bado, S. Williamson, G. Mourou, T. Baer, CLEO'84 (Anaheim, CA) paper PD3

10. I.N. Duling III, T. Norris, T. Sizer II, P. Bado, G.A. Mourou, J. Opt. Soc. Am. B 2 (1985) 616
11. P. Bado, M. Bouvier, J. Scott, Opt. Lett. 12 (1987) 319
12. M. Saeed, D. Kim, L.F. DiMauro, Appl. Opt. 29 (1990) 1752
13. M.D. Selker, R.S. Afzal, J.L. Dallas, A.W. Yu, Opt. Lett. 19 (1994) 551
14. M. Gifford, K.J. Weingarten, Opt. Lett. 17 (1992) 1788
15. L. Turi, T. Juhasz, Opt. Lett. 20 (1995) 154
16. D.R. Walker, C.J. Flood, H.M. Van Driel, U.J. Greiner, H.H. Klingenberg, CLEO/Europe'94 (Amsterdam) paper CThM1
17. T.E. Dimmick, Opt. Lett. 15 (1990) 177
18. S.J. Matthews, *Modulation in all things*, Laser Focus World 39 (8) (August, 2003) 110–115
19. R.O'B. Carpenter, J. Opt. Soc. Am. 40 (1950) 225
20. E. Treacy, IEEE J. QE-5 (1969) 454
21. D. Strickland, G. Mourou, *Compression of amplified chirped optical pulses*, Opt. Commun. 56 (3) (1985) 219–221
22. Sterling Backus, Charles G. Durfee, Gerard Mourou, Henry C. Kapteyn, Margaret M. Murnane, *0.2-TW laser system at 1kHz*, Optics Letters 22 (16) (1997) 1256–1258
23. M.D. Perry, D. Strickland, T. Ditmire, F.G. Patterson, *Cr:LiSrAlF*$_6$ *regenerative amplifier*, Opt. Lett. 17 (8) (1992) 604–606
24. F. Salin, J. Squier, G. Mourou, *Multikilohertz Ti:Al*$_2$*O*$_3$ *amplifier for high-power femtosecond pulses*, Optics Letters 16 (24) (1991) 1964–1966
25. C.P. Barty, C.L. Gordon, B.E. Lemoff, *Multiterawatt 30-fs Ti-sapphire laser*, Opt. Lett. 18 (1994) 1442–1444
26. M.E. Ferman, Opt. Lett. 23 (1) (1998) 52
27. G. Sucha, H. Endert, *Femtosecond fiber lasers hit power highs*, Laser Focus World 36 (8) (August, 2000) 133–136

– 7 –

The Measurement of Ultrashort Laser Pulses

Recent rapid progress in ultrafast laser technology is making the femtosecond and attosecond time scale accessible. Commercial laser systems can now operate at very short (sub–50 fs) pulse durations.

To be sure that an ultrashort pulse has been produced, we must have some way of measuring the temporal duration of the created pulse. Moreover, precise control of the output from our laser systems is essential in our research or applications to ensure accurate and reliable results. So, laser diagnostics and the ultrashort pulse characterization play a key role and are a part of the routine maintenance of the laser system.

However, we have to realize that for the ultrashort temporal duration of the femtosecond pulse there is generally no direct way of measuring its width or its complete temporal intensity profile. The traditional methods using photomultipliers (PMTs) or photodiode detectors are useless because the fastest PMTs on the market have characteristic rise times of less than 0.1 ns and the fastest photodiodes are still limited to approximately 10 ps. The best direct solution – streak cameras – can achieve sub-picosecond resolution, which is still insufficient for ultrashort measurements below 500 ps. How can we avoid the problems of the slow detectors?

Fortunately, we have a choice; we can use indirect methods for which the slow response of the detector no longer plays a role any. In fact, there are many commercial and non-commercial diagnostic tools available for examining ultrashort laser pulses. There are three main types of diagnostic techniques for the ultrashort pulse characterization:

1. autocorrelation techniques [1–4],
2. frequency-resolved optical gating (FROG) techniques [5–10],
3. the direct electric-field reconstruction (SPIDER) technique [11–12].

The first group – the autocorrelation techniques – is the oldest one. They offer virtually unlimited resolution and ease of use, measure the temporal pulse duration, but provide little information on pulse shape and phase.

The second group – the frequency-resolved optical gating (FROG) techniques – has blossomed in the 1990s. Several variations of the FROG technique are based on spectral resolution of the correlation functions for measuring the frequency-dependent intensity and phase, or the time-dependent (after Fourier transformation) temporal pulse shape and pulse duration. To extract the intensity $E(\omega)$ and phase $\phi(\omega)$ from the spectral trace, a sophisticated iterative inversion algorithm must be applied. The FROG techniques sacrifice simplicity in comparison with the autocorrelation techniques but obtain more information – frequency-dependent intensity, phase, temporal pulse shape and temporal pulse duration.

The third type, the SPIDER technique, is similar to the FROG phase-sensitive technique. In contrast to FROG, the recovery algorithm is not iterative, which is very attractive for real-time pulse characterization. Recently, however, real-time operation of the FROG technique has been reported and real-time commercial versions are already available. Another related pulse shape measurement method is known as the Spectrally and Temporally Resolved Upconversion Technique (STRUT) [13–14]. STRUT has been applied recently in telecommunication [15]. In the following, we will explain the fundamentals and basic principles for the autocorrelation and FROG groups of diagnostic techniques.

7.1. AUTOCORRELATION TECHNIQUES

The basic principle of the autocorrelation techniques is presented in Fig. 7.1. The autocorrelator consists of the Michelson interferometer, a nonlinear medium (for example the KDP crystal in which the second harmonic is generated) and a linear detector (PMT). We can remove the nonlinear medium and use a nonlinear detector (for example a regular photodiode in which a two-photon current is generated) instead of the PMT. The effect will be similar.

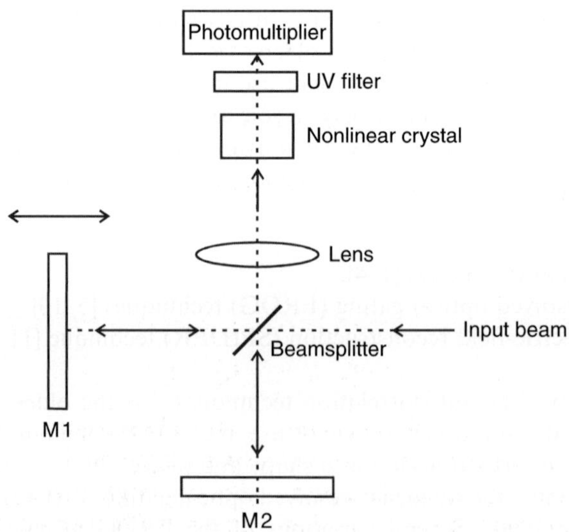

Fig. 7.1 Diagram of the autocorrelator for collinear beams.

The input pulse at the frequency ω is split in the beamsplitter into two identical pulses. Each of the pulses passes identical or different optical paths in the two arms of the interferometer. The optical path in one of the arms can be changed by moving the mirror (M1). The imbalance between the two paths generates the delay time τ between the replica pulses. Without the nonlinear medium, the interferometric output that reaches the linear detector in Fig. 7.1 is given by the intensity $I_1(\tau)$ as a function of the delay τ

$$I_1(\tau) = \int_{-\infty}^{+\infty} |\boldsymbol{E}(t) + \boldsymbol{E}(t-\tau)|^2 dt, \tag{7.1}$$

where $\boldsymbol{E}(t)$ is the electric field of the laser pulse. This can be expressed as a sum of the constant background contribution independent of the delay τ due to the intensities of the individual pulses and the first-order autocorrelation function $G_1(\tau)$

$$I_1(\tau) = \int I(t)dt + \int I(t-\tau)dt + 2G_1(\tau), \tag{7.2}$$

where

$$G_1(\tau) = \int_{-\infty}^{\infty} E(t)E(t-\tau)dt. \tag{7.3}$$

However, the first-order autocorrelation function $G_1(t)$ does not yield a unique solution for the temporal pulse duration. This is the reason why the autocorrelation technique uses higher-order autocorrelations generated by any instantaneous optical nonlinearity. The most commonly used nonlinearity is the second harmonic generation.

If the two replica pulses overlap in a nonlinear medium in collinear fashion, the SHG signal is

$$I_{2\omega}(\tau) \sim \int_{-\infty}^{+\infty} |(E(t) + E(t-\tau))^2|^2 dt. \tag{7.4}$$

Therefore

$$\begin{aligned} I_{2\omega}(\tau) &\sim \int_{-\infty}^{+\infty} |\{A(t)\exp i[\omega t + \phi(t)] + A(t-\tau)\exp i\omega[(t-\tau) + \phi(t-\tau)]\}^2|^2 dt \\ &= \int_{-\infty}^{+\infty} |2A^4(t) + 2A^2(t)A^2(t-\tau) + \\ &\quad + 4A(t)A(t-\tau)[A^2(t) + A^2(t-\tau)]\cos[\omega\tau + \phi(t) - \phi(t-\tau)] + \\ &\quad + 4A^2(t)A^2(t-\tau)\cos[2(\omega\tau + \phi(t)) - \phi(t-\tau)]|dt, \end{aligned} \tag{7.5}$$

where $A(t)$ and $\phi(t)$ are the time-dependent amplitude and phase, respectively. This expression contains the constant τ independent background $\int I^2(t)dt + \int I^2(t-\tau)dt$ and the part that is τ dependent. This part is still not so easy to analyse but it becomes much simpler after averaging over the delay time τ in the detector. The averaging in the detector is automatic because the delay τ is fast enough in comparison with the

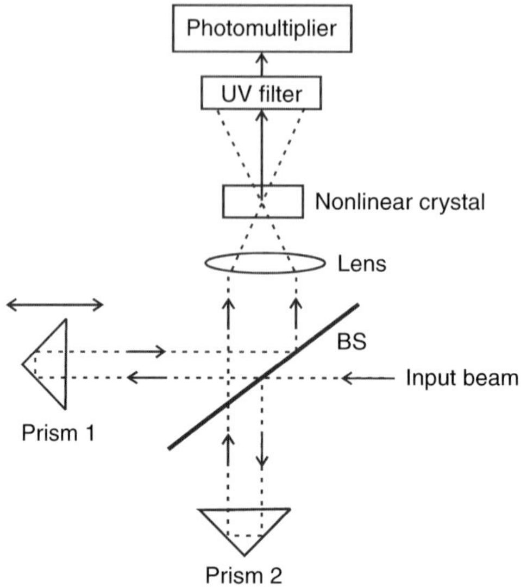

Fig. 7.2 Diagram of an autocorrelator for noncollinear beams.

detection time constant. Therefore, the autocorrelation signal becomes directly proportional to the second-order autocorrelation function $G_2(\tau)$ given by

$$G_2(\tau) = \frac{\int I(t)I(t-\tau)dt}{|I^2 dt|}. \qquad (7.6)$$

The constant background can be removed by non-collinear geometry of the Michelson interferometer presented in Figs. 7.2 and 7.3.

The typical autocorrelation signal is presented in Fig. 7.4.

Now, we will focus on practical configurations used in commercially available *autocorrelators*. There are two main groups of the autocorrelators:

1. *high-repetition-rate autocorrelator*,
2. *single-shot autocorrelator*.

The *high-repetition-rate autocorrelator* is used for measuring the temporal pulse duration of the pulses emitted by the high-repetition oscillators (for example, 82 MHz, 12 nJ, 80 fs of the Ti:Sapphire modelocked oscillator). *The single-shot*

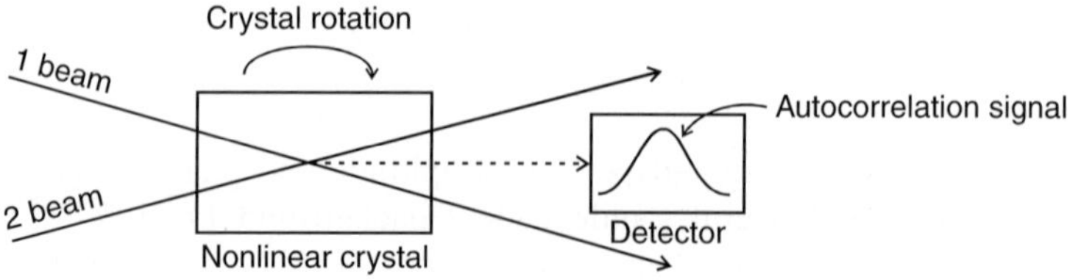

Fig. 7.3 Illustration of noncollinear beams leaving the autocorrelator and incident onto a nonlinear crystal and a linear detector.

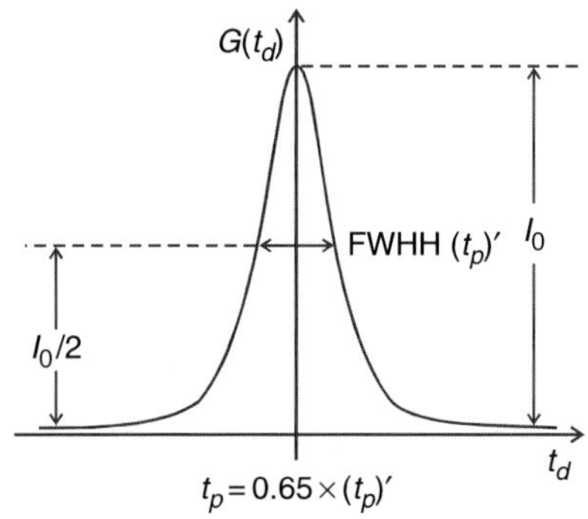

Fig. 7.4 Autocorrelation signal for $I(t) = \mathrm{sech}^2$ pulses.

autocorrelator is used to measure output from the high-peak-power amplified sources (for example, 1 kHz, 1 mJ, 100 fs, regenerative amplifier).

Figure 7.5 presents the configuration typical for commercially available autocorrelators.

The input laser beam, which is a sequence of temporal pulses, is split into two beams of the same intensity. Each beam passes through a slightly different optical path, and the difference of the optical paths is controlled and changed in a periodic manner. Subsequently, both beams are directed onto a nonlinear crystal, in which the second harmonic is generated. Depending on the difference of optical paths, the pulses of both beams incident onto the crystal overlap to a different degree. The

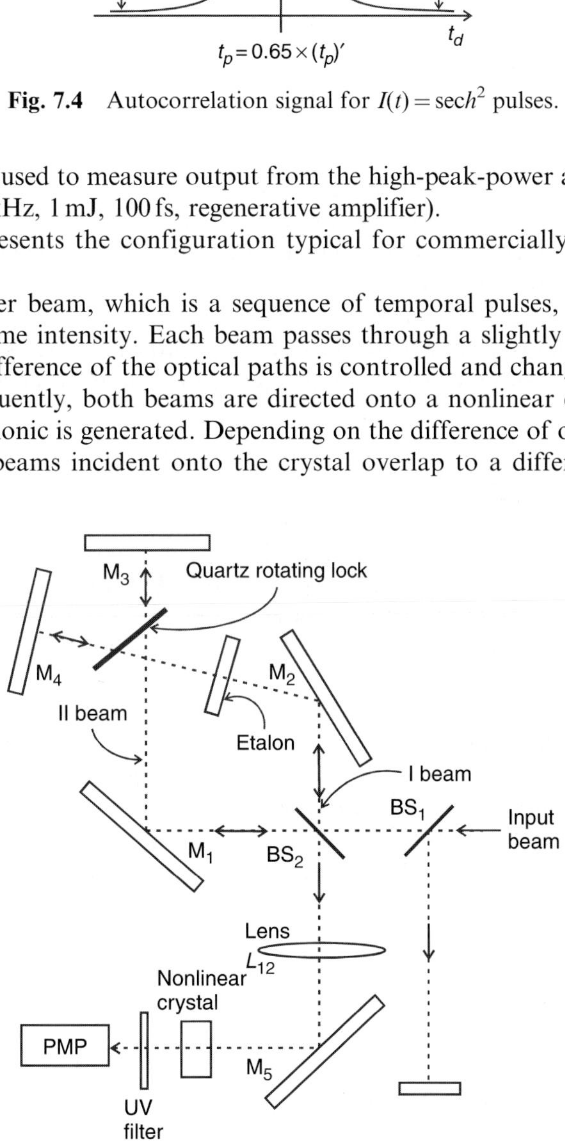

Fig. 7.5 Typical configuration for the commercially available high-repetition-rate autocorelator.

intensity of the second harmonic generated in the crystal depends upon the degree of the overlapping. When the difference of the paths is zero, the overlapping reaches its maximum. The signal produced in the nonlinear crystal, the *autocorrelation signal*, is sent to the detector. The signal is a function of the difference of the optical paths that controls the time delay between the pulses. The autocorrelation signal has a special name to distinguish it from the second harmonic signals generated separately by each of the beams. In the configuration proposed in Fig. 7.1 both beams are delayed together and the autocorrelation signal is collinear. The collinear configuration has many disadvantages, because the autocorrelation signal occurs at the top of the high background from the second harmonics of the separate beams. In order to avoid the background, the configuration presented in Fig. 7.2 is applied, where beams are incident on the crystal in a noncollinear manner and only the autocorrelation signal reaches the detector (Fig. 7.3). At first glace, the configuration in Fig. 7.5 looks like it is collinear. However, the beamsplitter BS_2 shifts the beams I and II due to the refraction and the lens L_{12} makes them noncollinear.

There are different tricks to delay the beams, e.g., moving mirrors (mirror M1 in Fig. 7.1) or prisms (prism 1 in Fig. 7.2) to increase the optical path of one of the beams. Another method applied in commercial devices is presented in Fig. 7.6. The beams I and II presented in Fig. 7.6 are incident onto the surface of a quartz block, which rotates at a given frequency (typically 50 or 60 Hz). As a consequence of its rotation, the beams I and II are incident into the surface of the quartz block at different angles (Fig. 7.6).

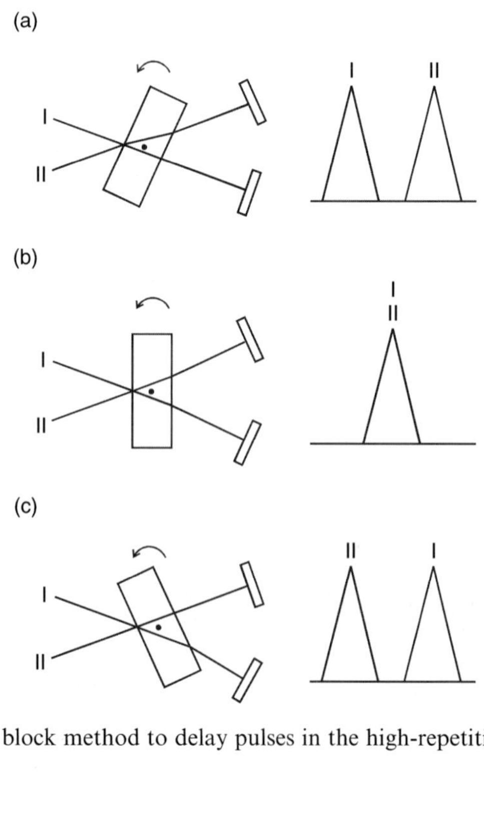

Fig. 7.6 Rotating block method to delay pulses in the high-repetition autocorrelators.

For the position of the quartz block denoted in Fig. 7.6 as (a), the beam I covers a shorter optical path than the beam II. The temporary pulses of the beam I reach the nonlinear crystal faster than those of the beam II, hence their signals do not overlap and no autocorrelation signal is sent to the detector. For the position of the block denoted as (b), the optical paths of both beams are the same, which means that the pulses reach the crystal at the same time and they effectively overlap, sending the autocorrelation signal to the detector. With further rotation of the block (position (c)) the beam II passes a shorter path than the beam I. The pulses again begin to diverge, resulting in the absence of any autocorrelation signal in the detector. Such a situation repeats with a frequency equal to half of the block rotation frequency. To generate a proper shape of the autocorrelation signal, the optical path difference must change in a linear manner with the rotation of the quartz block.

Figure 7.7 illustrates the optical paths of the beams passing through the rotating quartz block as a function of the incident angles Θ_1 and Θ_2 for the beams I and II onto the surface of the rotating block.

Using the Snell equation

$$\frac{\sin \Theta_1}{\sin \Theta_1'} = n, \tag{7.7}$$

the refraction index of quartz is n, and Θ_1' is the refraction angle of the beam I in quartz. It is easy to demonstrate that the optical path L_1 of the beam I passing twice through the quartz block of thickness l and incident at the angle Θ_1 is

$$L_1 = 2l\left(\sqrt{n^2 - \sin^2 \Theta_1} - \cos \Theta_1 + 1 - n\right).$$

We obtain an analogical expression for the beam II. Thus, the difference of optical paths $\Delta L = L_1 - L_2$ is

$$\Delta L = L_1 - L_2 = 2l\left[\left(\sqrt{n^2 - \sin^2 \Theta_1} - \cos \Theta_1\right) - \left(\sqrt{n^2 - \sin^2 \Theta_2} - \cos \Theta_2\right)\right]. \tag{7.8}$$

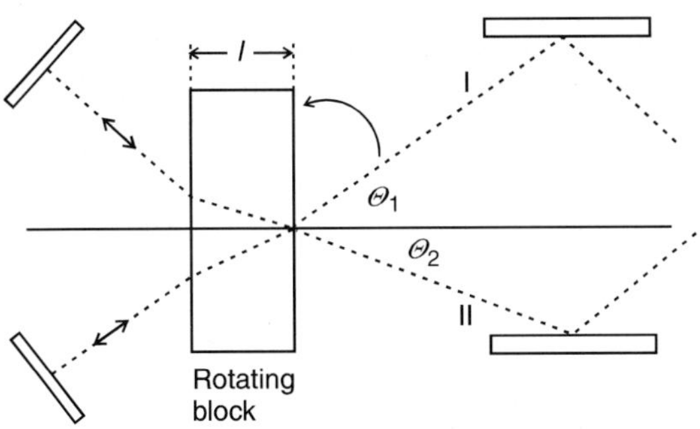

Fig. 7.7 Diagram illustrating optical paths of the beams passing through a rotating quartz block, as the functions of angles Θ_1 and Θ_2 of the beams I and II that are incident onto the surface of the block.

The typical dependence of the optical path difference, ΔL, on the incident angle Θ_1 of the beam I is presented in Fig. 7.8. One can see that in this example the difference of the optical paths, ΔL, changes in an almost linear manner for the rotation at the angles between 0° and 70°. As the sum of the angles $\Theta_1 + \Theta_2$ is the angle between the beams I and II (see Fig. 7.7) and it is constant for a given autocorrelator configuration, it should be about 70° for the configuration presented in the example.

The autocorrelation signal sent to the detector can be observed on the oscilloscope. The shape of the autocorrelation signal intensity profile $G(t_d)$ depends on the shape of the pulses $I(t)$ of the beam I and II and is a function of the delay time t_d

$$G(t_d) = \int_{-\infty}^{+\infty} I(t)I(t - t_d)dt. \tag{7.9}$$

The autocorrelation signal $G(t_d)$ presented in Fig. 7.4 corresponds to the pulses described by the sech^2 type function

$$I(t) = \sec h^2\left(\frac{1.76t}{\Delta t_{FWHH}}\right). \tag{7.10}$$

From the mathematical point of view, the relation described by eq. (7.9) is called convolution. The autocorrelation signal measured by the detector $G(t_d)$ and observed on the oscilloscope monitor is thus the convolution of the intensity profiles of the beam I ($I(t)$) and the beam II ($I(t - t_d)$). If we know $G(t_d)$ from the experiment, we may carry out a reverse procedure, called deconvolution, in order to determine $I(t)$. For some function types, $I(t)$, presented in Table 7.1, the deconvolution procedure leads to analytical solutions. It can be demonstrated that the full width at half height (FWHH) of the pulse profile ($I(t)$) corresponding to the pulse duration Δt_{FWHH} may be calculated from the full width at half height FWHH $(t_p)'$ of the autocorrelation signal $G(t_d)$ presented in Fig. 7.4 by multiplying it by a certain constant A. The values of these constants A are presented in Table 7.1 for several typical temporary pulses.

Finally, it should be explained why the *etalon*, which is the flat plate of a given length, is used in the configuration presented in Fig. 7.5. The etalon increases the

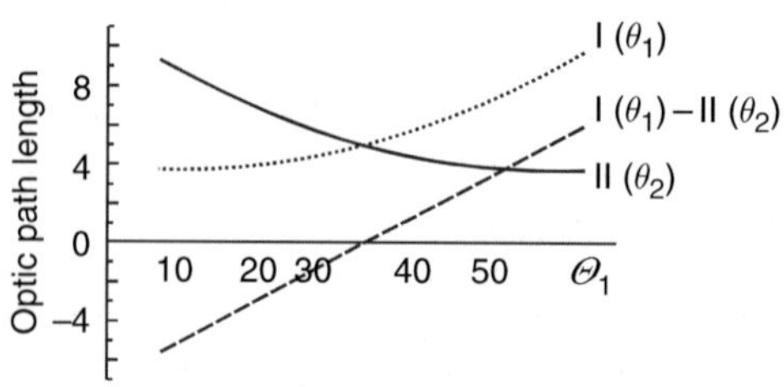

Fig. 7.8 Dependence of the length of optical paths ΔL on the angle Θ_1.

Table 7.1

Values of factors A for different pulse shapes

Function	$I(t)$	A
Square	$I(t) = 1;\ \lvert t\rvert \leq t_p/2$ $I(t) = 0;\ \lvert t\rvert > t_p/2$	1
Diffractive	$I(t) = \frac{\sin^2(t/\Delta t_{\mathrm{FWHH}})}{(t/\Delta t_{\mathrm{FWHH}})}$	0.751
Gaussian	$I(t) = \exp(-(4\ln 2)t^2/2\Delta t^2_{\mathrm{FWHH}})$	0.707
Hiperbolic Secant	$I(t) = sech^2\left(\frac{1.76t}{\Delta t_{\mathrm{FWHH}}}\right)$	0.648
Lorentzian	$I(t) = \frac{1}{1+(4t^2/\Delta t^2_{\mathrm{FWHH}})}$	0.500
Exponential	$I(t) = \exp(-(\ln 2)t/\Delta t_{\mathrm{FWHH}})$	0.413

time delay in one of the arms of the autocorrelator. This additional delay of a precisely specific magnitude serves to calibrate the time base of the oscilloscope. First, we generate the autocorrelation signal without etalon and observe it on the oscilloscope. We measure the number of divisions, N, on the oscilloscope monitor corresponding to its full width at half height (FWHH). The intensity profile of the autocorrelation signal obviously originates from a large number of single pulses, however the width of the autocorrelation signal is identical to that of a single pulse. Looking at the shape of the profile on the oscilloscope, we are unable to calculate the value of pulse duration, because we do not know the relation between the oscilloscope time base and the time delay. In order to calibrate the oscilloscope time base, we introduce the etalon, which is characterized by the precisely determined time delay, T. Etalon shifts the signal on the oscilloscope monitor by some number of divisions, ΔN. Thus, the autocorrelation signal width is FWHH $(t_p)' = NT/\Delta N$. Knowing the shape of temporal pulses $I(t)$, we can calculate the real pulse duration, Δt_{FWHH}, multiplying FWHH $(t_p)'$ by the factor A from Table 7.1 where the values of A are given for some typical shapes of laser pulses.

The pulse from the femtosecond oscillator can be aligned to achieve as short a pulse as possible that can be measured with a high-repetition-rate autocorrelator. The femtosecond seed pulse from the oscillator of energy a few nanojoules is often amplified. The pulse after amplification usually has a longer temporal duration because it undergoes various manipulations such as dispersing, stretching, amplifying, and eventually compressing. Unless all the devices are precisely aligned, the pulse suffers from many spatio-temporal distortions such as spatial chirp or pulse-front till, resulting in longer pulse duration. So, the pulse duration after the amplification should be measured once again. In contrast to oscillators the output from the amplifier is generally characterized by a low repetition rate. To measure the pulse duration, single-shot autocorrelators are usually needed.

The principle of a single-shot autocorrelator is presented in Fig. 7.9.

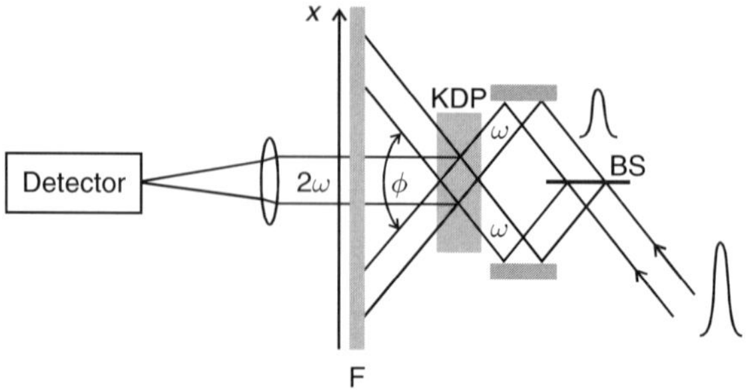

Fig. 7.9 The principle of a single-shot autocorrelator.

In the non-collinear configuration presented in Fig. 7.9, the delay τ is produced by geometry and is given by

$$\tau = \frac{nx \sin\frac{\phi}{2}}{c}. \tag{7.11}$$

where ϕ is an angle of intersection of the two beams, x denotes the axis along which a spatial intensity distribution of the autocorrelation signal is generated. As a result, the signal $\rho(x)$ from the linear detector is a function of the position x along the axis perpendicular to the direction of the SHG beam (Fig. 7.9)

$$\rho(x) = \int I(t+\tau)I(t-\tau)dt \tag{7.12}$$

and is directly proportional to the second-order correlation function $G_2(2\tau)$ (eq. (7.6)) from which we can determine the temporal pulse duration in the same way as it was explained for the high-repetition-rate autocorrelators.

The methods of pulse duration measurement discussed above are the traditional autocorrelation methods, contrary to the modern autocorrelation methods known as the *amphibious techniques*. The traditional methods deliver information about the width of "well behaved" pulses, they do not provide information about temporary changes of frequency and the subtle structure of the time profile. The most modern autocorrelation methods permit us to do the full temporary pulse analysis and they are known as *frequency-resolved optical gating – FROG* and *TADPOLE – temporary analysis by dispersing a pair of light E-field*. For details the reader is referred to the textbook on this subject [5].

7.2. FROG TECHNIQUES

To realize the main difference between the autocorrelation and FROG techniques, compare the Fig. 7.10 and Fig. 7.11. One can see that FROG is essentially a single-shot autocorrelator (SSA) with a signal imaging onto a spectrometer. The difference between SSA and FROG is that FROG measures the signal's spectral intensity versus delay rather than temporal intensity. FROG can use almost any optical

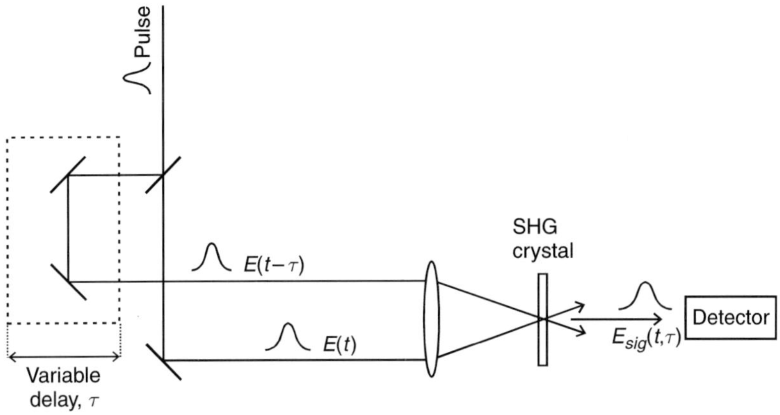

Fig. 7.10 The principle of an autocorrelator.

nonlinearity. The three most commonly used nonlinearities are: 1) second harmonic generation (SHG-FROG), 2) self-diffraction (SD-FROG), 3) polarization-gate (PG-FROG). In each case the nonlinearity generates the signal field $E_{sig}(t, \tau)$ which has a specific form for different nonlinearities.

This signal beam $E_{sig}(t, \tau)$ is then frequency resolved by a spectrometer, which generates the FROG trace

$$I_{FROG}(\omega, \tau) = \left| \int E_{sig}(t, \tau) e^{-i\omega t} dt \right|^2 . \tag{7.13}$$

The relation (7.13) uses the Fourier transform between the time-domain and the frequency-domain. The operation (7.13) is performed automatically by the spectrometer and $I_{\mathrm{FROG}}(\omega, \tau)$ represents the 2D experimental data – the frequency-dependent intensity as a function of the delay τ. Once the experimental FROG traces $I_{\mathrm{FROG}}(\omega, \tau)$ are obtained, a numerical algorithm must be applied to determine the intensity and the

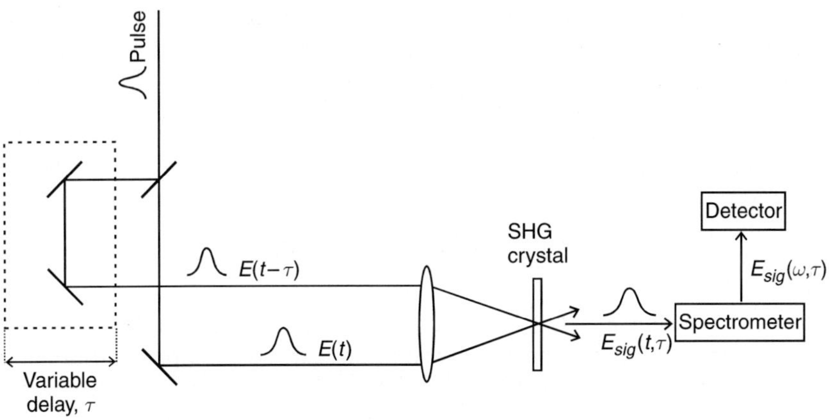

Fig. 7.11 The principle of the SHG FROG method.

phase of the pulse or, alternatively, the complete complex field envelope. This analysis is well known from the mathematical point of view and it is called the two-dimensional phase-retrieval problem.

The signal $E_{sig}(t, \tau)$ has a specific form for various nonlinearities. We have shown (eq. (7.6)) that for the SHG autocorrelation the signal is given by

$$E_{sig}(t,\tau) = |E(t)E(t-\tau)|^2. \tag{7.14}$$

For a polarization-gate the signal is the autocorrelation function of the third-order

$$E_{sig}(t,\tau) = E(t)|E(t-\tau)|^2. \tag{7.15}$$

In the two-dimensional phase-retrieval algorithm we have to guess the set of electric components with frequency-dependent amplitude $A(\omega)$ and phase $\phi(\omega)$

$$E(\omega) = A(\omega)e^{i(\omega t+\phi(\omega))}. \tag{7.16}$$

From the Fourier-transform we can calculate

$$E(t) = A(t)e^{i\omega t+i\phi(t)} \tag{7.17}$$

and

$$E(t-\tau) = A(t-\tau)e^{i\omega(t-\tau)+i\phi(t-\tau)} \tag{7.18}$$

that contains the intensity and the phase information on the measured pulse. From eq. (7.17) and (7.18) we calculate $E_{sig}(t, \tau)$ that is inserted into eq. (7.13). We can calculate $I_{\mathrm{FROG}}(\omega, \tau)$ that can be compared with the experimental intensity image obtained from the spectrometer $I_{FROG}^{\exp(\omega,\tau)}$. Then the algorithm is based on the iterative Fourier-transform (eq. (7.13)) used in the image phase retrieval problem. The convergence of the algorithm is determined as the difference of the FROG trace obtained at the nth iteration and the experimentally measured trace

$$\Delta(n) = \left\{\frac{1}{N^2}\sum_{i}^{N}\sum_{j}^{N}\left(I_{FROG}^{(n)}(\omega_i,\tau_j) - I_{FROG}^{\exp}(\omega_i,\tau_j)\right)^2\right\}^{\frac{1}{2}} \tag{7.19}$$

where $I_{FROG}^{(n)}$ is the recovered trace on n^{th} iteration.

Final error $\Delta < 0.04$ or less for an $N^x N$ array of points from the CCD camera is considered to be indicative of reasonable convergence of the algorithm. The FROG method is very accurate. Any known systematic error can be modeled in the algorithm, although this is not usually necessary, except for extremely short pulses (<10 fs). Another useful and important feature is that the 2D phase retrieval problem has an essentially unique solution and is a solved problem when certain additional information regarding $E_{sig}(t,\tau)$ is available. This is in contrast to the 1D problem autocorrelation techniques, where many solutions can exist, despite additional information. The two-dimensional phase-retrieval problem suffers only from "trivial" ambiguities, such as an absolute phase and a translation in time. So, the FROG solution is not totally unique, but essentially unique, which means that the solution is not perfect but good enough for practical measurements. In practice, it has been shown that FROG can measure temporal pulses from a few fs to many ps with energy in the range from a few fJ to mJ in the IR, visible, and UV spectral regions. There are attempts to extend FROG into the X-ray region.

The uniqueness of the FROG solutions can be verified by the presence of feedback. This feedback results from the fact that there are many more degrees of freedom in the FROG trace ($N^x N$ array of points) than we need to characterize the pulse – N intensity points and N phase points, that is, $2N$ points. So, a convergence of the FROG algorithm in eq. (7.19) for a different, randomly chosen combination of points virtually assures that the measured trace is free of systematic error. Conversely, non-convergence indicates the presence of systematic error. Indeed, the significant over-determination in the FROG trace guarantees a practical uniqueness of the full pulse characterization – intensity, shape, phase, temporal pulse duration. This feature is in contrast to information obtained from the autocorrelation technique. To recall, the intensity autocorrelation has only three constraints: a maximum at zero delay, zero for large delays, and even symmetry with respect to delay. These three constraints are not sufficient to produce an unique autocorrelation trace and temporal pulse profile.

If you are interested in reading more about FROG or other techniques used to measure ultrashort pulses you are referred to the excellent textbooks [5, 16], on this subject.

REFERENCES

1. C. Rullière, *Femtosecond Laser Pulses*, Principles and Experiments, Springer-Verlag (1998)
2. A. Laubereau, *Optical Nonlinearities with Ultrashort Pulses*, in: *Ultrashort Laser Pulses and Aplications*, W. Kaiser ed., Topics and Applied Physics, vol. 60, Springer-Verlag, Berlin Heidelberg (1988), p. 56
3. M. Maier, W. Kaiser, J.A. Giordmaine, Phys. Rev. Lett. 17 (1966) 1275
4. J.A. Armstrong, Apll. Phys. Lett. 10 (1967) 16
5. R. Trebino, *Frequency-Resolved Optical Gating: The Measurement of Ultrashort Laser Pulses*, Kluwer Academic Publishers, Boston (2002)
6. D.J. Kane, R. Trebino, Characterization of arbitrary femtosecond pulses using frequency resolved optical gating, IEEE J. Quantum Electron. 29 (1993) 571
7. R. Trebino, K.W. DeLong, D.N. Fittinghoff, J.N. Sweetser, D.J. Kane, Measuring ultrashort laser pulses in the time-frequency domain using frequency-resolved optical gating, Rev. Sci. Instr. 68 (1997) 3277
8. K.W. DeLong, R. Trebino, *Improved Ultrashort Pulse Retrieval Algorithm for Frequency-Resolved Optical Gating*, J. Opt. Soc. Am. A 11 (1994) 2429–2437
9. G. Taft, A. Rundquist, M.M. Murnane, H.C. Kapteyn, K.W. DeLong, R. Trebino, I.P. Christov, *Ultrashort optical waveform measurements using frequency-resolved optical gating*, Opt. Lett. 20 (7) (1995) 743
10. R. Trebino, D.J. Kane, *Using phase retrieval to measure the intensity and phase of ultrashort pulses: frequency-resolved optical gating*, J. Opt. Soc. Am. A 10 (1993) 1101–1111
11. C. Iaconis, I.A. Walmsley, *Spectral phase interferometry for direct electric-field reconstruction of ultrashort optical pulses*, Opt. Lett. 23 (1998) 792
12. C. Iaconis, I.A. Walmsley, *Self-referencing spectral interferometry for measuring ultrashort optical pulses*, IEEE J. Quantum Electron. 35 (1999) 501

13. J.L.A. Chilla, O.E. Martinez, *Direct measurement of the amplitude and the phase of femtosecond light pulses*, Opt. Lett. 16 (1991) 39
14. J.-P. Foing, J.-P. Likforman, J. Joffre, A. Migus, *Femtosecond pulse phase measurement by spectrally resolved up-conversion: application to continuum compression*, IEEE J. Quantum Electron. 28 (1992) 2285
15. M.R. Fetterman, J.C. Davis, H.-S. Tan, W. Yang, D. Goswami, J.-Krhee, W.S. Warren, *Fast-frequency hopping modulation and detection demonstration*, J. Opt. Soc. Am. B 18 (2001) 1372
16. J-C. Diels, W. Rudolph, *Ultrashort Laser Pulse Phenomena: Fundamentals, techniques and applications on a femtosecond time scale*, Academic Press, Boston (1996)

– 8 –

Selected Methods of Time-Resolved Laser Spectroscopy

In this chapter we shall discuss some of the most significant methods of time-resolved laser spectroscopy. This book does not discuss stationary methods of laser spectroscopy, selective excitation methods, molecular beam spectroscopy, high resolution Doppler spectroscopy and many other important fields of laser spectroscopy, which the reader may find in other textbooks [1].

Time-resolved spectroscopy focuses on the time evolution of emission, absorption or scattering processes to obtain information about the dynamics of chemical, physical or biological systems. From an historical perspective, time-resolved spectroscopy started on the millisecond time scale in the stopped-flow method (B. Chance, 1940) followed by flash photolysis (M. Eigen, R.G.W. Norrish, G. Porter, around 1950), reaching the microsecond and approaching the nanosecond time scale. The methods were used to provide kinetics of reactions occurring on these time scales. At that time these reactions were considered to be extremely fast and there was a feeling that this time resolution was the fastest that could be measured and sufficient for most relevant problems of chemistry. On the other hand, since 1935, when the *transition-state theory* was formulated by Eyring, Evans, and Polanyi, it was obvious that kinetics characterized by the rate constant does not provide a detailed molecular picture of the reaction. The actual atomic motions involved in chemical reactions are extremely fast and occur on the picosecond and femtosecond time scale. In order to study these motions we need a tool that provides a proper time resolution. Ultrafast pulsed lasers have made direct exploration of this temporal realm a reality, which had never been observed in real time in the past.

Molecular motions involved in chemical reactions–dynamics of the chemical bond at the atomic level such as the breaking and formation of bonds, ultrafast transformations leading to transition states, the redistribution of energy over different degrees of freedom–can now be monitored in real time. In this chapter we shall discuss the methods that can be employed to study dynamics at the molecular level.

Time-resolved laser spectroscopy methods provide information about the dynamics of various processes, such as:

1. dynamics of reaction coordinates, e.g., *cis* and *trans* isomerisation; excited state proton transfer, electron transfer, conformational changes, vibrational predissociation;
2. reorientational relaxation;
3. vibrational dephasing, T_2, in the ground and excited electronic state;
4. vibrational relaxation, T_1, in the ground and excited electronic state;
5. excess electron solvation.

Time-resolved spectroscopy may be classified using a variety of criteria such as the order of the response with respect to the applied fields, resonant and off-resonant techniques or temporal resolution. The simplest classification depends on the temporal resolution determined by the laser pulse duration: nanosecond (10^{-9} s), picosecond (10^{-12} s), and femtosecond spectroscopy (10^{-15} s).

Time-resolved laser spectroscopy methods can be carried out in resonant conditions monitoring phenomena such as absorption or emission or, in off-resonant conditions, monitoring phenomena such as Raman light scattering or wave mixing processes.

Time-resolved laser spectroscopy methods are related to linear or nonlinear polarization of various orders. *Fluorescence* is a linear phenomenon and depends on the first order $\chi^{(1)}$ susceptibility. The *pump-probe methods, coherent anti-Stokes Raman (CARS) scattering, photon echo*, and *transient grating* use the third order $\chi^{(3)}$ nonlinear phenomena. The Raman photon echo phenomena use non-linear processes of the seventh order $\chi^{(7)}$. *Quantum beats* are related to any of the above-mentioned spectroscopic methods and are based on certain quantum effects, which will be precisely defined and discussed in Chapter 8, section 5.1.

There is great variety of time-resolved spectroscopy methods. In this chapter we shall briefly survey the methods that are most commonly employed to probe dynamic processes of matter:

a) fluorescence decay;
b) pump-probe methods;
c) methods based on non-linear stimulated Raman scattering;
d) photon echoes.

In this chapter we focus on the physical principles employed in these methods, and in Chapter 9 we shall discuss some examples of their applications.

8.1. FLUORESCENCE DECAY

Conceptually, the simplest time-resolved method is fluorescence decay. A single laser beam (either pulsed or stationary) excites the system to a higher energy level and the emitted light is detected. The principle of emission is illustrated schematically in Fig. 8.1.

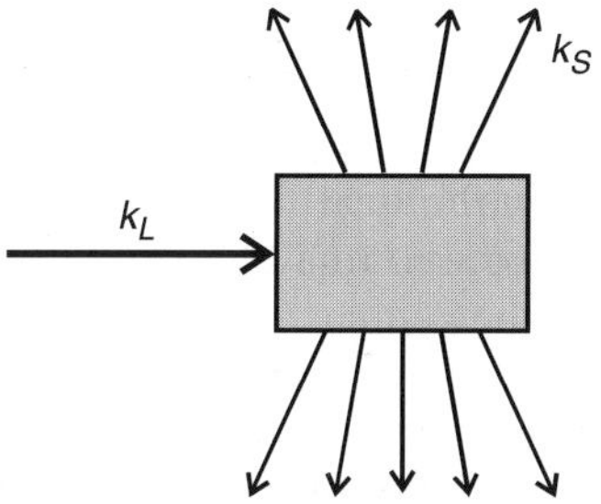

Fig. 8.1 Schematic representation of spontaneous light emission.

The spontaneous light emission signal consists of a spontaneous coherent Raman scattering and spontaneous, incoherent fluorescence. The spontaneous light emission is a linear process since the emitted light intensity is proportional to the incident light intensity. To avoid confusion, in this chapter we use the term spontaneous light emission to denote only the fluorescence.

Radiation interacts with a molecule in the singlet electronic ground state, S_0, absorption of a photon takes place, promoting the molecule to the excited electronic state S_1, when the energy difference of the specific electronic levels is equal to the energy of the photon. The molecule exists in the excited state for a short time, usually nanoseconds. The return to the ground state can be radiative or nonradiative in nature. Generally, the molecule gives back the excess energy as a result of the following processes:

a) radiative channel of energy dissipation, called fluorescence, followed by the molecule's return to the ground state, S_0;
b) nonradiative channel of energy dissipation to the surrounding molecules followed by the molecule's return to the singlet ground state, S_0, called *internal conversion*;
c) nonradiative transfer of energy from the excited singlet, S_1, to the triplet state, T_1, via *intersystem crossing*. This is the transition between the electronic states of different multiplicity (the singlet state, S_1, for which the total spin is 0 and the triplet state, T_1, of the total spin 1), which is forbidden by the selection rules within the dipole approximation;
d) radiative transfer of the energy from the excited triplet state, T_1, called *phosphorescence*, followed by the molecule's return to the ground singlet state, S_0. The lifetimes of the excited triplet states are much longer than the lifetimes of the excited singlet states.

The scheme of the channels for energy transfer was proposed for the first time in 1936 by Jabłoński, and is known in the literature as the *Jabłoński diagram*.

In this chapter we address the first channel of energy transfer, fluorescence. As the fluorescence lifetime, τ, of the molecule in the excited electronic state depends on molecular motions in its environment, it is important to monitor the time evolution of the emission spectrum. The lifetime measurements provide important information about the dynamics and structure, type of interactions, and molecular motions of the surrounding molecules. In contrast to the fluorescence spectrum in the frequency domain in stationary spectroscopy, which represents a picture in which all the objects are static, the measurement in the time domain, where the fluorescence decay is observed over time, can be compared to a film in which movement in time provides additional information about the system.

In the time-resolved measurements, fluorescence spectrometers use light from pulsed Q-switched lasers, pulsed flash lamps with filter or lasers working in the modelocking regime. The light exciting a sample causes fluorescence. The fluorescence radiation is focused onto the monochromator slit, which induces spatial separation of various fluorescence spectral components. The radiation of a definite wavelength falls onto a detector, usually a photomultiplier. The signal recorded in the detector is submitted to further analog or digital analysis, as a result of which one obtains the fluorescence decay as a function of time. In order to measure the fluorescence lifetime, τ, different detection methods may be used, such as: *streak camera detection*, the *time-correlated single photon counting* method, averaging by the *boxcar integrator, light-gating techniques*, the *up-conversion method*, and *phase-modulation methods*. The methods of detection may be classified by a variety of criteria such as the electronic, optical methods of gating or phase-modulation methods with heterodyne detection.

The streak camera detection, time-correlated single photon counting and averaging by the boxcar integrator techniques use electronic devices to obtain the time-resolved emission of the sample versus time. The light-gating techniques and the up-conversion method are optical methods that explore the correlation between the excitatory pulse, the gate-opening pulse, and the observed emission.

The techniques that include time-correlated single photon counting, the averaging by boxcar integrator, the light-gating techniques, and the up-conversion method measure the fluorescence decay as a function of time (Fig. 8.2b). In the case when fluorescence emission originates from only one type of molecules, the sample emission consists of a signal at t_0, corresponding to the excitatory laser pulse followed by the fluorescence decay signal described by the equation

$$I(t) = I_0 \exp(-(t - t_0)/\tau). \tag{8.1}$$

The lifetime τ, which characterizes the fluorescence signal decay by e times in comparison with the initial value of the fluorescence signal I_0 at time t_0, can be obtained from eq. (8.1). Such a simple determination is possible only when the fluorescence decay can be described by a single exponential function. However, the same principle applies to more complicated decays. Modern software allows us to determine the lifetimes, treating them as parameters in the standard procedures based on the root mean square method.

The second group, presented in Fig. 8.2c, comprises the *phase-modulation* methods, where the phase delay, $\Delta\Phi$, as well as the modulation of the fluorescence amplitude versus the temporary modulated light illuminating the sample are measured. Heterodyne detection combined with lock-in amplifiers allows us to determine directly the lifetime, τ.

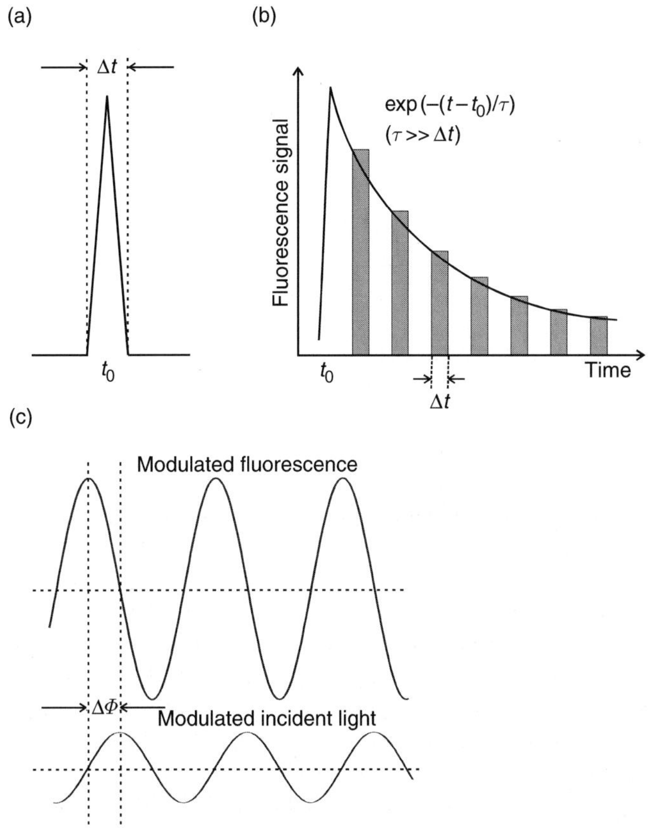

Fig. 8.2 (a) pulse of Δt duration excites the sample at the time t_0, (b) time-correlated single photon counting method, (c) phase-modulation method.

Figure 8.3 presents a typical experimental set up for time-correlated single photon counting measurements. A laser pulse of duration Δt, which is short in comparison with the measured signal decay times, excites a sample at $t = t_0$, inducing fluorescence. First, we need to clock the initial emission by defining its zero time. In order to determine the zero time t_0, the detector must receive information regarding when it should start the detection. For this purpose, the excitatory laser pulse is divided by a beamsplitter, BS (Fig. 8.3), into two beams: one is used to excite the sample, the second one of low intensity (called the *trigger pulse*) is sent to a fast photodetector coupled to a constant-fraction discriminator where the "start" electronic signal is generated. The "start" signal is sent to a device called a *time-amplitude converter* (*TAC*). The level of the trigger pulse is adjusted to minimize the jitter between the laser pulse and the electronic devices. The fluorescence emission is focused on the monochromator slit, which causes spatial separation of different fluorescence spectral components. The selected component that usually corresponds to the fluorescence maximum is detected by a photomultiplier or some other fast detector. The excitation level is adjusted in such a way that a single event is measured by the

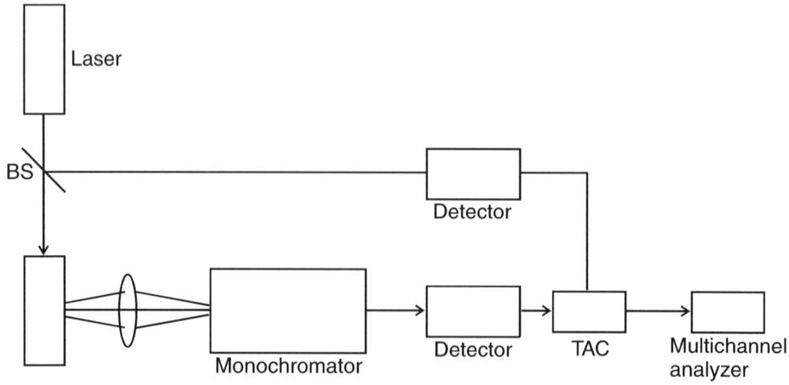

Fig. 8.3 Experimental set up for fluorescence lifetime measurement by the time-correlated single photon counting method, BS – beamsplitter, TAC – time-amplitude converter.

detector, generating a "stop" signal which is sent to the TAC. Thus, the TAC can easily record the time interval, Δt, between the absorption of a photon by the sample at $t = t_0$ and the emission of a photon. In the time-correlated single photon counting method, the photomultiplier records the number of photons. This method depends on counting individual photoelectrons emitted at a rate of n electrons per second, not the photocurrent $i = \int neGdt/\Delta t$ averaged over the interval Δt. Every time when an individual photon (e.g., from fluorescence emission) falls onto the surface of the photocathode, it causes an avalanche of photoelectrons with charge $q = Ge$, where G characterizes amplification in the photomultiplier. This charge then induces the voltage pulse $U = eG/C$ on the anode with the capacity C. The voltage pulse releases a fast discriminator, which sends the normalized TTL pulse to the counter or the digital–analog converter. A detailed description of the photon counting method can be found in the literature [1].

By measuring the number of pulses in a series of finished time intervals and storing each result in a multichannel analyzer, one obtains a histogram describing the fluorescence decay (Fig. 8.2b). The histogram is recorded by the TAC and the multichannel analyzer. The TAC generates a pulse with an amplitude proportional to the time interval, Δt, between the time when the emitted photon arrives at the detector and the zero time, t_0, when the excitatory pulse is absorbed by the sample. The multichannel analyzer counts the pulses with the corresponding amplitude and stores them in separate memory channels. The time resolution of the time-correlated single photon counting method is limited by the laser pulse duration and by the response time of the electronic devices.

All the methods based on fluorescence decay measure the fluorescence signal as a function of time delay. This corresponds to the measurement of the number of photons incident onto a detector (or a signal corresponding to this number) in definite time intervals, Δt, synchronized with the initial time, t_0. In the time-correlated single photon counting method, the device that controls the time intervals is the time-amplitude converter (TAC). However, the signal may be controlled with other detection methods, such as synchronized sampling by the boxcar integrator. The configuration applied in this method is similar to the one presented in Fig. 8.3, when the time-amplitude converter is replaced by the boxcar-integrator. In this electronic device, at a precisely determined moment correlated with the trigger pulse, an electronic entrance gate is opened for a certain short time, Δt, and the voltage corresponding to the fluorescence signal is measured.

In the methods discussed so far, different electronic types of gating in the detection analysis are applied. However, imagine that instead of gating of the detection system, the gating of the fluorescence emission can be used. Instead of a continuous fluorescence beam reaching a detector, one can provide discrete portions of fluorescent light to the detector at controlled time intervals. This idea has been applied in the light-gating method.

The principle of this method is presented in Fig. 8.4. The light emitted as a result of fluorescence illuminates carbon disulphide (CS_2) placed between two crossed polarizers. In this case the light does not reach the detector, because the crossed polarizers do not transmit the light to the detector. Carbon disulphide, as an isotropic liquid, does not affect the polarization of the incident light. However, if we place carbon disulphide in a strong electric field, the molecules begin to align in the direction of the electric field, and this ordering makes the liquid anisotropic and birefringent. The properties of the birefringent materials, including uniaxial crystals, were discussed in Chapter 5, which was dedicated to nonlinear optics. The birefringence generated in liquids by an electric field is known in the literature as the *Kerr effect*. As a result of birefringence, the incident light beam splits into two rays: the ordinary and the extraordinary rays, with mutually perpendicular polarizations, and the intensity of each depends on the angle between the polarization of the incident ray and the optical axis of the Kerr material. Thus, the Kerr effect changes the polarization of the beam that travels through the medium. If we apply an ultrashort laser pulse incident onto a liquid (Fig. 8.4) instead of a constant electric field, its electric field will induce an ultrashort Kerr effect. In this case the liquid placed between the crossed polarizers will act as an ultrafast shutter, which transmits

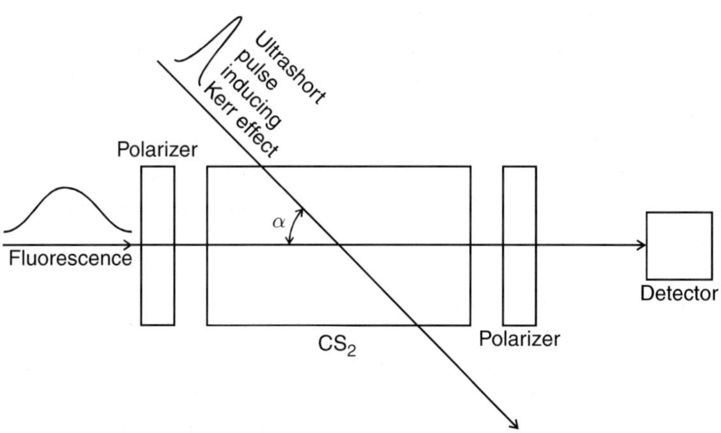

Fig. 8.4 Scheme illustrating the principle of the light-gating method.

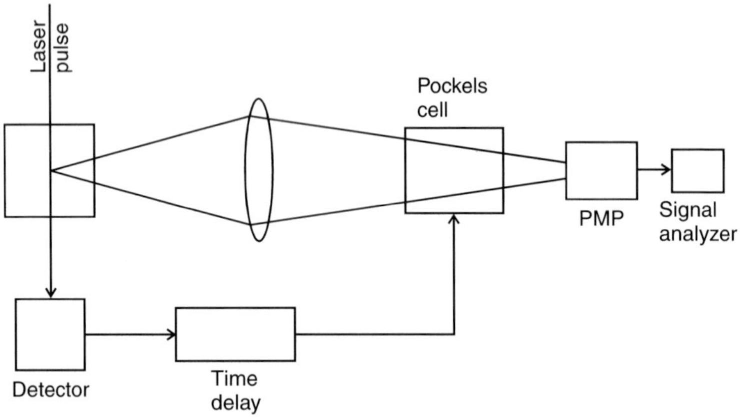

Fig. 8.5 Light-gating method using the Pockels cell, PMP – photomultiplier.

fluorescence as long as the liquid retains birefringence caused by the pulse, which is about 3.5 ps for CS_2. The fluorescence decay was initially measured using the echelon to produce time delays resulting from different optical paths on the echelon. In the currently applied light-gating methods, the time delay can be easily controlled with the Pockels cell, which plays the role of an ultrafast shutter like the Kerr cell. The Pockels cell was described in Chapter 6. The typical configuration employing the light-gating method is presented in Fig. 8.5.

Another version of the light-gating method is based on the nonlinear beam mixing phenomena discussed in Chapter 5. When a short laser pulse and the fluorescence beam are incident simultaneously on a nonlinear crystal at an angle allowing the phase matching condition to be fulfilled, the interaction of the beams produces a signal of a frequency equal to the sum frequency, called the *up-conversion signal*. The up-conversion signal is generated only when the beams overlap, that is only when the laser pulse illuminates the crystal. Thus, the crystal acts like an ultrafast shutter analogously to the Kerr or the Pockels cell. When the fluorescence intensity decreases with time, the up-conversion signal decreases as well, because the up-conversion signal intensity linearly depends on the intensity of each of the beams. The light-gating methods can be used to study fluorescence decays occurring at the picosecond scale.

The fluorescence decay method can be modified to determine lifetimes of short-living intermediates in a reaction. The time-correlated single photon counting method presented in Fig. 8.3 can be employed for this purpose with a slight modification that includes an additional non-monochromatic light source placed behind the sample (Fig. 8.6).

The idea of measuring the reaction intermediates is as follows:

1. a single, strong laser pulse induces a reaction that promotes molecules into transition states, resulting in the generation of intermediates ($t = 0$),
2. the sample is illuminated all the time with a continuous light from a source Z. The intermediates generated by the laser pulse absorb the light from the Z source

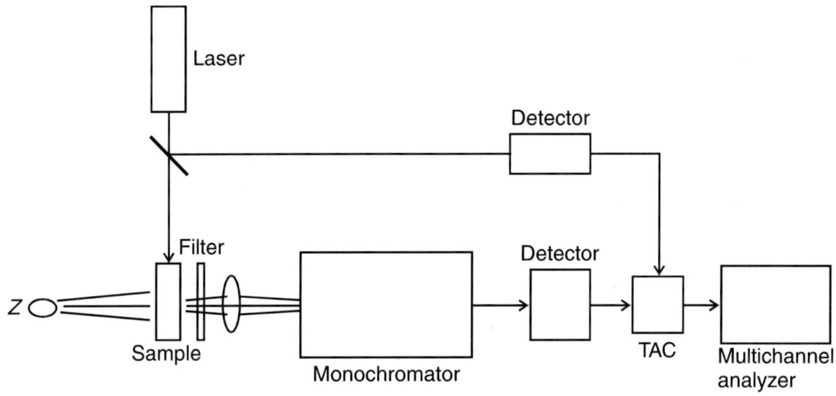

Fig. 8.6 Experimental set up for the fluorescence decay of reaction intermediates, Z – non-monochromatic light source, TAC – time-amplitude converter.

that is further emitted as fluorescence, if the intermediates dissipate the energy in a radiative manner,

3. separation of the fluorescence components takes place in a monochromator. The fluorescence components at specific wavelengths are recorded by the detection system,
4. when concentration of the intermediates decreases, the fluorescence decays as well.

Thus, the fluorescence signal decay is a measure of the intermediate lifetime. We may use this method only when other molecules of the sample, which are not intermediates, do not fluoresce, or when the wavelengths of their fluorescence significantly differ from that of the intermediates. In this case, employing a filter will eliminate background emission. This method can also be used when the decays of the intermediates and other molecules occur on a different time scale. Otherwise, the fluorescence decay signal is a complicated convolution of different contributions.

The fluorescence decay method has been used in a number of physical and chemical studies. This issue is addressed in a wide range of publications and interested readers are referred to refs. [2–5]. Some applications of the fluorescence decay method will be discussed in the next chapter.

8.2. THE PUMP-PROBE METHOD

The optoelectronic detection systems, such as fast photodiodes and stroboscopic oscilloscopes, only reach a time resolution of the order of 10^{-10} s. This is an insufficient resolution to examine dynamics of the numerous non-stationary processes that take place on a pico- or femtosecond time scale. One of the methods for extending the time resolution is to apply the streak camera as a detector to reach a resolution on the order of several picoseconds. However, so far no detector is able to measure the dynamics of femtosecond processes in real time. The insufficient time resolution of detectors was the reason why indirect methods that would go beyond the barriers connected with detection were needed. The most valuable method is the *pump-probe method.*

Application of the pump-probe method allows us to overcome the time resolution limitations of detectors. In this method a detector does not have to be characterized by the ultrafast time response to record dynamics of phenomena occurring on the pico- or femtosecond time scale. The concept of the pump-probe method is presented in Fig. 8.7. The beamsplitter divides the pulsed laser beam into two beams: the pump beam and the probe beam. These beams travel along different optical paths before they join up again. The delay of the probe beam with respect to the pump beam is $\Delta t = \frac{\Delta x}{c}$, where Δx is the optical path difference (Fig. 8.7). The signal in the detector depends on the delay of the probe beam with respect to the pump beam, Δt. Recording the signal for the different delays, we monitor the dynamics of the studied process. The time resolution of this method does not depend on the time response of the detector but on the pulse duration.

In the configuration presented in Fig. 8.7, the pump beam and the probe beam have the same wavelength, which considerably limits the number of possible applications. In most experimental configurations these beams have different frequencies (Fig. 8.8). Additionally, the frequency of the pump or/and the probe beams can be tuned over a broad spectral range by employing tunable light sources, such as tunable lasers, parametric generators (OPG), parametric oscillators (OPO), parametric amplifiers (OPA), and higher harmonic generators as described in Chapter 5, or by using white continuum sources (WC) emitting radiation in a broad spectral range. The origin of the WC generation is well described elsewhere [6–7]. The WC process is mainly governed by SPM (*self-phase modulation*) and *stimulated Raman emission*, described in Chapter 5. The most popular substances used to generate WC are water, deuterated water, quartz glass, and sapphire.

An ideal solution is to have independent tunability of the pump and the probe beams. There are many various modifications of the pump-probe method that differ according to the ways of excitation and probing. The most commonly used method is the technique that measures the modification of the absorption of the probe beam by the pump beam as a function of the time delay Δt. Let's consider a two-level system with energies E_1 and E_2. If the system is excited with the pump pulse of high intensity and frequency equal to $\omega = (E_2 - E_1)/\hbar$, some fraction of molecules will be promoted from the lower to the upper state. We observe a perturbation of the Boltzmann population distribution. As a consequence, at the lower level E_1 there are fewer

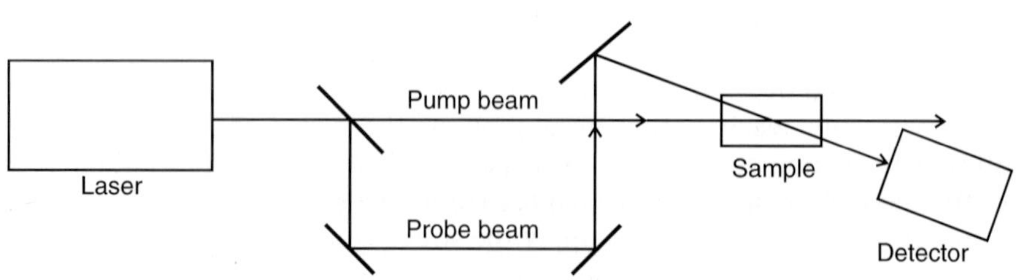

Fig. 8.7 Scheme illustrating the pump-probe method principle.

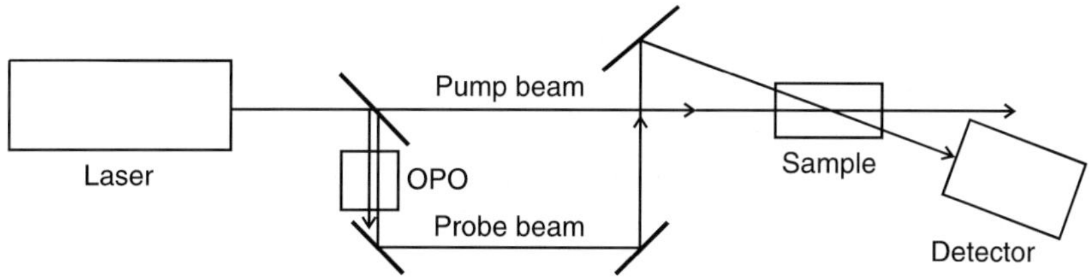

Fig. 8.8 Scheme illustrating the principle of the pump-probe method when the pump and the probe beams have different frequencies.

molecules than there were before excitation with the pump pulse in contrast to the upper level E_2, where there are more molecules. Therefore, the second probing pulse, delayed with respect to the pump pulse by Δt, monitors the transient absorption $A(E_1 \rightarrow E_2)$ that is smaller than the absorption A_0 recorded for the same probing pulse without employing the pumping pulse. When the time delay, Δt, of the probe beam increases and exceeds the lifetime of the excited level T_1 ($\Delta t > T_1$), one observes the return of the excited molecules to the lower level and recovery of the initial absorption value observed before the action of the pump pulse. A typical signal of the absorption change, $\Delta A = A - A_0$, as a function of the time delay Δt, is presented in Fig. 8.9. The changes in the absorption spectrum may involve the spectral shifts and/or the optical density changes for a given wavelength. Analysis of the spectral changes provides information about new species or transitions created by the pump excitation, while the temporal analysis for a given wavelength provides information about the population dynamics.

In order to achieve a significant change in the energy level population, the pumping pulse must have the intensity close to the saturation intensity I_s (Chapter 1, eq. (1.40)) $I_s = \frac{h\nu}{2\sigma t_p}$, where ν is the frequency of radiation emitted by the laser beam, σ is the absorption cross-section of the sample, and t_p is the pulse duration.

The pump-probe methods can be applied for various transitions between two quantum levels, E_1 and E_2, in all the spectral ranges. The absorption pump-probe method is more effective, the larger the absorption across section σ and the longer

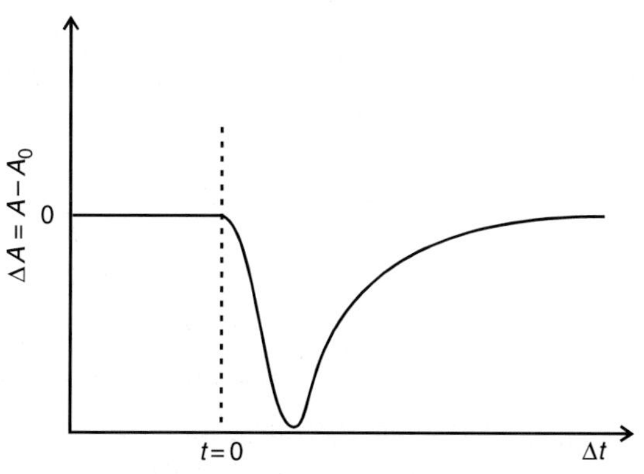

Fig. 8.9 A typical signal obtained in the pump-probe absorption method.

the lifetimes T_1 are. The pump-probe absorption is often used to study electronic transitions to monitor new electronic states of intermediates or photochemical products created by the pump pulse.

The pump-probe methods are especially useful in monitoring the dynamics of populations of vibrational states. In the infrared range the absorption cross section for vibrational modes is usually small ($\sigma \cong 10^{-20}$ cm^2), which requires large pumping intensities on the order of $I_s \cong 10^{11}-10^{12}$ W/cm^2.

The effect of the pumping pulse on the sample can be analyzed not only by the transient absorption signal $\Delta A = A - A_0$ versus the time delay. Other methods such as spontaneous anti-Stokes scattering or two-pulse fluorescence can sometimes be very useful. Fig. 8.10 compares the different pump-probe methods that can be employed for probing the instantaneous population of vibrational states.

In Fig. 8.10a the pump beam has a frequency ν_1, corresponding to the vibrational transition in the IR range. The pump beam excites some molecules from the vibrational ground level ($\upsilon = 0$) to the excited vibrational level ($\upsilon = 1$). The probe beam of the ν_2 frequency causes *spontaneous anti-Stokes scattering* at frequency $\nu_{as} = \nu_2 + \nu_1$. If redistribution of vibrational energy has taken place within the time delay, Δt, that passed between the pulses ν_1 and ν_2 and, as a consequence of radiationless transitions, another vibrational mode ν_i has been populated, an additional anti-Stokes scattering at frequency $\nu_{as} = \nu_2 + \nu_i$ should also be observed. This technique provides information on the dynamics of populations of different vibrational states and on the rate of the energy redistribution between the different vibrational modes.

In Fig. 8.10b the pump pulse ν_1 (IR) excites a fraction of the molecules at the vibrational ground level ($\upsilon = 0$) to the excited vibrational level ($\upsilon = 1$). The energy redistribution makes some of the excited molecules transfer energy to another vibrational mode of frequency ν_i and populate the vibrational level of the mode ν_i. The probe beam at frequency ν_2 that is smaller than the transition frequency between

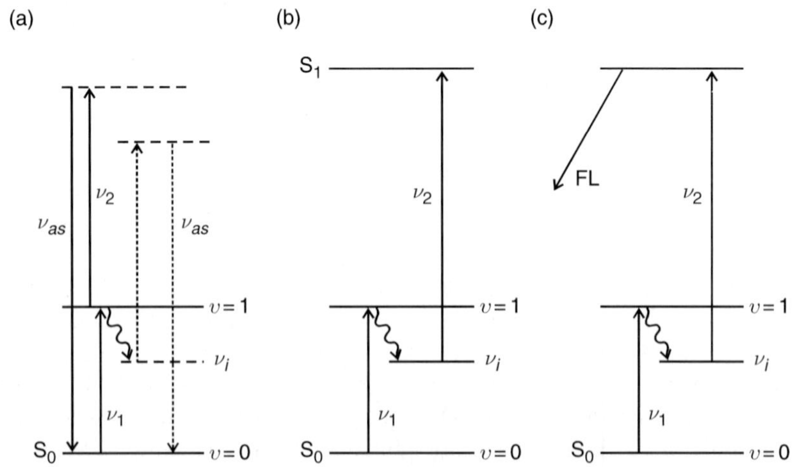

Fig. 8.10 Techniques applied in monitoring population dynamics of vibrational states: (a) spontaneous anti-Stokes Raman scattering, (b) absorption change, (c) two-pulse fluorescence.

the electronic states $S_0(v = 0) \rightarrow S_1(v = 0)$ monitors the population changes at the vibrational level of the mode ν_i.

Two-pulse fluorescence is another configuration to study population changes of vibrational levels (Fig. 8.10c). The beam at the frequency ν_1 pumps the system to the excited vibrational state, which dissipates energy via energy exchange, causing the population of another vibrational mode ν_i. The probe beam excites the molecules from the ground electronic state $S_0(\nu_i)$ to the excited electronic state S_1. By measuring fluorescence, FL, one can get information about the population dynamics of the vibrational mode ν_i since the fluorescence signal is directly proportional to the population of the probed vibrational state. All the methods mentioned above can be used for probing ultrafast dynamics of vibrational energy transfer and for measuring lifetimes, T_1, of vibrational levels.

Most of the pump-probe methods require tuning of the pump beam ν_1 and the probe beam ν_2. The development of laser technology has provided stable pico- and femtosecond pulses from solid-state lasers (vibronic lasers described in Chapter 4) that are tunable over a wide range. These lasers work in the modelocking regime at high repetition rate (on the order of tens of MHz). The time-resolved measurements with these lasers are likely to contain fewer mistakes than earlier measurements based on *Q*-switch lasers of lower repetition rates. The pulses emitted by the modelocking lasers are much more reliable, stable and the high repetition rate assures proper statistics of experiments. The typical pulse energy, which is 10–20 nJ for the lasers working at the repetition rate of 80 MHz, is several orders lower than the energy of the low-repetition-rate *Q*-switched lasers. This is often regarded as an advantage because energy that is too high may lead to undesirable nonlinear effects or even damage a sample. On the other hand, in many cases energies on the order of 10 nJ are insufficient and pulses must be amplified in the amplifiers that were described in Chapter 6. In commercially available regenerative amplifiers, the energy of the 10 nJ pulse is amplified to around 1 mJ at a repetition rate of 1 kHz. High intensities of the output beam from the amplifier permit us to generate higher harmonics, which considerably extends the spectral range. An additional way to extend the spectral range is to generate a white continuum (WC). Moreover, the spectral range can be extended by using parametric generators, parametric oscillators, and parametric amplifiers as described in Chapter 6.

Figure 8.11a shows a typical experimental laser system for the pump-probe absorption measurements. The configuration consists of the femtosecond Ti:sapphire laser, which is pumped with the *cw* solid state laser Nd:YVO_4 (yttrium vanade crystal doped with neodymium). Its second harmonic (532 nm, power 5.5 W) is used as a pumping source for the Ti:sapphire femtosecond laser working in the modelocking regime. The femtosecond Ti:sapphire laser is tunable in a broad range. Usually, the wavelength at 796 nm is used for further amplification. The pulse with an energy of 12 nJ, duration of 80 fs and a repetition rate of 82 MHz is amplified in the regenerative amplifier. The small part of the seed pulse is sent by the beamsplitter to the spectrum analyzer in the frequency domain to monitor the quality of the modelocking. The regenerative amplifier is pumped by the solid state laser (Nd:YLF) working with a repetition of 1 kHz, and a pulse duration of 250 ns. The pulsed mode is achieved by *Q*-switching with an optoacoustic device. The laser emits at 527 nm

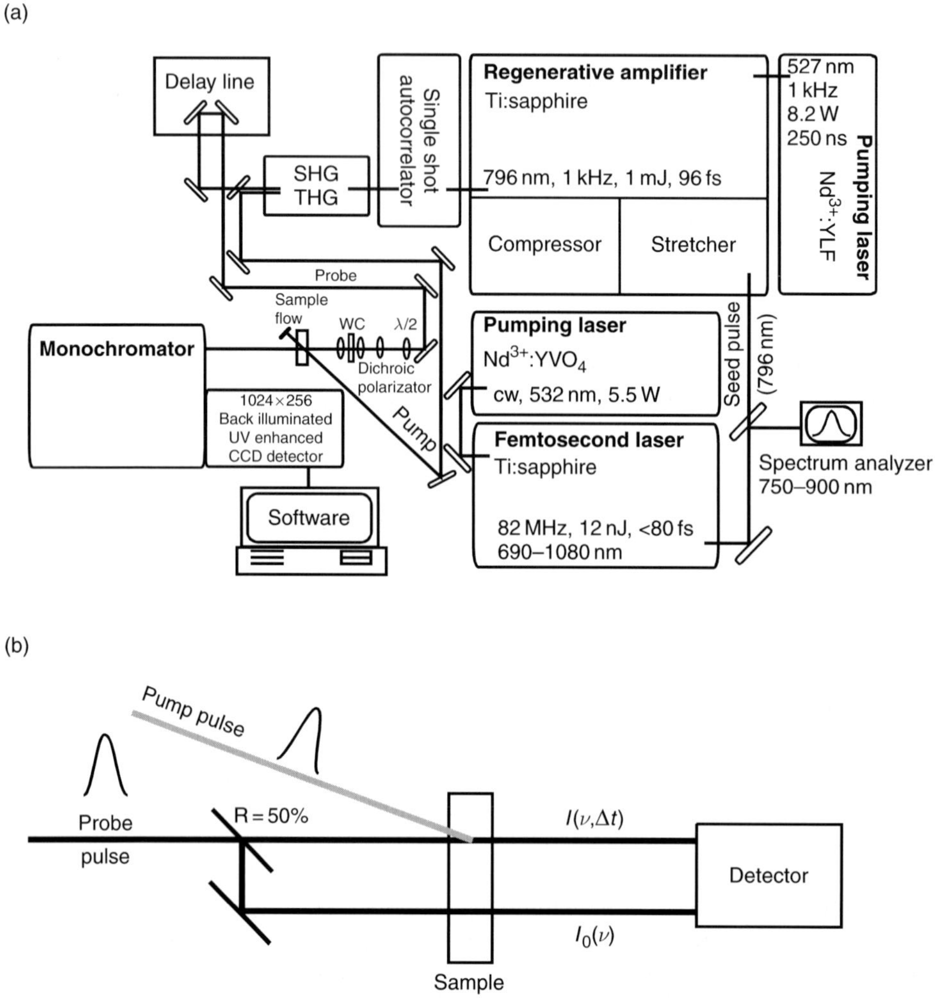

Fig. 8.11 (a) Typical experimental laser system for pump-probe absorption measurements, (b) pump-probe transient absorption with the reference probe beam.

(second harmonic from LBO nonlinear crystal). After the amplification the energy of the pulse is 1 mJ, the repetition 1 kHz, and the wavelength 796 nm. The pulse duration is measured with the single shot autocorrelator. The amplified pulse interacts nonlinearly with KDP crystals to generate SHG and THG which are used as a pump beam. The fundamental 796 nm beam passes through the delay line, the half wave plate, the dichroic polarization analyzer, the lens, and the sapphire plate when the white continuum (WC) is generated in the broad 398–1000 nm range and focused onto the flowing sample. The WC beam is used as a probe beam. The probe and a pump beams are focused onto the flowing sample to ensure complete replacement of the sample between each laser shot. Care is taken to compensate for the group velocity dispersion. The time delay is provided by an optical delay line, which is controlled by a motor-driven translational stage. The radiation from the sample enters the entrance slit of a spectrometer. The dispersed radiation is detected by a multichannel charge-coupled device detector, which is attached to the exit slit of the spectrometer. The data acquisition and full computer control is provided by the software. The difference absorption signal, $\Delta A = A - A_0$, is measured, where A_0 is the absorption without pump pulse, and A the absorption when both the pump and pulses excite the sample. ΔA is measured as a function of a time delay between the pump and the probe pulses for a given wavelength.

To improve the sensitivity of the pump-probe transient absorption, it is better to record the spectral density of the probe beam with respect to the reference probe beam (8.11b). This can be performed by splitting the probe beam into two equal beams: the first beam crosses the area perturbed by the pump beam while the other probe beam crosses an unperturbed part of the sample.

8.3. CARS AS A TIME-RESOLVED METHOD

In Chapter 5, section 5.2 we discussed the CARS method, which applies coherent anti-Stokes Raman scattering. This method is based on two beams simultaneously incident onto a sample: at the frequencies ω_L and ω_S, the difference between them, $\omega_L - \omega_S = \omega_{vib}$, is equal to a specific vibrational mode frequency, ω_{vib}, of a molecule. If the incident beams fulfill the phase matching condition $2\boldsymbol{k}_L - \boldsymbol{k}_S = \boldsymbol{k}_{as}$, stimulated coherent anti-Stokes scattering is observed at frequency $2\omega_L - \omega_S$ in the $\boldsymbol{k}_{as}$ direction. It results from the polarization induced by interaction between the incident beam at frequency ω_L and the vibrational frequency ω_{vib}

$$\omega_L + \omega_{vib} = \omega_L + (\omega_L - \omega_S) = 2\omega_L - \omega_S \tag{8.2}$$

As long as the sample interacts with the electric field of the beams at frequencies ω_L and ω_S, all molecules vibrate with the same frequency and the same phase. When the electric field of the laser beams is removed, the oscillators lose the coherence because they start to vibrate with different frequencies and phases due to interaction with surrounding molecules. Processes causing oscillators to dephase are called *phase relaxation* whereas the time characterizing the dynamics of the changes is called the *dephasing time* T_2.

If the probe beam at frequency ω_L is delayed by t_D with respect to the moment when the pump beams at the frequencies ω_L and ω_S were applied simultaneously, the interaction described by eq. (8.2) between the probe beam ω_L and the coherent polarization at ω_{vib} that was generated by the pump pulses will again produce anti-Stokes scattering at a frequency of $2\omega_L - \omega_S$. However, the anti-Stokes signal will be weaker than at $t = 0$ because the coherent polarization at the vibrational frequency ω_{vib} begins to disappear (Fig. 8.12) as a result of vibrational dephasing. The longer the time delay t_D, the weaker is the CARS signal.

The CARS signal decay is a measure of the vibrational coherence decay in a sample and it permits us to determine the vibrational dephasing time T_2. The vibrational dephasing time T_2 can also be estimated from the width of the vibrational

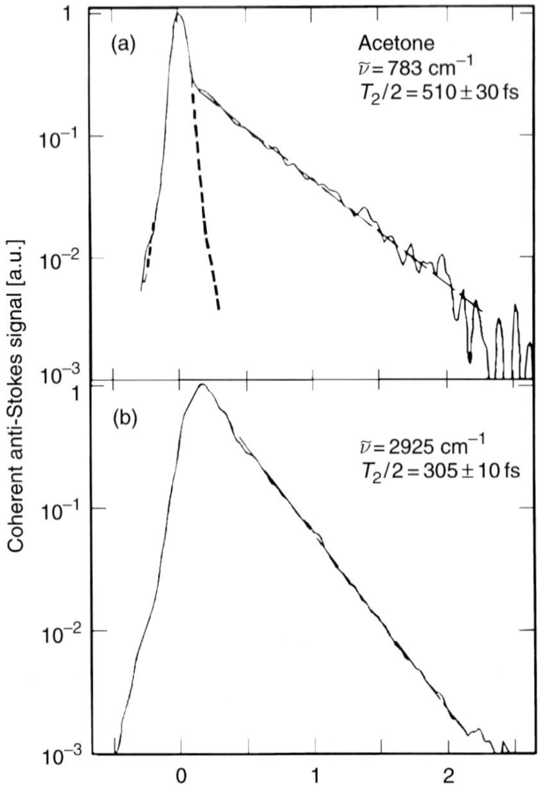

Fig. 8.12 The CARS signal decay for acetone, (a) mode of 783 cm^{-1}, (b) mode of 2925 cm^{-1} [8].

line in stationary spontaneous linear Raman spectroscopy. The vibrational coherence in condensed phases decays through pure dephasing and vibrational energy (population) relaxation [9–11]. Pure dephasing occurs through fluctuations in vibrational frequencies and due to differences in vibrational frequencies of molecules in different molecular environments. The portion of dephasing occurring due to static or slowly relaxing disorder gives rise to inhomogeneous broadening in the vibrational lineshape. The amount of static disorder can be strongly dependent on the thermodynamic state of the sample and is expected to increase considerably in going from the liquid to the glassy state [12–14]. The decay time, T_2, obtained from the homogeneous vibrational lineshape or the CARS measurements, is related to *the pure dephasing time* T_2^* and population relaxation time T_1 by

$$\frac{1}{T_2} = \frac{1}{2T_1} + \frac{1}{T_2^*}. \tag{8.3}$$

Those readers who are more interested in the energy and phase relaxation processes should read Chapter 9, section 8.

Both the CARS method and stationary Raman spectroscopy determine the dephasing time T_2, which depends on homogeneous line broadening processes in

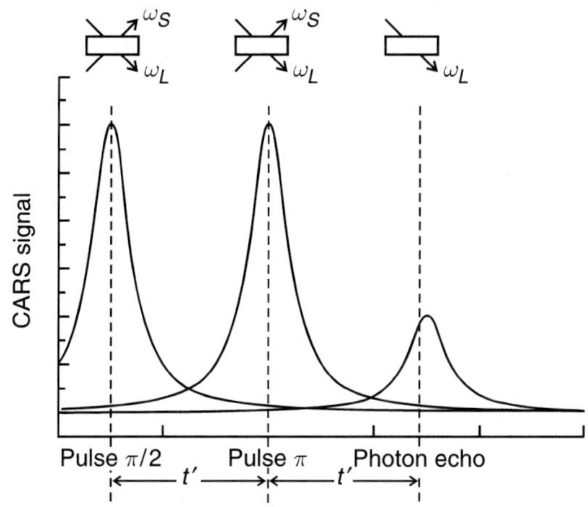

Fig. 8.13 Photon echo in CARS.

the frequency domain (fast processes) and inhomogeneous line broadening processes (slow processes). These methods are not able to separate contributions originating from the homogeneous and inhomogeneous vibrational dephasing processes. In order to separate these, the CARS photon echo method (Fig. 8.13) or IR photon echo method must be applied. The photon echo method, which employs pulse sequences (called $\pi/2$ and π pulses), is based on optical resonance phenomena. The photon echo method will be described in the section 8.4.2. The details can also be found in ref. [15–19].

8.4. PHOTON ECHO

Photon echo is an analog of the spin echo observed in nuclear magnetic resonance (NMR) [20]. In NMR spectroscopy the spin echo method is used to measure the spin-lattice relaxation time, T_1, and the spin-spin relaxation time, T_2. In order to understand the principles of the photon echo, let us first focus on the spin echo in NMR. The spin echo in NMR can be explained in a classical way, which will help us to get an intuitive insight into the optical echo effect. Although there is no classical analogy for the optical echo, the classical Bloch formulae for the magnetization vector $\boldsymbol{M}$ have their equivalent for the polarization vector $\boldsymbol{P}$. Feynman, Vernon and Hellwarth [21] have established the exact isomorphism between the Bloch equations and the evolution of two level systems interacting with an optical field, known as the *optical Bloch equations*. A theoretical description of the photon echo can be found in ref. [22].

8.4.1. Spin Echo in NMR

An atomic nucleus can have spin, just like an electron has. Spin is a fundamental property of nature. Not only electrons but also protons and neutrons possess a spin of 1/2. According to the shell model the nucleons in the nucleus, just like electrons, fill orbitals. Because nucleons have spin, their spins can pair up when the orbitals of the nucleus are filled and cancel out. That is why some nuclei have non-zero nuclear spin, while the others do not manifest any spin. The net spin depends on the total number of unpaired protons and unpaired neutrons in the nucleus.

If the nucleus has spin $\boldsymbol{I}$, it also possesses the nuclear magnetic moment $\boldsymbol{\mu}$, which can be expressed in terms of the nuclear spin $\boldsymbol{I}$ in the form

$$\boldsymbol{\mu} = g_N \mu_N \boldsymbol{I}, \tag{8.4}$$

where g_N is the gyromagnetic ratio or g-factor, μ_N denotes a *nuclear magneton* ($\mu_N = e\hbar/2mc$). The magnitude of the magnetic moment $|\boldsymbol{\mu}|$ can be written as

$$|\boldsymbol{\mu}| = g_N \mu_N \sqrt{\boldsymbol{I}(\boldsymbol{I}+1)}. \tag{8.5}$$

For more details the reader is referred to textbooks on NMR spectroscopy [23].

If the nucleus is placed in the magnetic field $\boldsymbol{B}_0$, the nuclear magnetic moment interacts with the magnetic field, tending to line up with the magnetic field. The potential energy associated with the interaction is given by

$$E = g_N \mu_N \boldsymbol{I}\boldsymbol{B}_0 \tag{8.6}$$

Taking into account the quantization of nuclear spin $\boldsymbol{I}$, the measured quantity is the interaction energy proportional to the z–component of the magnetic moment $\mu_z = g_N \mu_N m_I$ along the direction of an applied magnetic field B_0

$$E_I = g_N \mu_N m_I B_0, \tag{8.7}$$

where m_I is a nuclear spin quantum number and it describes the projection of the spin $\boldsymbol{I}$ on the direction of the magnetic field. The quantum number m_I is equal to one of the $(2I+1)$ values: $-I, -(I-1), -1, 0, 1, \ldots, (I-1), I$.

For a system consisting of N molecules with nuclear spin $\boldsymbol{I}$ placed in the magnetic field $\boldsymbol{B}_0$, the energy levels E_I (Fig. 8.14) are populated according to the Boltzmann distribution

$$\frac{N_I}{N_{(I-1)}} = \exp\left(-\frac{\Delta E}{kT}\right), \tag{8.8}$$

where N_I and N_{I-1} are the number of molecules occupying the levels E_I and E_{I-1}, respectively, ΔE is the energy difference between the spin states, k is Boltzmann's

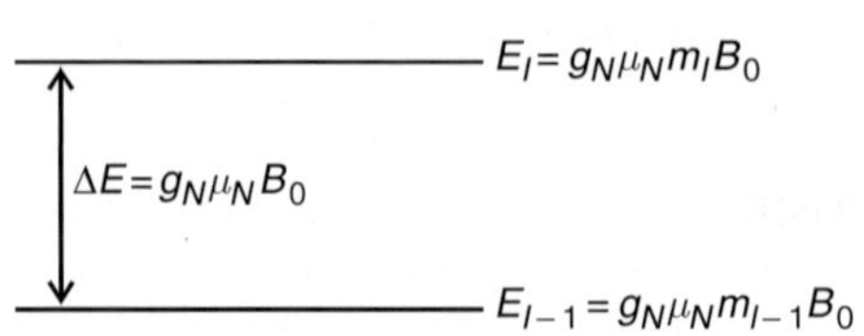

Fig. 8.14 Energy levels of a nucleus with spin $\boldsymbol{I}$ in the magnetic field B_0.

constant, T is the temperature in degrees Kelvin. From eq. (8.7) one can calculate the energy separation between the two spin states

$$\Delta E = g_N \mu_N B_0. \tag{8.9}$$

A molecule can undergo a transition between the two spin states by the absorption of a photon when the energy of the photon matches the energy gap between the initial and final spin states, $\Delta E = h\nu$. The resonance frequency is called the *Larmor frequency*. In NMR spectroscopy, the frequency of the photon is in the radiofrequency (RF) range. The exact value depends on the type of nuclei and the magnitude of the applied magnetic field–for hydrogen nuclei the typical values are between 60 and 800 MHz.

Eq. (8.8) describes the system of N spins in the magnetic field B_0 at equilibrium. When the system is illuminated with photons precisely matching the energy gap between the initial and final states, $\Delta E = h\nu$, absorption of photons occurs, stimulating transitions to the upper state, and the population distribution deviates from the Boltzmann equation

$$\frac{N_I}{N_{I-1}} = e^{-g_N \mu_N B_0/kT} \tag{8.10}$$

The system, which is deviated from equilibrium, tends to return to equilibrium. The energy excess is dissipated to the environment and the system returns to equilibrium after some characteristic time, T_1. The time T_1 characterizes the lifetime of the upper level. The process of giving the nuclear spin excess energy back to the surrounding (lattice) is named the *longitudinal relaxation* T_1 (or the *spin-lattice relaxation*) and governs the population of the spin states

$$\frac{N_0 - N}{N_0 - N_a} = \exp\left(-\frac{t}{T_1}\right), \tag{8.11}$$

where $N_0 = n_{I-1} - n_I$ is the population difference between the lower and upper states at equilibrium, $N_a = n_{I-1} - n_I$ is the population difference between the lower and upper states at $t = 0$, $N = n_{I-1} - n_I$ is the population difference between the lower and upper states at $t > 0$, n_{I-1} and n_I denote the number of molecules in the lower and upper states, respectively. The moment of absorption of a photon of energy $\Delta E = h\nu$ is taken as the initial moment $t = 0$. The longitudinal relaxation times T_1 can be relatively long, even of the order of 10^4 s. The question arises as to what mechanisms cause the relaxation of the system to equilibrium and how the excited molecules give the excess energy back to the surroundings. First, spontaneous emission does not contribute to the relaxation process since it is negligible in this spectral range. Indeed, one can see from eq. (1.16) that the Einstein coefficient, A_{mn}, depends on the third power of the radiation frequency and for the radiofrequency range the spontaneous emission is negligibly small. The radiationless transitions are mainly responsible for the relaxation processes. They occur when the resonance condition $\Delta E_{spin} = \Delta E_{lattice}$ is fulfilled, that is when the difference of the energy between the lower and upper spin states matches the difference of the lattice energy levels (e.g., rotational, translational levels). The main type of coupling between the spin and the lattice that is responsible for the energy relaxation T_1 originates from interactions between the magnetic moments of the spins and the magnetic moments of the lattice molecules. The magnetic field of the lattice fluctuates in time due to chaotic translational, rotational and vibrational motions of molecules. Time varying fields at the Larmor frequency cause transitions between the spin states. However, fluctuating fields due to the random motions of the molecules perturb also the energy levels of the spin states, leading to dephasing similar to that observed in vibrational spectroscopy and discussed in the previous section, 8.3. This type of relaxation is known as *spin–spin relaxation* and is characterized by the time constant T_2.

The relaxation processes occurring on a microscopic scale can be studied macroscopically, which is more convenient. At any instant in time, the magnetic field originating from the molecular spins can be represented by a magnetization vector $\boldsymbol{M} = \sum_i^N \boldsymbol{\mu}_i$, which is the vector sum of the magnetic moments of individual spins. We will show that time-varying fields at the Larmor frequency which cause transitions between the spin states (T_1 processes) are responsible for a change in the z component of magnetization whereas T_2 dephasing processes govern the return to equilibrium of the transverse magnetization. The time constant T_1 describes the rate of recovery along the z-axis whereas T_2 describes the decay of the transverse magnetization.

Here we describe the time evolution of the magnetization vector $\boldsymbol{M} = \sum_i^N \boldsymbol{\mu}_i$ of a sample that absorbs radiation, followed by the relaxation processes T_1 and T_2. The equations describing the time evolution of the magnetization vector are known as the *Bloch equations* [24].

Recall that the magnetic moment of a molecule $\boldsymbol{\mu}$ is expressed in terms of the total angular momentum $\boldsymbol{J}$ in the form

$$\boldsymbol{\mu} = \gamma \boldsymbol{J}. \tag{8.12}$$

If one assumes the total angular momentum of a molecule $\boldsymbol{J}$ is equal to the nuclear spin $\boldsymbol{I}$, the γ factor is given by

$$\gamma = g_N \mu_N. \tag{8.13}$$

The torque $\boldsymbol{N}$ on the magnetic moment $\boldsymbol{\mu}$ in the magnetic field $\boldsymbol{B}_0$ is given by

$$\boldsymbol{N} = \boldsymbol{\mu} \times \boldsymbol{B}_0. \tag{8.14}$$

The equation of motion for the total angular momentum $\boldsymbol{J}$ in the laboratory frame of reference can be written in the form

$$\frac{d\boldsymbol{J}}{dt} = \boldsymbol{N} \Rightarrow \frac{d\boldsymbol{J}}{dt} = \gamma \boldsymbol{J} \times \boldsymbol{B}_0. \tag{8.15}$$

On the other hand, the linear velocity $\boldsymbol{\upsilon}$ is given by

$$\boldsymbol{\upsilon} = \frac{d\boldsymbol{r}}{dt} = \boldsymbol{r} \times \boldsymbol{\omega}_0, \tag{8.16}$$

and

$$\frac{d\boldsymbol{J}}{dt} = \boldsymbol{J} \times \boldsymbol{\omega}_0. \tag{8.17}$$

Comparing (8.15) and (8.17) one sees that in the magnetic field $\boldsymbol{B}_0$ the magnetic vector $\boldsymbol{\mu}$ rotates about $\boldsymbol{B}_0$ with an angular velocity equal to $\boldsymbol{\omega}_0 = \gamma \boldsymbol{B}_0$ (Fig. 8.15). It is worth noting that the frequency is equal to the resonance Larmor frequency (eq. 8.9). So far, we have just looked at the behavior of spins in the laboratory frame of reference. It is convenient to define a rotating frame of reference, which rotates about the z-axis at the Larmor frequency (Fig. 8.15). In such a frame of reference, $\boldsymbol{B}_0 = 0$, with the magnetization vector appearing stationary, the z-component of magnetization looks the same as it did in the laboratory frame, and a transverse magnetization will appear stationary in the rotating frame. In Fig. 8.15 we distinguish this rotating system from the laboratory system by primes on the x- and y-axes, x' and y'.

To study relaxation processes by NMR we must apply an additional pulsed magnetic field of a proper sequence and pulse duration. The pulsed magnetic field along the x'-axis can be provided by a coil of wire passing an alternating current at the Larmor frequency $\boldsymbol{\omega}_0 = \gamma \boldsymbol{B}_0$. Indeed, a coil of wire passing a direct current produces a magnetic field along the x-axis in the laboratory frame of reference. An alternating current will produce a magnetic field, which alternates in direction along the x-axis. In a frame rotating about the z-axis at the frequency equal to that of the alternating current the magnetic field along the x'-axis will be constant (Fig. 8.15), just as in the direct current in the laboratory frame. When the alternating current through the coil is turned on and off, it creates a pulsed magnetic field along the x'-*axis* which is called the $\boldsymbol{B}_1$ magnetic field

$$\boldsymbol{B}_1 = \boldsymbol{B}_{10} \exp(i\omega t - \boldsymbol{k} \cdot \boldsymbol{r}). \tag{8.18}$$

When the frequency ω of the field $\boldsymbol{B}_1$ reaches the Larmor frequency $\omega_0 = \gamma \boldsymbol{B}_0$, the $\boldsymbol{B}_1$ vector is directed along x'. The magnetization vector $\boldsymbol{M}$ responds to this pulse by

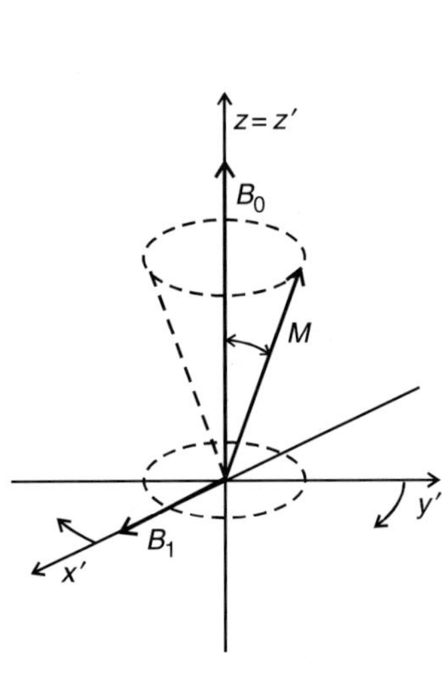

Fig. 8.15 Rotating frame of reference.

rotation about the direction of the applied $\boldsymbol{B}_1$ field, because the torque $\boldsymbol{N} = \boldsymbol{M} \times \boldsymbol{B}_1$ appears. The rotation angle θ about the x'-axis is given by

$$\theta = \omega_1 t = \gamma B_1 t, \tag{8.19}$$

where t is the duration of the magnetic pulse $\boldsymbol{B}_1$. When the pulse duration is such that it causes clockwise rotation by 90 degrees, it is called the $\pi/2$ *pulse*. The pulse causing the rotation by $\theta = \pi$ is called the π *pulse*.

Figure 8.16 illustrates the time evolution of the magnetization vector $\boldsymbol{M} = \sum_i \boldsymbol{\mu}_i$ of the sample after the application of the $\pi/2$ pulse followed by the π pulse.

Figure 8.16a shows the situation at equilibrium (before the application of the $\pi/2$ pulse) in the frame of reference rotating at $\boldsymbol{\omega}_0 = \gamma \boldsymbol{B}_0$. The magnetization vector $\boldsymbol{M}$ has the following components

$$M^0_{z'} = N_0 \mu_z = (n_I - n_{I-1}) \mu_z, \tag{8.20}$$

$$M^0_{x'} = 0, \tag{8.21}$$

$$M^0_{y'} = 0, \tag{8.22}$$

where μ_z is a z-component of the magnetic moment of a single spin.

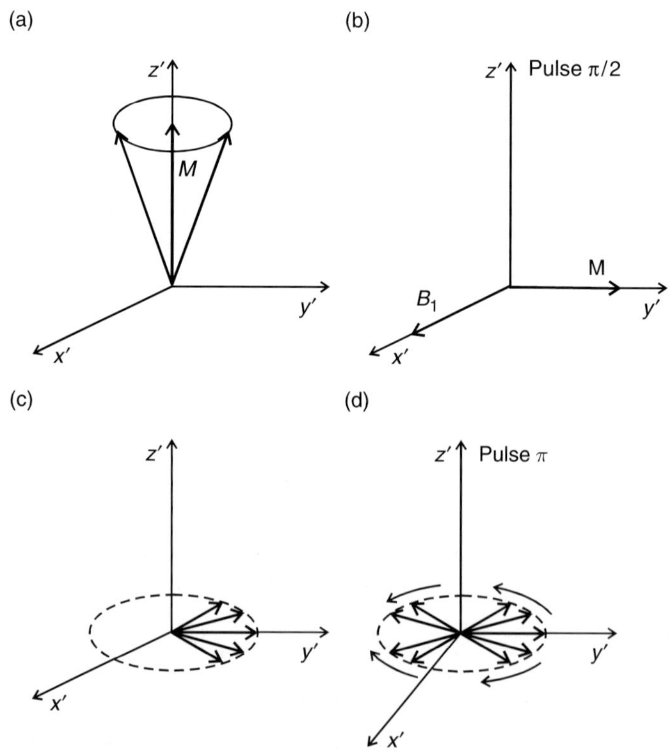

Fig. 8.16 Time evolution of the magnetization vector in the spin echo experiment.

Fig. 8.16b presents a situation at $t=0$ (just after the action of the $\pi/2$ pulse). As a result of absorption of an RF photon provided by the $\boldsymbol{B}_1$ magnetic field, a change of population number takes place as well as the rotation of the magnetization vector $\boldsymbol{M}$ by the angle $\pi/2$ about the x'-axis. The vector of magnetization $\boldsymbol{M}$ has the following components

$$M_{z'} = 0, \tag{8.23}$$

$$M_{y'} = M_z^0, \tag{8.24}$$

$$M_{x'} = 0. \tag{8.25}$$

Figure 8.16c illustrates the situation at $t>0$. As a consequence of the fact that the magnetic field $\boldsymbol{B}_0$ is not perfectly homogeneous and due to molecular interactions, the frequencies of the magnetic moments $\boldsymbol{\mu}$ of different spins differ a little from the resonance frequency ω_0. This indicates that in the reference system rotating at the angular velocity ω_0 the magnetic moments of individual molecules are not directed only along the y-axis. Some moments rotate more quickly, the others more slowly than ω_0. The "fan" of the magnetic moments draws aside. The longer the time t, the larger are the differences. Dephasing processes tend to vanish the transverse magnetization $M_{x'} \to 0$ and $M_{y'} \to 0$ for long times. Simultaneously, as a result of relaxation processes T_1 (spin-lattice) the magnitude of the component M_z begins to recover. The time constant T_1 describes how M_z returns to its equilibrium value. Thus, the time evolution of the magnetization vector $M(t)=(M_x, M_y, M_z)$ after the use of the $\pi/2$ pulse can be described by the equation

$$\frac{d\boldsymbol{M}}{dt} = \boldsymbol{M} \times \boldsymbol{\omega}_0 - \left\{ \frac{M_x}{T_2}; \frac{M_y}{T_2}; \frac{M_z}{T_1} \right\}. \tag{8.26}$$

The vector equation (8.26) can be written in the form of the equations for the x, y, z components:

$$\frac{dM_x}{dt} = \omega_0 M_y - \frac{M_x}{T_2} \tag{8.27}$$

$$\frac{dM_y}{dt} = -\omega_0 M_x - \frac{M_y}{T_2}, \tag{8.28}$$

$$\frac{dM_z}{dt} = -\frac{M_z - M_z^0}{T_1}. \tag{8.29}$$

The eq. (8.26) results from eq. (8.17) after multiplying both sides of the equation by the factor γ. The decay of the transverse components M_x and M_y is called FID (*Free Induction Decay*).

Figure 8.16d presents the following sequence of pulses:

1. at $t=0$ the pulse $\pi/2$ is applied (Fig. 8.16b),
2. after some time t the pulse π is applied which causes the rotation of the transverse components of the magnetic moments by the angle π. The pulse π reverses the temporary evolution of dephasing (Fig. 8.16c). The magnetic

moments begin to coincide again. After time $2t$ the "fan" of the magnetic moments disappears and the transverse magnetization lies along the y'-axis again, but in the opposite direction. The magnitude of the signal corresponding to the sample magnetization after the sequence: pulse $\pi/2$–t–pulse $\pi - t$ is presented in Fig. 8.17. The signal observed after the pulse sequence $\pi/2 - t - \pi - t$ is called the *spin echo* or *NMR echo*.

8.4.2. Optical Resonance

Echo in NMR can be quite easily understood using classical analogies. For photon echo it is difficult to find a classical analogy. The polarization vector $\boldsymbol{P}$, which is the sum of molecular dipole moments, corresponds to the magnetization vector $\boldsymbol{M}$ for the molecular magnetic moments. The dipole moment, however, cannot be expressed in terms of angular momentum, as can the magnetic moment and it is impossible to derive Bloch equations for the polarization vector $\boldsymbol{P}$ from the classical equations of motion. Although there is no classical description for the optical echo, the classical Bloch formula for the magnetization vector $\boldsymbol{M}$ has its analog for the polarization vector $\boldsymbol{P}$ [22]

$$\frac{d\boldsymbol{P}}{dt} = \boldsymbol{P} \times \boldsymbol{\omega}_0 - \left\{ \frac{P_x}{T_2}; \frac{P_y}{T_2}; \frac{P_z}{T_1} \right\}. \tag{8.30}$$

The photon echo experiment can be described qualitatively (Fig. 8.18) applying the Bloch equation for the polarization vector $\boldsymbol{P}$.

At equilibrium, before the excitation of the sample with a laser pulse, the polarization vector $\boldsymbol{P}$ is directed along the z'-axis (Fig. 8.18a). At $t = 0$ the sample is excited with a strong, coherent, and short $\pi/2$ pulse of a laser light E_0, which "rotates" the polarization vector $\boldsymbol{P}$ by 90 degrees. In the quantum description this means that some molecules of the sample are coherently excited from the lower state E_1 to the

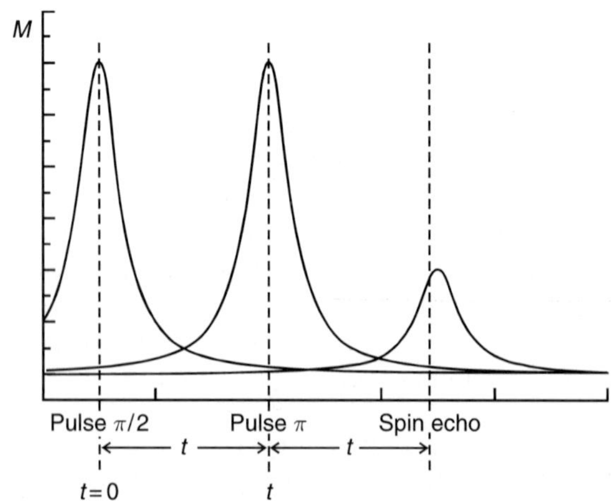

Fig. 8.17 The sequence of pulses in the NMR spin echo method.

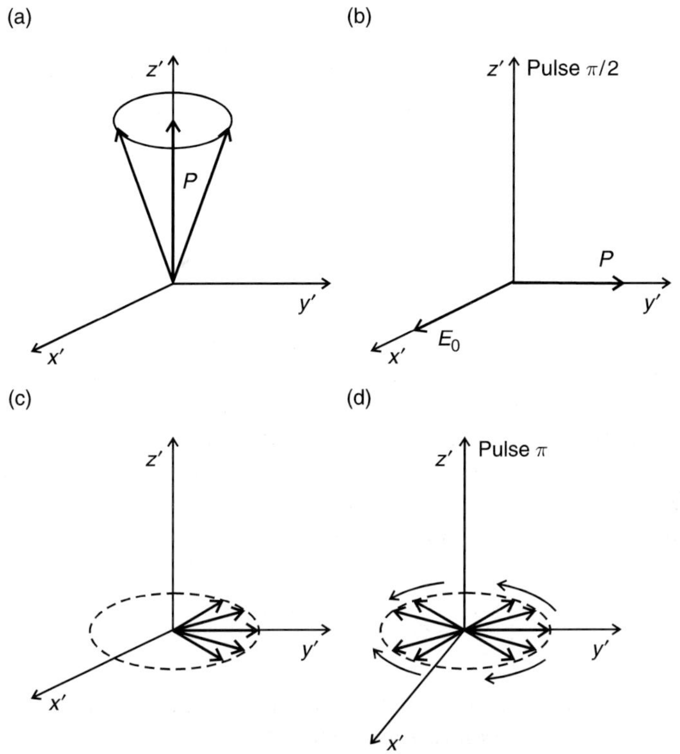

Fig. 8.18 Illustration of the photon echo.

upper state E_2. The pulse duration has to be chosen in such way in order to change the probability of the population of the states E_1 and E_2 from $|a_1|^2 = 1$; $|a_2|^2 = 0$ to $|a_1|^2 = |a_2|^2 = 0.5$. At $t = 0$, just after the pulse $\pi/2$, all the induced dipole moments are in the same phase and the polarization vector $\boldsymbol{P}$ lies along the y'-axis (Fig. 8.18b). At $t = t'$ ($t' > 0$), and $t' > T_2$ after switching off the $\pi/2$ pulse, the phases begin to distribute randomly (Fig. 8.18c). When the sample is excited with the second laser pulse (pulse π) at $t = t''$ ($t'' < T_1$), the time evolution of the phases of the molecular dipole moments will be reversed (Fig. 8.18d). This means that after time $t = 2t''$ all the excited molecules will be again in the same phase with the transverse polarization component along the y'-axis, but in the opposite direction. Then the sample sends a signal called the *photon echo* (which is N times stronger than the intensity of incoherent fluorescence) (Fig. 8.19).

The $\pi/2$ pulse duration, $t_{\pi/2}$ for the spin echo can be determined from the condition

$$\theta = \frac{\pi}{2} = \gamma B_1 t_{\pi/2}. \tag{8.31}$$

Knowing γ ($\gamma = g_N \mu_N$; $\mu_N = \frac{e\hbar}{2mc}$) and B_1 one can determine $t_{\pi/2}$ from the eq. (8.31). The NMR pulses durations are on the order of 1–100 microseconds, which can easily be generated in the radiofrequency range.

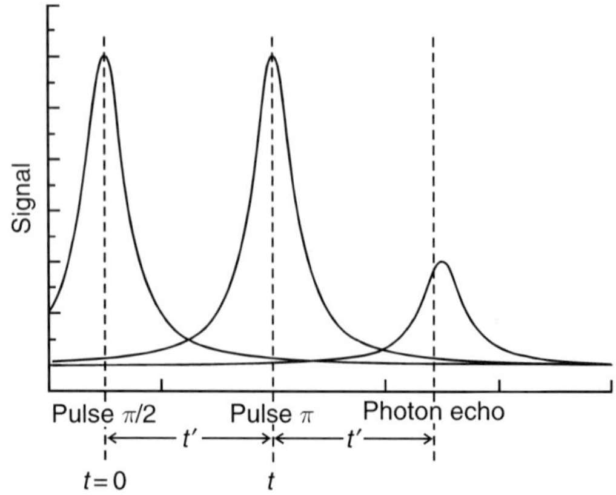

Fig. 8.19 The sequence of pulses in the photon echo method.

For the photon echo the condition of the optical resonance requires

$$\frac{1}{t_{\pi/2}} = eR_{12}E_0 > \pi\hbar\left(\frac{1}{2T_1} + \frac{1}{T_2^*}\right). \tag{8.32}$$

In eq. (8.32) E_0 is the electric field intensity of the laser beam, R_{12} is the dipole moment of the quantum transition $1 \to 2$. In next section we will derive the condition (8.32). The condition (8.32) allows us to estimate the laser peak power, which should be used in the photon echo experiments. Because the times T_1 and T_2^* change in the range 10^{-6}–10^{-12} s depending on spectral range, the condition (8.32) requires laser powers in the range from kilowatts to megawatts. With the advances in laser technology it is now easy to reach such powers in the Q-switching or the modelocking lasers. The condition (8.32) permits us to understand why the photon echo experiments, although theoretically predicted many years ago, have been performed much later when a proper level of laser technology was available. For these reasons the spin echo was a mature method thirty years ago, while the IR and Raman photon echo began to develop at the beginning of the 1990s.

The photon echo permits to eliminate inhomogeneous processes of phase relaxation T_2^{inh} delivering information about homogeneous processes T_2^h. The inhomogeneous processes of phase relaxation T_2^{inh} occurring in the period of time between $t=0$ and $t=2t''$ are usually caused by the Doppler effect in gases or inhomogeneous distribution in condensed matter. The inhomogeneous phase relaxation is a slow process and it has no influence on the photon echo signal as long as the condition $2t'' < T_2^{inh}$ is fulfilled. The pulse π reverses the temporal phase evolution of the excited molecules between $t=0$ and $t=t''$ and the signal after time $t=2t''$ should be equal to the signal at $t=0$, when there are no other factors leading to the phase decoherence. This can be intuitively explained by the analogy of runners who start with different velocities (phase inhomogeneous distribution T_2^{inh}). After some time the distances between the runners increase. However, if after time $t=t''$ the signal (the pulse π) makes them come back, all runners will be in the same place again–at the start line (photon echo) after time $t = 2t''$. However, if the velocity of the runners undergoes rapid fluctuations during running (homogeneous dephasing T_2^h), they do not meet at the start line simultaneously after time $t = 2t''$. Similarly, the photon echo signal after time $t = 2t''$ decays due to rapid homogeneous dephasing T_2^h. This denotes that only the homogeneous processes of phase relaxation have an influence on the magnitude of the photon echo signal. The change of the photon echo signal occurring after the time $2t''$ caused by the homogeneous dephasing T_2^h can be described by

$$N_2(2t'') = N_2(0)e^{-2t''/T_2^h}. \tag{8.33}$$

8.4.3. Quantum-Classical Description of the Photon Echo

So far we have used rather intuitive analogies to the spin echo in NMR to understand the photon echo. Now we will introduce the quantum-classical description of this phenomenon. The description of a quantum system interacting with the electromagnetic field can be found in every textbook on quantum mechanics [25–26]. The description of the photon echo presented here is based on ref. [1].

Let us consider a system of N dipole moments, $\boldsymbol{\mu} = e\boldsymbol{r}$, interacting with the electromagnetic field, $\boldsymbol{E} = \boldsymbol{A}_0 \cos(\omega t)$. The time-dependent Schrödinger equation describing the quantum system is given by

$$\hat{H}\Psi = i\hbar\frac{\partial \Psi}{\partial t}. \tag{8.34}$$

The energy operator of the system, the Hamiltonian $\hat{H}$ can be written as

$$\hat{H} = \hat{H}_0 + \hat{V}, \tag{8.35}$$

where $\hat{H}_0$ describes the energy of the system unperturbed by the electromagnetic field, $\hat{V}$ is the semiclassical Hamiltonian of the radiation-matter interaction describing the interaction between the dipole moment $\boldsymbol{\mu}$ of a molecule with the field $\boldsymbol{E}$

$$\hat{V} = \boldsymbol{\mu} \cdot \boldsymbol{E} = \boldsymbol{\mu}\boldsymbol{A}_0 \cos(\omega t). \tag{8.36}$$

The solution of the Schrödinger equation (8.34) can be written as a linear combination of the unperturbed solutions

$$\Psi(\boldsymbol{r}, t) = \sum_{n=1}^{\infty} C_n(t) u_n(\boldsymbol{r}) \exp(-iE_n t/\hbar), \tag{8.37}$$

where the spatial part of the wavefunction $u_n(\boldsymbol{r})$ obeys the stationary Schrödinger equation

$$H_0 u_n(\boldsymbol{r}) = E_n u_n(\boldsymbol{r}). \tag{8.38}$$

For a two-level system with the quantum states $|a\rangle$ and $|b\rangle$ and energies E_a and E_b, eq. (8.37) can be rewritten as a sum of two terms:

$$\Psi(\mathbf{r}, t) = a(t)u_a e^{-iE_a t/\hbar} + b(t)u_b e^{-iE_b t/\hbar}. \tag{8.39}$$

The coefficients $a(t)$ and $b(t)$ are the time-dependent amplitudes determining the probability of finding the system in the state a and b, respectively. They satisfy the relation ($|a(t)|^2 + |b(t)|^2 = 1$). Substituting (8.39) into (8.34), one gets

$$i\hbar \frac{da(t)}{dt} u_a e^{-iE_a t/\hbar} + i\hbar \frac{db(t)}{dt} u_b e^{-iE_b t/\hbar} = a(t)\hat{V} u_a e^{-iE_a t/\hbar} + b(t)\hat{V} u_b e^{-iE_b t/\hbar}, \tag{8.40}$$

when the relation $\hat{H}_0 u_n = E_n u_n$ has been applied. Multiplying both sides of equation (8.40) by the conjugated function $u_n^*(n = a, b)$ and integrating over the space coordinates one obtains

$$\frac{da(t)}{dt} = -(i/\hbar)\left[a(t)V_{aa} + b(t)V_{ab} e^{i(E_a - E_b)t/\hbar}\right], \tag{8.41}$$

$$\frac{db(t)}{dt} = -(i/\hbar)\left[b(t)V_{bb} + a(t)V_{ba} e^{i(E_a - E_b)t/\hbar}\right], \tag{8.42}$$

where

$$V_{ik} = \int u_i^* \hat{V} u_k d\tau = -eE \int u_i^* r u_k d\tau = -eER_{ik}; \tag{8.43}$$

eR_{ik} is the matrix element of the dipole moment of a molecule. The V_{aa} and V_{bb} terms disappear, because $\boldsymbol{r}$ is an odd function and finally one obtains

$$\frac{da(t)}{dt} = i/2 \frac{eR_{ab}A_0}{\hbar}\left[e^{i(\omega_{ab} - \omega)t} + e^{i(\omega_{ab} + \omega)t}\right] b(t), \tag{8.44}$$

$$\frac{db(t)}{dt} = i/2 \frac{eR_{ab}A_0}{\hbar}[e^{i(\omega_{ab} - \omega)t} + e^{i(\omega_{ab} + \omega)t}] a(t) \tag{8.45}$$

where

$$\omega_{ba} = (E_b - E_a)/\hbar = -\omega_{ab}. \tag{8.46}$$

Weak Field Approximation

Let's assume that the amplitude A_0 is weak and perturbations in the system due to the interaction with the electromagnetic field are small. In this approximation, the equations (8.44) and (8.45) can be solved with an iterative method, assuming that at $t = 0$, $a(0) = 1$ and $b(0) = 0$. This means that at $t = 0$ all molecules are in the lower energy state E_a. In the weak field approximation the first iteration result is sufficiently accurate

$$a(t) \cong a(0) \cong 1, \tag{8.47}$$

$$b(t) = \frac{eR_{ab}A_0}{2\hbar}\left[\frac{e^{i(\omega_{ba}-\omega)t}-1}{\omega_{ba}-\omega}+\frac{e^{i(\omega_{ba}+\omega)t}-1}{\omega_{ba}+\omega}\right], \tag{8.48}$$

with the b coefficient obeying the condition $b(t) \ll 1$.

Rotating-Wave Approximation

The second term of the sum in eq. (8.48) can be omitted as it is much smaller than the first term in the resonance when $\omega \to \omega_{ba}$. The neglection of the second term is called the *rotating-wave approximation*, because we keep only the term for which the phase factors for the wave function and the wave of the field, $\exp(-i\omega_{ab}t)$ and $\exp(-i\omega t)$, have the same sign. In eq. (8.48) we used the relation (8.46). Thus, in the rotating-wave approximation the coefficient $b(t)$ can be written in the form

$$b(t) = \frac{eR_{ab}A_0}{2\hbar}\frac{e^{i(\omega_{ba}-\omega)t}-1}{\omega_{ba}-\omega}. \tag{8.49}$$

The probability of finding a molecule in the state E_b after time t of interaction with the radiation is

$$|b(t)|^2 = \left(\frac{eR_{ab}A_0}{2\hbar}\right)^2\left(\frac{\sin(\omega_{ba}-\omega)t/2}{(\omega_{ba}-\omega)/2}\right)^2. \tag{8.50}$$

When $\omega \to \omega_{ba}$, $\frac{\sin^2 xt}{x^2} \to t^2$, because

$$x = \frac{\omega_{ba}-\omega}{2} \to 0. \tag{8.51}$$

Therefore, under the resonance condition the expression for $|b(t)|^2$ is

$$|b(t)|^2 = \left(\frac{eR_{ab}A_0}{2\hbar}\right)^2 t^2. \tag{8.52}$$

In the weak field approximation

$$|b(t)|^2 = \left(\frac{eR_{ab}A_0}{2\hbar}\right)^2 t^2 \ll 1, \tag{8.53}$$

so the period of interaction must be much shorter than

$$t \ll T = \frac{2\hbar}{eR_{ab}A_0}. \tag{8.54}$$

This condition is not obviously fulfilled for the monochromatic radiation (in the Fourier limit $\Delta\omega$ is inversely proportional to $\frac{1}{T}$, so for the monochromatic wave $\Delta\omega \to 0$ and $T \to \infty$). The weak field approximation is valid only for short laser pulses corresponding to a broad frequency distribution $\Delta\omega$. To take into account the distribution of the incident radiation frequency, one should replace the amplitude A_0 with the spectral density of the energy $\rho(\omega)$. The intensity of a plane wave

$E = E_0 \cos(\omega t - kx)$ traveling in vacuum in the x direction should be replaced by the intensity of the radiation I [W/m^2] of the broadband excitation

$$I = c\int \rho(\omega)d\omega = c\varepsilon_0 E^2 = c\varepsilon_0 E_0^2 \cos^2(\omega t - kx), \tag{8.55}$$

where

$$\begin{aligned} E &= A_0 e^{i(\omega t - kx)} + A_0^* e^{-i(\omega t - kx)} \\ &= 2A_0 \cos(\omega t - kx) = E_0 \cos(\omega t - kx). \end{aligned} \tag{8.56}$$

The intensity must be averaged over time because detectors have response times that are much longer than the fast wave oscillations, which occur on the order of 10^{13}–$10^{15}\,\mathrm{s}^{-1}$ in the visible and the near infrared range

$$\bar{I} = \overline{c\varepsilon_0 E_0^2 \cos^2(\omega t - kx)} = \frac{1}{2} c\varepsilon_0 E_0^2 = 2c\varepsilon_0 A_0^2. \tag{8.57}$$

From the comparison between eqs. (8.55) and (8.57) we get

$$A_0^2 = \frac{\int \rho(\omega)d\omega}{2\varepsilon_0}. \tag{8.58}$$

The equation describing the probability of finding the system in the state E_b takes the form

$$P(t) = \int |b(t)|^2 d\omega = \frac{(eR_{ab})^2}{4\hbar^2} \int \frac{\rho(\omega)}{2\varepsilon_0} \left(\frac{\sin(\omega_{ba} - \omega)t/2}{(\omega_{ba} - \omega)/2}\right)^2 d\omega. \tag{8.59}$$

Because $\rho(\omega)$ changes very slowly in comparison with $\frac{\sin^2 xt}{x^2}$ and is practically constant in the area in which $\frac{\sin^2 xt}{x^2}$ is large (when $\omega = \omega_{ba}$) one can write

$$\begin{aligned} P(t) &= \int |b(t)|^2 d\omega = \frac{(eR_{ab})^2}{4\hbar^2 2\varepsilon_0} \rho(\omega_{ba}) \int_0^{2\pi} \left(\frac{\sin(\omega_{ba} - \omega)t/2}{(\omega_{ba} - \omega)/2}\right)^2 d\omega \\ &= \frac{(eR_{ab})^2}{4\hbar^2 2\varepsilon_0} \rho(\omega_{ba}) 2\pi 2^2 t^2 = \frac{\pi e^2}{\varepsilon_0 \hbar^2} R_{ab}^2 \rho(\omega_{ab}) t^2. \end{aligned} \tag{8.60}$$

Strong Field Approximation. Rabbi Frequency

In the *weak field approximation*, when $|b(t)| \ll 1$, the population of the initial state stays nearly constant and it does not undergo significant change as a result of the interaction with the electromagnetic field. Now we will consider the case in which the radiation field is strong enough to make the assumption invalid. The *theory of the strong field*, which has been developed by *Rabbi*, describes the probability of finding a molecule in the upper state E_b or the lower state E_a as a function of time.

Let us apply the equations (8.44) and (8.45) for the amplitudes of probability in the rotating-wave approximation for a monochromatic field at frequency ω

$$\frac{da(t)}{dt} = \frac{1}{2} i \left(\frac{eR_{ab}A_0}{\hbar} \right) e^{-i(\omega_{ba}-\omega)t} b(t), \tag{8.61}$$

$$\frac{db(t)}{dt} = \frac{1}{2} i \left(\frac{eR_{ab}A_0}{\hbar} \right) e^{-(\omega_{ba}-\omega)t} a(t) \tag{8.62}$$

Putting a trial solution $a(t) = \exp(i\mu t)$ into (8.61) one obtains

$$b(t) = \frac{2\mu\hbar}{eR_{ab}A_0} e^{i(\omega_{ba}-\omega+\mu)t}. \tag{8.63}$$

Inserting (8.63) into (8.62), one gets

$$2\mu(\omega_{ba} - \omega + \mu) = \frac{1}{2} \left(\frac{eR_{ab}A_0}{\hbar} \right)^2, \tag{8.64}$$

which is a quadratic equation with respect to the unknown μ parameter. The solutions of the equation (8.64) are

$$\mu_{\pm} = -\frac{1}{2}(\omega_{ba} - \omega) \pm \frac{1}{2} \left[(\omega_{ba} - \omega)^2 + \left(\frac{eR_{ab}A_0}{\hbar} \right)^2 \right]^{1/2}. \tag{8.65}$$

Thus, the trial solutions for $a(t)$ and $b(t)$ get the form

$$a(t) = C_1 e^{i\mu_+ t} + C_2 e^{i\mu_- t}, \tag{8.66}$$

$$b(t) = \hbar / eR_{ab}A_0 \cdot e^{i(\omega_{ba}-\omega)t} \left(C_1 \mu_+ e^{i\mu_+ t} + C_2 \mu_- e^{i\mu_- t} \right). \tag{8.67}$$

The coefficients C_1 and C_2 can be determined from the initial conditions $a(0) = 1$; $b(0) = 0$ from which we obtain the following relations $C_1 + C_2 = 1$, $C_1\mu_+ = -C_2\mu_-$ or

$$C_1 = -\mu_- / (\mu_+ - \mu_-), \tag{8.68}$$

$$C_2 = +\mu_+ / (\mu_+ - \mu_-). \tag{8.69}$$

Introducing $\mu = \mu_+ - \mu_-$, we obtain the following expression for the probability amplitude $b(t)$

$$b(t) = i \left(\frac{eR_{ab}A_0}{\hbar\mu} \right) e^{i(\omega_{ba}-\omega)t/2} \sin \mu t/2. \tag{8.70}$$

The probability of finding the system in the state E_b, $|b(t)|^2 = b^*(t)b(t)$ is given by

$$|b(t)|^2 = \left(\frac{eR_{ab}A_0}{\hbar\mu} \right)^2 \frac{\sin^2(\mu t/2)}{\mu^2}, \tag{8.71}$$

where the frequency μ is expressed as

$$\mu = \sqrt{(\omega_{ba} - \omega)^2 + (eR_{ab}A_0/\hbar)^2}. \tag{8.72}$$

The frequency is known as the *Rabbi frequency*.
The probability of finding the system in the state E_a is given by

$$|a(t)|^2 = 1 - |b(t)|^2 = 1 - \left(\frac{eR_{ab}A_0}{\hbar}\right)^2 \frac{\sin^2(\mu t/2)}{\mu^2}. \quad (8.73)$$

Thus, the system illuminated with strong monochromatic light oscillates between the levels E_a and E_b at the Rabbi frequency μ. The Rabbi frequency depends on the difference between the incident radiation frequency and the resonance frequency, $\omega_{ba} - \omega$, the field amplitude A_0 and the matrix element of the dipole moment eR_{ab}. Under the resonance conditions, $\omega_{ba} = \omega$, the Rabbi frequency is given by the formula

$$\mu = \frac{eR_{ab}A_0}{\hbar}, \quad (8.74)$$

and the probabilities $|b(t)|^2$ and $|a(t)|^2$ are given by

$$|b(t)|^2 = \sin^2[eR_{ab}A_0 t/2\hbar], \quad (8.75)$$

$$|a(t)|^2 = \cos^2[eR_{ab}A_0 t/2\hbar]. \quad (8.76)$$

When the electromagnetic field pulse interacts with the sample for the time duration of $t = T = \pi\hbar/eR_{ab}A_0$, one can see from eqs. (8.75) and (8.76) that the probabilities become $|b(T)|^2 = 1$ and $|a(T)|^2 = 0$. It indicates that the initial probabilities $|b(0)|^2 = 0$, $|a(0)|^2 = 1$ have been reversed. The pulse $T = \frac{\pi\hbar}{eR_{ab}A_0}$ is called the π *pulse* because it changes the phases of the wave functions by π. The eqs. (8.75) and (8.76) permit us to understand the idea of the π pulse in the optical resonance.

The relations derived in eqs. (8.75) and (8.76) are illustrated in Fig. 8.20 for the cases without damping and with damping. The damping is caused by relaxation processes to energy levels different from the levels a and b.

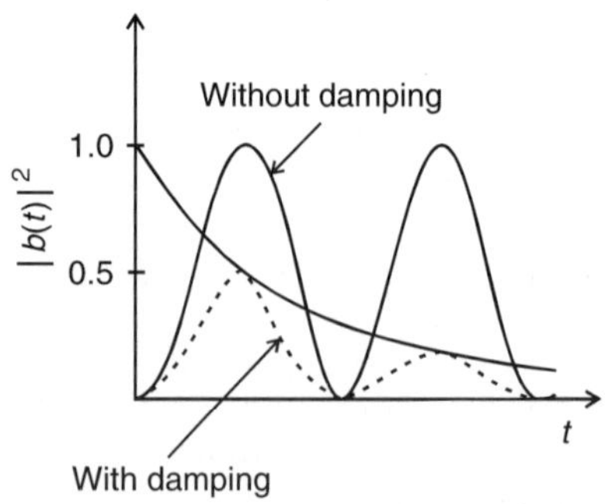

Fig. 8.20 Oscillation of the population of the levels E_a and E_b at the Rabbi frequency due to the interaction with a strong field. The damping is caused by relaxation processes to energy levels different from the levels a and b.

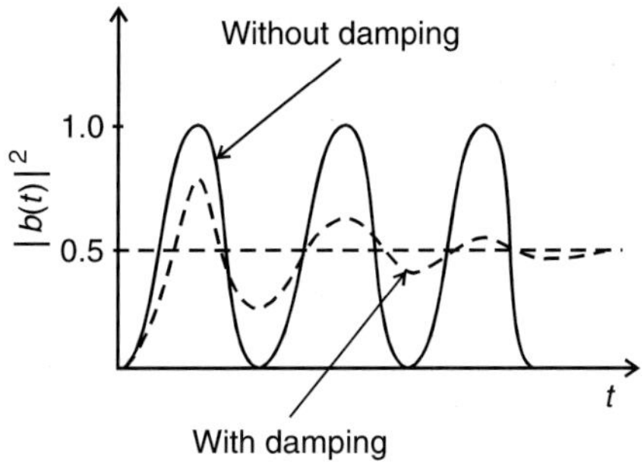

Fig. 8.21 Oscillation of the population of the levels E_a and E_b at the Rabbi frequency due to the interaction with a strong field. The damping occurs via relaxation processes between the levels E_a and E_b.

Figure 8.21 illustrates the situation when the damping occurs via relaxation processes between the levels E_a and E_b.

In the previous section we introduced the photon echo in a qualitative way using the analogy with the spin echo in NMR. Quantum formalism allows to understand the photon echo in a quantitative way.

The wave function of a two-level system E_1 and E_2 is expressed by

$$\Psi(t) = \sum_{n=1}^{2} a_n(t) u_n e^{-iE_n t/\hbar}. \tag{8.77}$$

Before using the first pulse ($\pi/2$) all molecules are in the state E_1, which indicates that $|a_1| = 1$ and $|a_2| = 0$. The system perturbed by the pulse of duration t can be described by a new wave function (see the equations (8.37), (8.75), (8.76))

$$\Psi(t) = \cos\left(\frac{eR_{ab}A_0 t}{2\hbar}\right) u_1 e^{-iE_1 t/\hbar} + \sin\left(\frac{eR_{ab}A_0 t}{2\hbar}\right) u_2 e^{-iE_2 t/\hbar}. \tag{8.78}$$

If the pulse duration t is equal to T as determined from

$$\frac{eR_{ab}A_0 T}{\hbar} = \frac{\pi}{2}, \tag{8.79}$$

the pulse is called the $\pi/2$ *pulse*, because it changes the phase of the amplitudes by $\pi/2$. The wave function from eq. (8.78) take the form

$$\Psi(t) = \frac{1}{\sqrt{2}}\left(u_1 e^{-iE_1 t/\hbar} + u_2 e^{-iE_2 t/\hbar}\right). \tag{8.80}$$

After time $t = t'$ following the excitation with the $\pi/2$ pulse, the phases are equal to $E_1 t'/\hbar$ and $E_2 t'/\hbar$. If the π pulse is applied at time $t = t'$, which indicates that the laser pulse duration T obeys the condition

$$\frac{eR_{ab}A_0 T}{\hbar} = \pi, \tag{8.81}$$

the population is reversed from ($|a(0)| = 1$) and ($|b(0)| = 0$) to ($|a(0)| = 0$) and ($|b(t')| = 1$). This means that the wave functions of the upper and the lower states simply change with each other in such a manner that we have at $t \geq t'$

$$e^{-iE_1t'/\hbar}u_1 \rightarrow e^{-iE_1t'/\hbar}u_2e^{-iE_2(t-t')/\hbar}, \tag{8.82}$$

$$e^{-iE_2t'/\hbar}u_2 \rightarrow e^{-iE_2t'/\hbar}u_1e^{-iE_1(t-t')/\hbar}. \tag{8.83}$$

The total wave function of a molecule at $t \geq t'$ following the use of the π pulse is given by

$$\Psi(t) = \frac{1}{\sqrt{2}}[u_2e^{-i\omega_k(t-2t')/2} - u_1e^{i\omega_k(t-2t')/2}], \tag{8.84}$$

where $\omega_k = (E_2 - E_1)/\hbar$. The transition dipole moment is given by

$$\mu_{21} = \langle -\Psi^*|er|\Psi\rangle = -\langle u_2^*|er|u_1\rangle e^{-i\omega_k(t-2t')}. \tag{8.85}$$

The intensity of the macroscopic fluorescence I is directly proportional to the square of the sum of microscopic dipole moments μ_{21}

$$I \propto \left|\sum_N \mu_{21}\right|^2 \propto N\mu_{21}^2. \tag{8.86}$$

Because the individual molecules have somewhat different frequencies of absorption ω_k due to a different environment, the phase factors of the individual molecules differ from one another. The intensity of the macroscopic fluorescence (8.86) is an incoherent superposition of contributions from all molecules. However, at $t = 2t'$ the phase factor for all molecules equals zero, which indicates that all the dipole moments μ_{21} have the same phase, leading to the enhanced signal of the coherent fluorescence, which is called the photon echo

$$I \propto \left|\sum_N \mu_{21}\right|^2 = |N\mu_{21}|^2 = N^2\mu_{21}^2. \tag{8.87}$$

Comparing the signal of the incoherent (8.86) and the coherent superposition (8.87), one can see that the coherent fluorescence is N times larger.

The photon echo can be applied in all optical spectral ranges. This method was applied for the first time in the visible range in 1964 (14 years after the discovery of the spin echo in NMR). The photon echo in the infrared range was applied only in the 1990s. This was related to difficulties in obtaining tunable short-pulse laser sources in this spectral range. The development of parametric amplifiers removed obstacles and allowed the extension of the photon echo method into the infrared range [27]. Raman photon echo was also observed for the first time in the 1990s [28], although the theoretical background had been given much earlier.

8.4.4. Practical Advantages of Photon Echo Applications

What practical advantages result from the application of the photon echo method? What information about the system does the photon echo deliver, which cannot be obtained by the other, less sophisticated spectroscopic methods? Let's concentrate on the IR and Raman echoes.

Stationary vibrational spectroscopy allows us to record the spectrum (Fig. 8.22). Every line corresponds to a certain quantum transition. Every line has a definite position, intensity, width and shape. The width and the shape of the line depend on the vibrational dynamics characterized by the processes of energy relaxation, T_1, and phase relaxation, T_2.

In linear optics the band shape $I(\omega)$ and the system dynamics are related through the Fourier transform of the dipole correlation function $\langle\mu(0)\mu(t)\rangle$

$$I(\omega) \propto \int \langle\mu(0)\mu(t)\rangle e^{i\omega t} dt. \tag{8.88}$$

If we record the shape of the band $I(\omega)$ in the stationary experiment, it is possible to calculate the inverse Fourier transform to determine vibrational dynamics that are contained in the dipole correlation function $\langle\mu(0)\mu(t)\rangle$

$$\langle\mu(0)\mu(t)\rangle \propto \int I(\omega)^{-i\omega t} d\omega. \tag{8.89}$$

Thus, in linear optics the relationship between the frequency domain and the time domain is very simple and permits us to obtain information about T_1 and T_2 processes occurring in the pico- and femtosecond time scale from the stationary measurements of the shape and the width of the spectroscopic bands (Fig. 8.23). When the band shape $I(\omega)$ is described by a Lorentzian function, the dipole correlation function decays exponentially.

Nonlinear methods of time-resolved spectroscopy can also determine the vibrational dephasing time, T_2, directly in the time domain. In the previous section we

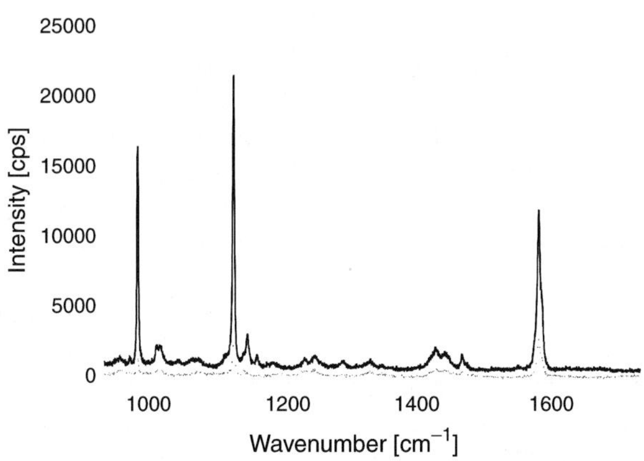

Fig. 8.22 Raman spectrum of diphenylacetylene.

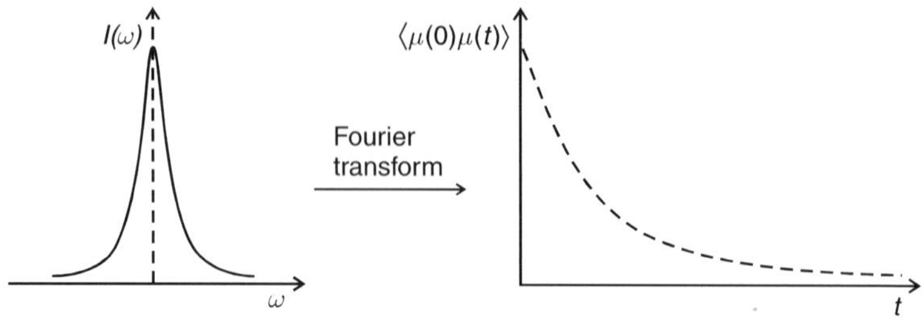

Fig. 8.23 Relationship between the frequency domain and the time domain.

described the CARS method, which allows us to determine the vibrational dephasing time, T_2. If we determine the time T_1 in an independent pump-probe experiment, one can get information about the pure dephasing characterized by T_2^* from eq. (8.3).

The question arises why the vibrational photon echo is used to determine the dephasing, when a simple conventional vibrational spectroscopy or CARS do the same. The problem is that these methods cannot separate dephasing processes occurring on a different time scale. As long as dephasing processes are very rapid, leading to homogeneous band broadening, conventional linear vibrational spectroscopy, nonlinear CARS, and nonlinear vibrational echo methods are equivalent. However, the pure dephasing time T_2^* depends on various mechanisms, which can be divided into homogeneous (fast) T_2^h and inhomogeneous T_2^{inh} (slow) processes. The inhomogeneous processes contribute to *inhomogeneous broadening*, the homogeneous processes to *homogeneous band broadening*. The stationary vibrational methods and time-resolved CARS are unable to distinguish between contributions from homogeneous and inhomogeneous broadening in contrast to the photon echo method (Fig. 8.24).

Fig. 8.24 summarizes the principle of the photon echo. Photon echoes are standard tools of nonlinear spectroscopy for determining the dephasing time, T_2. In photon echo experiments a short laser pulse ($\pi/2$ *pulse*) initially excites the population at a

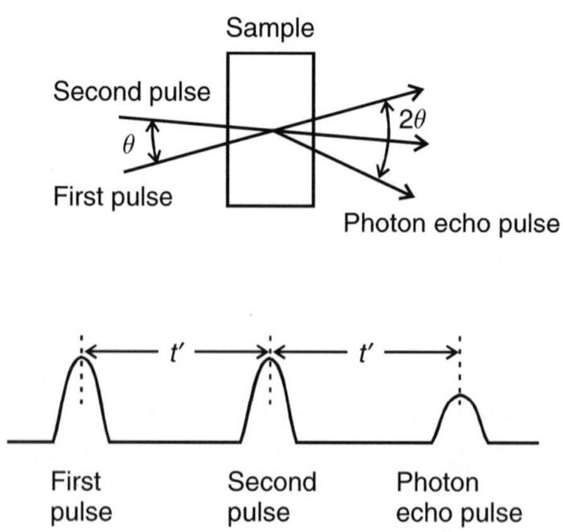

Fig. 8.24 The photon echo allows to eliminate inhomogeneous relaxation processes.

frequency corresponding to a quantum transition, followed by another pulse (π *pulse*) from the same source. If the time interval, $\Delta t = t'$, between the two pulses is less than the dephasing time, a photodetector will detect an optical signal at a time Δt after the second pulse: the *photon echo*. The photon echo rapidly diminishes as Δt becomes larger than the dephasing time T_2.

The principles of the Raman echo are generally the same as presented in Fig. 8.24. The only difference is that the first pulse ($\pi/2$ pulse) consists of two pulses of visible light of different frequency applied simultaneously, followed by the second excitation (π pulse) again consisting of two visible light pulses physically identical to the first one. The difference between the two visible frequencies matches the vibrational frequency. By means of this stimulated Raman process, a low frequency vibration (IR region) is excited with high frequency, visible light. A crucial property of the stimulated Raman process is that the vibrational excitation is coherent, i.e., all the molecules in the sample vibrate with the same phase, which is equal to the phase of the driving difference frequency of the two light fields, and therefore all the vibrational phases match one another. As for incoherent excitation, the presence of vibrationally-excited molecules can be detected by anti-Stokes Raman scattering of another pulse of light. If we detect the scattering signal without applying additional excitation via a π pulse, we measure just the FID (free induction decay) discussed in section 8.3. Recall that the coherence introduces important differences between the spontaneous anti-Stokes scattering and the coherent anti-Stokes scattering. These differences were discussed in Chapter 5. The coherent vibrations cause the anti-Stokes light field to be emitted by each molecule with a definite phase relationship. Along one particular direction, all the fields add constructively, and there is a tremendous amplification of the anti-Stokes light. The coherent anti-Stokes light is emitted as a highly directional beam in a unique direction determined by the appropriate phase-matching equation, $2\boldsymbol{k}_L - \boldsymbol{k}_S = \boldsymbol{k}_{as}$ (see 5.54). The intensity of the CARS signal (or FID signal) is directly proportional to the square of the vibrational coherence. The dephasing time, T_2, is measured from the CARS signal decay as a function of delay time between the excitation and scattering pulses. The Raman echo method differs from the CARS method in that between the excitation and scattering pulses, additional second excitation (π pulse) is applied. During the period between the two excitations, the vibrations start to dephase. The second excitation sets up the vibrations to rephrase. The rephrased signal is measured as the coherent signal in the Raman echo. This is not perfect coherence recovery, because the solvent forces and resulting frequency perturbations change between the two excitation pulses.

The quantum-classical description of the photon echo, presented in section 8.4.3, and the theoretical background on the phase matching conditions in the coherent Raman scattering provided in Chapter 5, will help to answer the questions that may arise on this issue. Why does the second excitation have a different effect from the first excitation? Why does it create a rephrasing rather than a single excitation as the first pulses did? The mechanism of rephrasing is explained by the eqs. (8.85) and (8.87). In fact, both rephrasing and a single excitation with the second pulse and their various combinations do occur in the sample. However, a proper control of the phase matching conditions described in Chapter 5, by correctly arranging the angles of the incident pulses, induces the constructive interference needed to generate coherent anti-Stokes scattering, and can limit interactions to only the combination responsible for the Raman echo.

8.5. QUANTUM BEATS

The quantum beats method is a high-resolution spectroscopic method that can be employed with the use of relatively inexpensive equipment. For example, quantum beats can be observed in the fluorescence decays on the nanosecond time scale (Fig. 8.25). The Fourier transform of the decay presented in Fig. 8.25 provides the spectrum in the frequency domain characterized by a high resolution.

Quantum beats spectroscopy allows us to obtain spectral resolution larger than the Doppler width. This resolution may be larger by a number of orders of magnitude than the width of incident radiation. Quantum beats may be related to any of the time-resolved spectroscopic methods mentioned in this chapter and are based on certain quantum effects rather than on nonlinear effects.

8.5.1. Quantum Description

Quantum beats are the quantum-mechanical interference effect, which occurs during the time evolution of a molecular system. Let us consider the molecular system with the energy levels presented in Fig. 8.26.

Let a photon from a laser pulse of energy $\hbar\omega = E_a - E_p$ and time duration ΔT excite the system presented in Fig. 8.26. The pulse duration, ΔT, is related to a certain spectral width, $\Delta\omega$. Let the difference of the energy levels $|a\rangle$ and $|b\rangle$ be equal to $\Delta E = E_a - E_b$. When $\hbar\Delta\omega > \Delta E$, the photon will excite the molecule from the initial state $|p\rangle$ to the state $|a\rangle$ or $|b\rangle$, because the *Heisenberg principle* says that the states $|a\rangle$ and $|b\rangle$ cannot be distinguished from each other. Therefore the wave

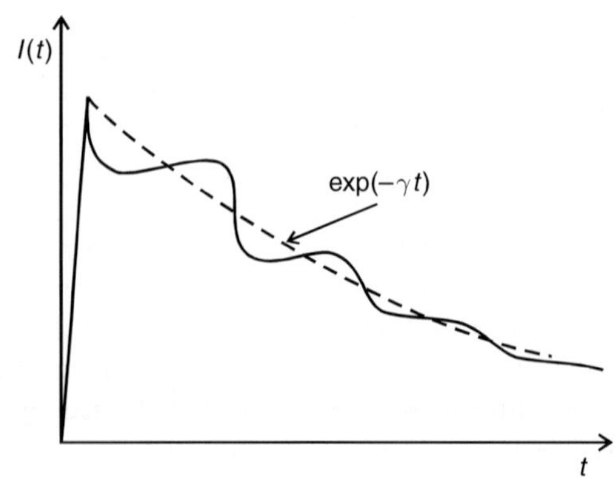

Fig. 8.25 Fluorescence decay signal modulated by quantum beats.

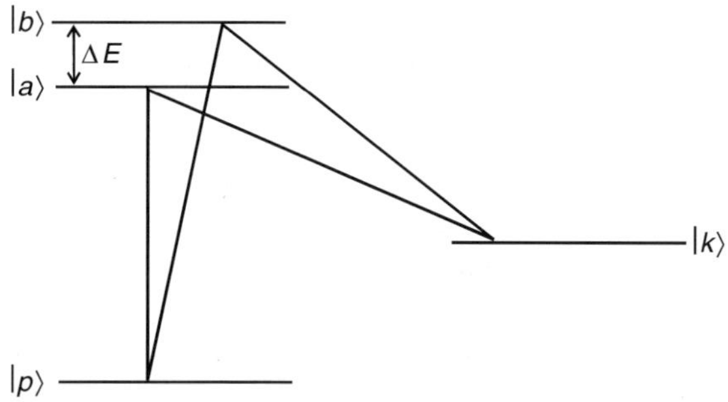

Fig. 8.26 Diagram illustrating the origin of the quantum beats.

function for the excited state immediately after the laser pulse ($t=0$) is expressed in the following way

$$\Psi(0) = C_a|a\rangle + C_b|b\rangle, \tag{8.90}$$

where

$$C_a = \langle a|\mu|p\rangle = \mu_{ap} \qquad C_b = \langle b|\mu|p\rangle = \mu_{bp}. \tag{8.91}$$

C_a^2 and C_b^2 are the probabilities of finding a molecule in the $|a\rangle$ or $|b\rangle$ states, μ_{ap} and μ_{bp} are the matrix elements of the dipole moment μ for the $p \rightarrow a$ and $p \rightarrow b$ transition, respectively. The wave function $\Psi(0)$ describes a non-stationary state, which evolves in time according to the time-dependent Schrödinger equation

$$\hat{H}\Psi(t) = i\hbar\frac{\partial\Psi(t)}{\partial t}. \tag{8.92}$$

The solution of the eq. (8.92) is given by the following wave function $\Psi(t)$

$$\Psi(t) = \mu_{ap}e^{-(iE_a/\hbar+\gamma_a/2)t}|a\rangle + \mu_{bp}e^{-(E_b/\hbar+\gamma_b/2)t}|b\rangle, \tag{8.93}$$

where E_a and E_b are the energies of the $|a\rangle$ and $|b\rangle$ states. The radiative and radiationless relaxation processes are described by the phenomenological parameters γ_a and γ_b.

Subsequently, the photon from the $|a\rangle$ or $|b\rangle$ states is emitted to the final state $|k\rangle$. Different methods are used to monitor the time evolution of the system. One of the most popular methods is the fluorescence decay, for which the signal intensity is expressed as

$$I(t) \propto (\langle k|\mu|\Psi(t)\rangle)^2. \tag{8.94}$$

Inserting (8.93) into (8.94), one obtains

$$I(t) \propto |\mu_{ap}|^2|\mu_{ka}|^2e^{-\gamma_a t} + |\mu_{bp}|^2|\mu_{kb}|^2e^{-\gamma_b t} + \\ + 2|\mu_{ap}\mu_{bp}\mu_{ka}\mu_{kb}|e^{-(\gamma_a+\gamma_b)/2}\frac{\cos(E_b - E_a)t}{\hbar}. \tag{8.95}$$

One can see from eq. (8.95) that the fluorescence signal decays exponentially as a result of relaxation processes. However, there is an additional term that modulates the signal with the frequency $\omega = \frac{E_b - E_a}{\hbar}$. As a consequence, the detector will record the modulated fluorescence signal presented in Fig. 8.25. After applying the Fourier transform to the fluorescence decay presented in Fig. 8.25 we obtain the spectrum in the frequency domain with a high resolution. For decays of the order of 10^{-6} s the resolution is $10^{-6}\,\mathrm{cm}^{-1}$. This resolution is much greater than that obtained from the methods of conventional spectroscopy. The phenomenon of quantum beats is observed only when the channels $|a\rangle$ and $|b\rangle$ are indistinguishable due to the spectral distribution of the excitation pulse. When any of the channels is blocked, the effect of the quantum beats disappears. This effect has a quantum origin, because the total probability amplitude of two indistinguishable processes is the sum of the relevant amplitudes, and the observed signal intensity is the square of this sum. The interference phenomenon of the quantum beats is analogous to the interference of light on two slits in the Young experiment.

8.5.2. Examples of Quantum Beats Applications

Quantum beats may be observed for any spectroscopic method and for any time resolution, including the pico- and femtosecond time scale. However, the shorter the pulse, the broader is the excitation, and the larger the number of indistinguishable energy levels. This results in producing many different quantum beats that may be averaged to zero. The selection of the pulse duration depends on the phenomena we want to examine. If we are concerned with high-resolution measurements in the frequency domain to determine certain structural constants, we should choose lasers emitting longer pulses on the order of nanoseconds, for which is possible to obtain a spectral resolution on the order of $10^{-6}\,\mathrm{cm}^{-1}$or more. In contrast, the application of picosecond lasers to examine quantum beats allows us to obtain much lower resolution of about $1\,\mathrm{cm}^{-1}$. Therefore, in many cases there is no reason to apply expensive, sophisticated pico- or femtosecond methods, because this resolution may easily be obtained by conventional spectroscopy methods. However, there are cases when the quantum beats with pico- and femtosecond resolution provide very important information. For example, studying quantum beats with pico- and femtosecond resolution allows us to monitor the dynamics of chemical reactions, transition states, and vibrational dynamics in real time [29]. On the other hand, the application of nanosecond lasers permits the determination of molecular structure parameters, such as the spin-orbital coupling constant, the interaction constants in the Zeeman effect, the dipole moments in excited states, the rotation constants and the anharmonic constants [30].

Let us discuss, for example, the spin-orbital coupling for the case where the excited electron singlet $|s\rangle$ and triplet $|t\rangle$ states of a molecule have similar energies (Fig. 8.27). The spin-orbital coupling is the interaction between an electron spin and its orbital moment. Let us assume that the spin-orbital interaction is strong enough to mix the singlet and the triplet states, and consequently to produce splitting of the states $|k\rangle$ and $|k'\rangle$ (Fig. 8.27)

$$|k\rangle = C_{sk}|s\rangle + C_{tk}|t\rangle, \tag{8.96}$$

$$|k'\rangle = C_{sk'}|s\rangle + C_{tk'}|t\rangle. \tag{8.97}$$

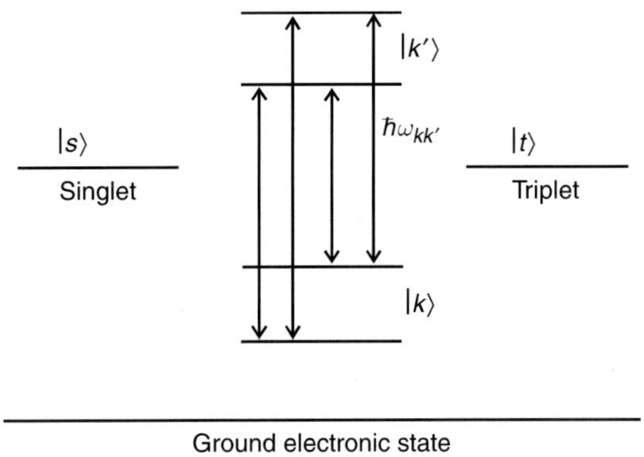

Fig. 8.27 Energy levels for the case where the spin-orbital coupling causes the mixing of the singlet $|s\rangle$ and triplet $|t\rangle$ states.

One can show [30] that in this case the energy difference between the $|k\rangle$ and $|k'\rangle$ states, $\Delta E = \hbar\omega_{kk'}$, is expressed by

$$\hbar\omega_{kk'} = \left(\Delta E_{st}^2 + 4|V_{st}|^2\right)^{1/2}, \tag{8.98}$$

where

$$\Delta E_{st} = E_s - E_t. \tag{8.99}$$

E_s and E_t are the energies of the singlet and triplet states in the absence of the spin-orbit coupling, V_{st} is the spin-orbital coupling constant. Analysis of the quantum beats from the fluorescence decay allows us to calculate the energy difference, $\hbar\omega_{kk'}$, between the states $|k\rangle$ and $|k'\rangle$. Knowing the term $\hbar\omega_{kk'}$ one can determine the spin-orbital coupling constant V_{st} from eq. (8.98).

The spectral resolution obtained by the quantum beats method is so high that we can go one step further. If the system presented in Fig. 8.27 is placed in the magnetic field, the rotational quantum states characterized by the total angular momentum $\boldsymbol{F}$ split into $2\boldsymbol{F}+1$ closely-spaced lines. First observed by Zeeman, this splitting is attributed to the interaction between the magnetic field and the magnetic dipole moment associated with the orbital angular momentum. The quantum beats make it possible to observe the subtle structure of the rotational states after the *Zeeman splitting*.

REFERENCES

1. W. Demtröder, *Laser spectroscopy*, Basic Concepts and Instrumentation, 3rd ed., Springer-Verlag, Berlin (2003)
2. *Ultrashort Light Pulses*. Picosecond Techniques and Applications, 2nd Ed., S.L. Shapiro (ed.), *Topics in Applied Physics*, vol. 18, Springer, Berlin (1977)

3. *Picosecond Phenomena III*, C.V. Shank, E.P. Ippen, S.L. Shapiro (eds.), Springer Series in Chemical Physics 4, Springer Verlag, Berlin (1978)
4. *Picosecond Phenomena II*, R.M. Hochstrasser, W. Kaiser, C.V. Shank (eds.), Springer Series in Chemical Physics 14, Springer, Berlin (1980)
5. *Ultrashort Laser Pulses and Applications*, W. Kaiser (ed.), *Topics in Applied Physics*, vol. 60, Springer-Verlag, Heidelberg (1988)
6. R.R. Alfano, S.L. Shapiro, Phys. Rev. Lett. 24 (1970) 592
7. *Ultrafast Phenomena VIII*, F. Salin, J. Watson, J.F. Cormier, P. Georges, A. Brun (eds.), Springer Series in Chemical Physics 55, Springer, Berlin Heidelberg (1992), p. 306
8. *Ultrashort Laser Pulses and Applications*, W. Kaiser (ed.), *Topics in Applied Physics*, vol. 60, Springer-Verlag, Heidelberg (1988), p. 249
9. W.G. Rothschild, *Dynamics of Molecular Liquids*, Wiley, New York (1984)
10. D.W. Oxtoby, Adv. Chem. Phys. 47 (1981) 487
11. M. Kolodziejski, G. Waliszewska, H. Abramczyk, *Vibrational dephasing in bromocyclohexane: how to separate contributions from different mechanisms*, Chem. Phys. 213 (1996) 341–356
12. H. Abramczyk, K. Paradowska-Moszkowska, *The correlation between the phase transitions and vibrational properties by Raman spectroscopy: liquid – solid β and solid β – solid α acetonitrile transitions*, Chem. Phys. 265 (2001) 177–191
13. H. Abramczyk, K. Paradowska-Moszkowska, *Crystallization and glass formation processes in methylcyclohexane. Vibrational dynamics as a possible molecular indicator of the liquid-glass transition*, J. Chem. Phys. 115 (24) (2001) 11221–11227
14. H. Abramczyk, K. Paradowska-Moszkowska, G. Wiosna, *Premelting Structure: Vibrational dynamics of liquid, undercooled liquid, glassy and crystal states in methylcyclohexane and deuterated methylcyclohexane*, J. Chem. Phys. 118 (9) (2003) 4169–4175
15. A. Tokmakoff, M.D. Fayer, *Infrared Photon Echo Experiments: Exploring Vibrational Dynamics in Liquids and Glasses*, Acc. Chem. Res. 28 (1995) 437–445
16. *Vibrational Dephasing in Liquids: Raman Echo and Raman Free-Induction Decay Studies*, M.A. Berg, in *Ultrafast Infrared and Raman Spectroscopy*, M.D. Fayer (ed.), Marcel Dekker, New York (2001), pp. 383–436
17. K.D. Rector, M.A. Berg, D. Sengupta, M.D. Fayer, *Effects of Solvent Viscosity on Protein Dynamics: Infrared Vibrational Echo Experiments and Theory*, J. Phys. Chem. B 105 (2001) 1081
18. K.A. Merchant, Qing-Hua Xu, D.E. Thompson, M.D. Fayer, *Frequency Selected Ultrafast Infrared Vibrational Echo Studies of Liquids, Glasses, and Proteins*, J. Phys. Chem. A 106 (38) (2002) 8839–8849
19. Qing-Hua Xu, M.D. Fayer, *Temperature-dependent vibrational dephasing: Comparison of liquid and glassy solvents using frequency-selected vibrational echoes*, J. Chem. Phys. 117 (6) (2002) 2732–2740
20. A. Abragam, *The Principles of Nuclear Magnetism*, Oxford University Press, London (1961)
21. R.P. Feynman, F.L. Vernon, R.W. Hellwarth, J. Appl. Phys. 28 (1957) 49–52
22. L. Allen, J.H. Eberly, *Optical Resonance and Two-Level Atoms*, Dover Publications (December 1, 1987)
23. A. Carrington, A.D. McLachlan, *Introduction to Magnetic Resonance*, Chapman and Hall, London (1967)
24. F. Bloch, *Nuclear Induction*, Phys. Rev. 70 (7–8) (1946) 460–474
25. A.S. Davydov, *Quantum Mechanics*, 2nd Edition, Pergamon, New York (1976)
26. L.I. Schiff, *Quantum Mechanics*, McGraw–Hill, New York (1968)
27. D. Zimdars, A. Tokmakoff, S. Chen, S.R. Greenfield, M.D. Fayer, T.I. Smith, H.A. Schwettman, *Picosecond infrared vibrational photon echoes in a liquid and glass using a free electron laser*, Phys. Rev. Lett. 70 (18) (1993) 2718–2721

28. D. Vanden Bout, L.J. Muller, M. Berg, *Ultrafast Raman echoes in liquid acetonitrile*, Phys. Rev. Lett. 67 (26) (1991) 3700–3703
29. J.C. Polanyi, A.H. Zewail, *Direct Observation of the Transition State*, Acc. Chem. Res. 28 (1995) 119–132
30. H. Bitto, J.R. Huber, *Molecular quantum beats. High-resolution spectroscopy in the time domain*, Acc. Chem. Res. 25 (1992) 65–71

– 9 –

Ultrafast Chemical and Physical Processes

The development of laser technologies has permitted the generation of ultrashort picosecond and femtosecond pulses, opening new research possibilities by ultrafast spectroscopic methods. Vibrational motions take place on the femtosecond timescale. The period of a single molecular oscillation is a few femtoseconds. By probing with femtosecond pulses we can control the breaking of chemical bonds during chemical reactions, and obtain "snapshots" of how atoms in a molecule oscillate and of exact reaction paths. This new branch, approximately fifteen years old, is a rapidly developing field known as *femtochemistry*. This is an area of chemistry, biology, and physics, dealing with ultrafast chemical reactions by using experimental methods based on the use of femtosecond laser pulses. A number of different ultrafast spectroscopic methods has been developed for studying various aspects of the dynamics in molecular systems. These methods have been discussed in Chapter 8. Scientists would, of course, like to investigate and monitor even shorter times and distances. Recent progress in the generation of attosecond pulses extends spectroscopic access to electron motions. The periods of electrons revolving in atomic orbitals are on the attoseconds timescale. Spectroscopic access to this timescale would allow us to see directly the configuration of electrons in an atom. So, we are participating in an exciting race from femtochemistry to attophysics to elucidate the molecular world of atomic processes during unprecedentedly short time intervals.

One may ask why the investigation of ultrafast processes is so important, when practical effects occur on a macroscopic scale? We shall give some examples that illustrate how basic ultrafast research yields information of direct importance to the real world, whether it be photoisomerization in vision, or proton pumping through a biological membrane, proton-transfer relating to acid catalysis, or studies of clusters which relate to nanotechnology. The properties of matter in the surrounding world depend on the reorganization of chemical bonds and their mutual transformations. Although the phenomena observed on the macroscopic time and spatial scale seem to be static or slow, an unexpectedly large number of macroscopic phenomena is directly determined by ultrafast processes at a molecular level, occurring on the picosecond, femtosecond, or attosecond timescale. A chemical reaction observed on a macroscopic scale depends on elementary processes representing a sequence of chemical bonds forming and breaking that occur on a femtosecond timescale and angstrom spatial scale. The development of laser technologies has made it possible to monitor the dynamics of ultrafast processes. By using ultrashort multiple laser pulses, one can not only precisely start reactions but also monitor the decay of the reactants or the formation of the products by using a second probing laser pulse with a frequency set to an appropriate spectroscopic transition. With a given delay between the pump and the probe laser beams, a snapshot of the populations of the reactants and products can be recorded at that instant along the reaction pathway. By recording numerous snapshots, with incrementally increasing delay-times between the laser pulses—a "movie" of the reaction at the molecular level is produced with an extremely high time-resolution of femtoseconds, that is sufficient to see bonds breaking, new bonds forming, and the reorganization of the atoms' geometry.

Ahmed Zewail has pioneered the use of ultrafast lasers in studies of femtosecond transition-state dynamics, and received the Nobel Prize in Chemistry in 1999 for investigations in the field of ultrafast spectroscopy.

To study femtosecond events we need proper tools. Ultrafast chemical and physical processes are studied with the time-resolved methods described in Chapter 8. For example, pump-probe spectroscopy is often used to monitor dynamics of chemical reactions. The first pumping pulse initiates a chemical process by excitation of a molecule to a reactive state. The second pulse monitors changes in the population of reactants, intermediates, or products. Femtosecond spectroscopy provides access to transition-state dynamics that in the past were the subject of theoretical speculations.

Experimental investigations in this field started from the most elementary reactions, such as the bond rupture in diatomic molecules as observed during dissociation dynamics in NaI [1]. The femtosecond pump pulse promotes an NaI molecule to the covalent excited electronic state in which NaI may dissociate into neutral Na and I atoms ($NaI \rightarrow Na + I$) when the distance between atoms reaches 6.93 Å. However, the excited NaI may choose another reaction-coordinate path at longer internuclear distances. The excited state at longer internuclear distances can be represented by an ionic potential-energy curve. The covalent and ionic potential curves cross at 6.93 Å. The stretching oscillation between Na and I atoms leads to the internuclear separation, and at the crossing-point the excited NaI may either dissociate or transform into the ionic form ($Na^+ \cdots I^-$). Therefore, the reaction coordinate for the excited complex NaI changes character, from being covalent at short distances to being ionic at larger distances. A series of pulses, delayed in time with respect to the pump pulse , monitors the system—probing absorption corresponding to either the excited complex NaI, or the free atom Na. The oscillatory character of the signal for the excited complex NaI in Fig. 9.1. illustrates oscillations between atoms in the excited complex. One can see that the oscillations are damped, providing valuable information about dephasing of the wave packet at the crossing point. The wave packet theory will be described in section 9.1.

There are many examples of chemical reactions or physical processes directly affecting reaction rates that could not be understood without ultrafast laser spectroscopic methods. The time-resolved spectroscopic studies involve the understanding of chemical and physical processes such as:

1. isomerization, proton transfer, vibrational predissociation, electron transfer, recombination, dissociation,
2. processes determining a solvent's role on a reaction coordinate, singlet – singlet and triplet – triplet energy transfer,

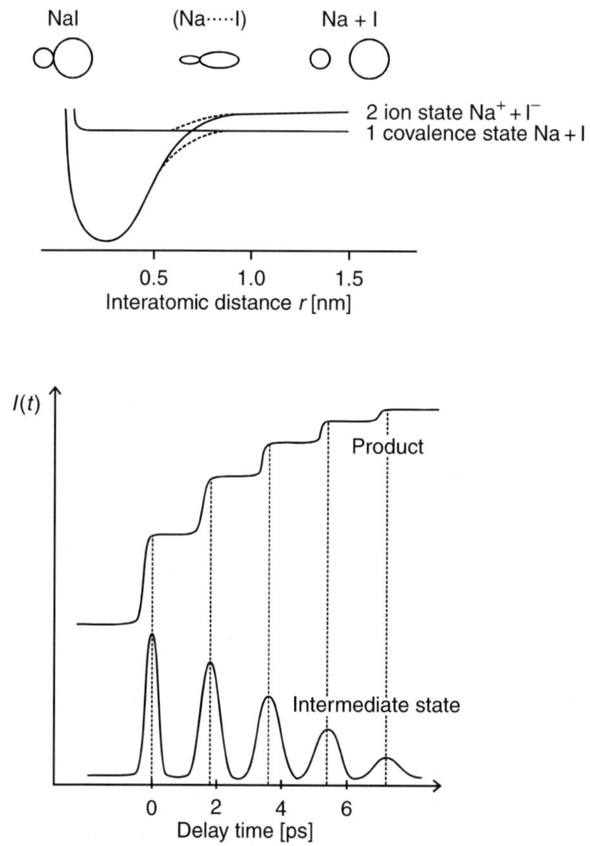

Fig. 9.1 Dynamics of reaction of dissociation of the NaI molecule. [1]

3. vibrational relaxation processes T_1,
4. vibrational dephasing processes T_2,
5. solvation of an excess electron.

Some of these topics are discussed in this chapter. For details the reader is referred to [2–8].

9.1. FEMTOCHEMISTRY. WAVE PACKET DYNAMICS. THEORY

The idea of the wave packet has been known for many decades to all who are familiar with the fundamentals of quantum mechanics. The wave function $Y(x, t)$ must obey Schrödinger's equation

$$-\frac{\hbar^2}{2m}\frac{\partial^2 \Psi(x,t)}{\partial x^2} + U(x)\Psi(x,t) = i\hbar\frac{\partial \Psi(x,t)}{\partial t} \tag{9.1}$$

where $U(x)$ is the potential well.

The superposition principle states that any linear superposition of wavefunctions is also the solution of Schrödinger's equation. Thus, the most general solution to Schrödinger's equation is an infinite sum of the stationary states

$$\Psi(x,t) = \sum_{n=1}^{\infty} C_n \Psi_n(x,t) = \sum_{n=1}^{\infty} C_n \exp\left(-\frac{iE_n t}{\hbar}\right)\Psi_n(x). \quad (9.2)$$

Each term in the sum corresponds to a specific energy E_n of the system, $|C_n|^2$ represents the probability of being found in the E_n state. This sum in eq. (9.2) represents the *wave packet* and contains information about all the molecular states involved. Now consider the general solution of Schrödinger's equation at time $t_0 = 0$,

$$\Psi(x,0) = \sum_{n=1}^{\infty} C_n \Psi_n(x). \quad (9.3)$$

When the system described by the wave function $\Psi(x,0)$ is disturbed with the laser pulse at $t_0 = 0$, the molecular system evolves in time, and can be described mathematically by the evolving wave packet $\Psi(x,t)$ from eq. (9.2). For many decades the time-evolution of the wave packet represented only a theoretical idea used in quantum mechanics that could not be tested directly in an experiment. Historically, the idea of the mystical "transition state" reflected our lack of knowledge about the whole trip from reagents to products

$$\mathrm{A} + \mathrm{BC} \rightarrow (\mathrm{ABC})^{\#} \rightarrow \mathrm{AB} + \mathrm{C}. \quad (9.4)$$

The nature of the transition state $(\mathrm{ABC})^{\#}$ had a very speculative character, and was based on empiricism and many severe approximations. It is not surprising that the region of the $(\mathrm{ABC})^{\#}$ state was not accessible by experiment because the femtosecond time-scale required for monitoring the electronic processes of bond rupture and bond formation was immeasurably short at that time.

With the development of ultrafast lasers the time evolution of the wave packet can be monitored in real-time measurements. Now, the time-dependent wave packets are widely used in nonlinear laser spectroscopy to study and model various phenomena in physics, chemistry and biology [2–5]. These branches of laser spectroscopy are often called "femtochemistry" and "femtobiology". With ultrashort laser pulses, one is able to follow the elementary dynamics of transition states, reaction rates, and energy distribution.

So far, the wave packet in eq. (9.2) represents any superposition of quantum molecular states. In practice, for most photophysical and photochemical reactions, electronic and vibrational degrees of freedom decide the elementary dynamics along the reaction coordinate. Therefore the most commonly studied wave packet is the vibrational wave packet.

We shall look at a very typical situation in the pump-probe measurements presented in Fig. 9.2. With the femtosecond pump pulse, the molecules from the ground electronic state are promoted to the excited electronic state. It is obvious that the

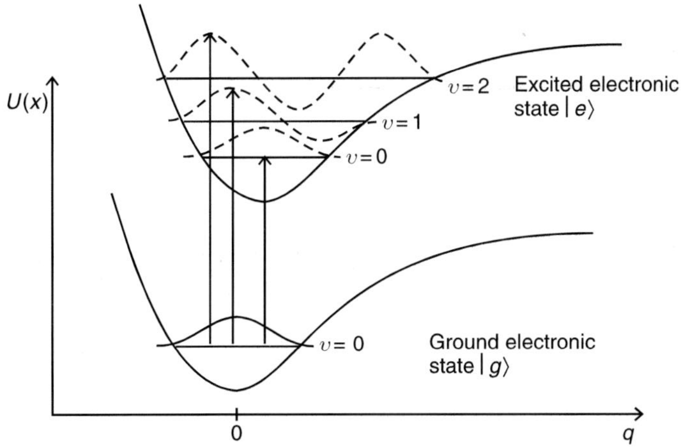

Fig. 9.2 A particle with two electronic states, a ground state $|g\rangle$ and an electronic excited state $|e\rangle$, which are coupled to nuclear degrees of freedom characterized by a nuclear coordinate q; the dashed lines represent the wave functions of the harmonic oscillator for the quantum numbers $\upsilon = 0$, $\upsilon = 1$, $\upsilon = 2$.

vibrational wave packet in the ground state at $t_0 = 0$, shortly before the pulse excitation

$$\Psi(q, 0) = \sum_{n=1}^{\infty} C_n \Psi_n(q) \tag{9.5}$$

is not the same as that in the excited electronic state. First, the Franck-Condon vibrational excited states are populated during the electronic transition, owing to the different geometries of the potential wells. Secondly, the femtosecond pulse is not monochromatic, and contains a distribution of various wavelengths that excite different vibrational levels, and different vibrational oscillators. For example, the femtosecond pulse of 6 fs covers the frequency spectrum of 3700 cm^{-1}. Thus, for such short pulses it is hard to define any single oscillator coordinate, the pump pulse prepares a coherent state called the vibrational wave packet (eq. (9.6)) that contains information about the collective molecular state, instead of a collection of separate eigenstates. In contrast, if we use a pulse longer than the characteristic period of a single oscillation—as we do in *the hole-burning techniques*—we can get more information about the individual vibrational normal modes. Therefore, the pump pulse in Fig. 9.2 prepares the coherent state described by the vibrational wave packet that evolves in time and can be monitored with the second pulse, delayed with respect to the pump pulse,

$$\Psi(q, t) = \sum_{n=1}^{\infty} C_n \exp\left(-\frac{iE_n t}{\hbar}\right) \Psi_n(q). \tag{9.6}$$

To understand how we can monitor the time-evolution of the vibrational wave packet we have to know how to describe the interaction between the electric field of the laser $\boldsymbol{E}(\boldsymbol{r}, t)$ and the molecular system, and how to measure the response of the molecular system. Here, we will present only the basic ideas of the formalism of the nonlinear response theory that is commonly applied in nonlinear laser spectroscopy. The detailed analysis of the nonlinear response functions goes beyond the present treatment. Further details of the theoretical aspects of nonlinear spectroscopy are given in Mukamel's excellent book on the subject [6].

Polarization plays a key role in optical spectroscopy; it controls the radiation– matter interaction as well as the intermolecular forces. Consequently, a complete knowledge of the optical polarization is sufficient for the interpretation of any time-domain or frequency-domain spectroscopic measurements. The response of the electronic and nuclear degrees of freedom, as well as the time-evolution via relaxation processes can be observed in optical measurements only through their effects on the polarization.

The polarization operator of the material system is given by a summation over all molecules

$$\hat{P}(\mathbf{r}) = \sum_m P_m(\mathbf{r}) \tag{9.7}$$

where $P_m(\mathbf{r})$ is the polarization operator of molecule, m,

$$P_m(\mathbf{r}) = \sum_\alpha \int_0^1 du\, q_{m\alpha}(\hat{r}_{m\alpha} - R_m)\delta[\mathbf{r} - R_m - u(\hat{r}_{m\alpha} - R_m)]. \tag{9.8}$$

Here, $\hat{r}_{m\alpha}$ is the position operator, $q_{m\alpha}$ is the electric charge of particle α (electron or nucleus) belonging to molecule m, R_m is a molecular center of mass (or charge), and the u–integration is a number integration that ensures the correct coefficients of the multipolar expansion of the polarization operator [6].

Most commonly we shall invoke the dipole approximation, where we keep only the first electric term in the multipolar expansion. The eq. (9.8) is then simplified to the form

$$\hat{P}(\mathbf{r}) = \sum_m \mu_m \delta(\mathbf{r} - \boldsymbol{R}_m) \tag{9.9}$$

where μ_m is the dipole-moment operator for the mth molecule, and is given by

$$\mu_m = \sum_\alpha q_{m\alpha}(\hat{r}_{m\alpha} - R_m). \tag{9.10}$$

The $\delta(r - R_m)$ is the Dirac delta function

$$\delta(r - R_m) = \begin{cases} \infty & r = R_m \\ 0 & r \neq R_m \\ \int f(r)\delta(r - R_m)dr = f(R_m). \end{cases} \tag{9.11}$$

The electric field of the laser, $\boldsymbol{E}(\mathbf{r}, t)$ interacts with the matter through the polarization, $\boldsymbol{P}(\mathbf{r})$. The radiation–matter interaction is given by the semi-classical Hamiltonian

$$H_{\text{int}} = -\int E(\mathbf{r}, t)P(\mathbf{r})d\mathbf{r}. \tag{9.12}$$

The radiation–matter interaction described by eq. (9.12) can be simplified if we assume the dipole approximation (eq. (9.9)). The dipole approximation denotes that a single particle has a size much smaller than the optical wavelength, and may be represented by a point dipole. When the dipole approximation is made we can consider just the temporal response of a single particle to the electric laser field, and the Hamiltonian H_{int} from eq. (9.12) can be written as

$$H_{\text{int}}(t) = -\boldsymbol{E}(\boldsymbol{r}, t)\boldsymbol{V}, \tag{9.13}$$

with the dipole operator

$$\boldsymbol{V} = \sum_{\alpha} q_{\alpha}(\boldsymbol{r} - \boldsymbol{r}_{\alpha}) \tag{9.14}$$

where the sum runs over all the electrons, and the nuclei, a, with charges, q_a, at positions $\boldsymbol{r}_{\alpha}$.

In the semi-classical description of the interaction the material system is treated quantum mechanically, whereas the transverse radiation field $\boldsymbol{E}(\boldsymbol{r}, t)$ is considered to be classical.

In the Schrödinger picture, the semi-classical approximation results in the Maxwell–Liouville equations:

$$\nabla \times \nabla \times E(\boldsymbol{r}, t) + \frac{1}{c^2}\frac{\partial^2}{\partial t^2}E(\boldsymbol{r}, t) = -\frac{4\pi}{c^2}\frac{\partial^2}{\partial t^2}P(\boldsymbol{r}, t) \tag{9.15}$$

$$P(\boldsymbol{r}, t) = Tr[\hat{P}(\boldsymbol{r})\rho(t)] \tag{9.16}$$

$$\frac{\partial \rho(t)}{\partial t} = -\frac{i}{\hbar}[H_T, \rho(t)]. \tag{9.17}$$

The dynamics of the system are calculated by solving coupled equations for the electric field and the polarization. As we have said, the optical polarization $P(\boldsymbol{r}, t)$ is the primary goal of any theory of optical spectroscopy because it plays a key role in interpreting optical measurements. To calculate the polarization, $P(\boldsymbol{r}, t)$, which is physically observable, we have to take the average over the statistical ensemble (eq. (9.16)) with the non-equilibrium density operator, $\rho(t)$, by taking a trace Tr. The density operator $\rho(t)$ can be obtained by solving the Liouville equation, (9.17), where H_T is given by

$$H_T = H + H_{\text{int}} \tag{9.18}$$

and describes the Hamiltonian H of the molecular system (matter) and the Hamiltonian of the interaction of matter with radiation, H_{int}, given by eq. (9.12).

Unfortunately, eq. (9.17) is easily solved only for the thermal equilibrium ($t = t_0$) when the electric field of the laser does not disturb the system ($H_{\text{int}} = 0$), and the canonical density operator is given by

$$\rho(t_0) = \rho(-\infty) = \frac{\exp(-\beta H)}{Tr\exp(-\beta H)} \tag{9.19}$$

where $\beta = (k_B T)^{-1}$, k_B is the Boltzmann constant, and T is the temperature.

For a non-equilibrium situation, when the laser beam begins to interact with the molecular system, the eq. (9.17) is not easily solved. Usually, a perturbative order by order expansion of the response in the fields is applied, which indicates that the time-dependent density operator can be expanded in terms of the powers of the electric fields

$$\rho(t) = \rho^{(0)}(t) + \rho^{(1)}(t) + \rho^{(2)}(t) + \rho^{(3)}(t) + \dots \tag{9.20}$$

where $\rho^{(n)}$ denotes the nth -order contribution in the electric field, $\rho^{(0)}(t) = \rho(-\infty)$.

By substituting eqn. (9.20) into (9.16) we obtain the Taylor expansion of the polarization in powers of the radiation field $E(\mathbf{r}, t)$,

$$P(\mathbf{r}, t) = P^{(1)}(\mathbf{r}, t) + P^{(2)}(\mathbf{r}, t) + P^{(3)}(\mathbf{r}, t) + \dots \tag{9.21}$$

As we explained in Chapter 5, this expansion corresponds to the linear and non-linear optical processes. The linear term $P^{(1)}$ is responsible for linear optics, $P^{(2)}$ is responsible for second-order processes such as SHG or frequency-sum generation, and $P^{(3)}$ is the third-order polarization that is responsible for THG and many other processes, measured by a broad variety of laser techniques such as four-wave mixing, stimulated Raman scattering, pump-probe spectroscopy, or polarization-gating. It can be shown [6] that the nth-order polarizability, $P^{(n)}(\mathbf{r}, t)$, is expressed by the nth-order non-linear response functions $S^{(n)}$ that provide complete information about the time-evolution of the system measured by the optical spectroscopy of the nth order, and is given [6] by

$$P^{(n)}(\mathbf{r}, t) = \int_0^{\infty} dt_n \int_0^{\infty} dt_{n-1} \dots \int_0^{\infty} dt_1 S^{(n)}_{t_n, t_{n-1}, \dots, t_1} E(\mathbf{r}, t - t_n) \times$$
$$\times E(\mathbf{r}, t - t_n - t_{n-1}) \dots E(\mathbf{r}, t - t_n - t_{n-1} - \dots t_1) \tag{9.22}$$

where the response function $S^{(n)}$ is given by

$$S^{(n)}(t_n, t_{n-1}, \dots, t_1) = \left(\frac{i}{\hbar}\right)^n \theta(t_1)\theta(t_2)\dots\theta(t_n)$$
$$\times \langle V(t_n + \dots + t_1)[V(t_{n-1} + \dots + t_1),$$
$$[\dots[V(t_1), [V(0), \rho(-\infty)]]\dots]]\rangle \tag{9.23}$$

Here, $\theta(\tau)$ is the Heavyside step-function [$\theta(\tau) = 1$ for $t > 0$ and $\theta(\tau) = 0$ for $t < 0$], $\rho(-\infty) = \rho_0$ represents the equilibrium-density operator, which only does not evolve with time when the molecular Hamiltonian, H, does not interact with the electric field of the laser. The operator $V(t)$ from eq. (9.23) can be expressed as

$$V(\tau) = \exp\left(\frac{i}{\hbar} H\tau\right) V \exp\left(-\frac{i}{\hbar} H\tau\right) \tag{9.24}$$

and represents the dipole operator from eq. (9.14) in the interaction picture. One can see from eqs. (9.22–9.24) that the polarization response, $P^{(n)}$, is expressed by the n-time points correlation functions.

For the linear response, $P^{(1)}$, that controls all linear-spectroscopy optical measurements (for example absorption, spontaneous Raman scattering) the response function, $S^{(1)}$, is simplified to the form

$$S^{(1)}(t_1) = \frac{i}{\hbar}\theta(t_1)\langle[V(t_1), V(0)]\rho(-\infty)\rangle \tag{9.25}$$

and represents the well-known *two-points correlation functions*. These functions describe the dynamics of *vibrational relaxation* in IR and Raman spectroscopy through the *vibrational correlation function* $\langle Q(t)Q(0)\rangle$ and the dynamics of *reorientational relaxation*. The higher-order response functions are a little more complicated, but in some approximations they can be factorized and expressed as a product of two time-point correlation functions. The exact form of various responses for a specific nonlinear spectroscopy, including pump-probe, polarization-getting, echoes, and coherent Raman, are to be found in Mukamel's book [6].

The expression derived so far for the responses $S^{(n)}$ and $P^{(n)}$ do not seem to be directly related to *wave packet dynamics*. However, we will show now that this relationship is very close.

The present approach is based on the propagation in a Hilber-space for the general case, without any specification of the molecular system. Let us consider the situation presented in Fig. 9.2. Consider a particle with two electronic states, a ground state $|g\rangle$ and an electronically excited state $|e\rangle$, which are coupled with nuclear degrees of freedom characterized by q. The Hamiltonian, H (from eq. (9.24)), for such a molecular system is given by

$$H = |g\rangle H_g(q)\langle g| + |e\rangle H_e(q)\langle e| \tag{9.26}$$

$$V = V_{ge}(q)|g\rangle\langle e| + V_{eg}|e\rangle\langle g|. \tag{9.27}$$

The quantities V_{ge} and V_{eg} are no longer dipole operators, with respect to the electronic degrees of freedom, but they are still operators in the nuclear space. After averaging over the nuclear space they become the transition dipole moments.

To calculate the response functions, $S^{(n)}$ (eq. (9.23)) we need $\rho(-\infty)$ which represents the equilibrium-density operator in the absence of the electric field. As we consider the electronic transition, the electronic energy gap, ΔE, is much larger than k_bT, and most molecules occupy the ground electronic state. So, we can assume that initially ($t \to -\infty$), the system is described by the equilibrium ground-state nuclear density operator, ρ_g

$$\rho(-\infty) = |g\rangle\rho_g\langle g| \tag{9.28}$$

$$\rho_g = \frac{\exp(-\beta H_g)}{Tr\exp(-\beta H_g)}. \tag{9.29}$$

Using (9.26) and (9.27) in (9.24) we obtain the time-evolution of the dipole operator V_{ge},

$$V(t_1) = \exp\Big(\frac{i}{\hbar}H_g t_1\Big)V_{ge}\exp\Big(-\frac{i}{\hbar}H_e t_1\Big) \tag{9.30}$$

which can be used to calculate the response function $S^{(n)}$ (eq. 9.23). For example, for the first-order response $S^{(1)}$ (eq. 9.25) we obtain

$$S^{(1)}(t_1) = \frac{i}{\hbar}\theta(t_1)[J(t_1) - J^*(t_1)] \tag{9.31}$$

where,

$$J(t_1) = \left\langle \exp\left(\frac{i}{\hbar}H_g t_1\right) V_{ge} \exp\left(-\frac{i}{\hbar}H_e t_1\right) V_{eg}\rho_g \right\rangle. \tag{9.32}$$

The equation can be interpreted as follows. The system is initially at thermal equilibrium in the ground electronic state, and is represented by its nuclear density operator ρ_g (eq. 9.29). Then the dipole operator V_{eg} (generated by a laser beam at $t=0$) acts on it, converting it to the electronic excited state $|e\rangle V_{eg}\langle g|$ This electronically excited system is now in an optical coherence $|e\rangle\langle g|$. The time evolution of the coherence at t_1 is given by the term

$$\exp\left(\frac{i}{\hbar}H_g t_1\right) V_{ge} \exp\left(-\frac{i}{\hbar}H_e t_1\right). \tag{9.33}$$

Finally, averaging $\langle\ \rangle$ over the nuclear coordinates gives the expected value of the dipole. The Fourier transform of the time response, $S^{(1)}$ (eq. 9.31) gives the linear response function in the frequency domain and provides information about the dynamics through band broadening.

We can re-cast eq. (9.32) in a different form that is more suitable for clarifying the idea of the nuclear wave packet's evolution [6]. By using the nuclear eigenstates $|a\rangle,|b\rangle$ in eq. (9.32) for the H_g and H_e Hamiltonians

$$H_g|a\rangle = E_a|a\rangle \tag{9.34}$$

$$H_e|b\rangle = E_b|b\rangle \tag{9.35}$$

we obtain

$$J(t_1) = Tr[V_{ge}\rho_1(t_1)] \tag{9.36}$$

where,

$$\rho_1(t_1) = \sum_{a,\,b} P(a)\mu_{ba}\exp -i\omega_{ba}t_1|b\rangle\langle a| \tag{9.37}$$

represents the time-dependent nuclear wave packet when the electronic system is in an optical coherence.

The above picture can be extended to the nonlinear response functions. The expression for $S^{(n)}$ is more complicated but the idea is similar. The response $S^{(n)}$ provides information about the time-evolution of the nuclear wave packet. Details can be found in Mukamel's book [6].

9.2. FEMTOCHEMISTRY. SPECTROSCOPIC APPLICATION OF WAVE PACKET DYNAMICS

In the previous section we provided the background for understanding what "wave packet dynamics" means. Now, we will show how the wave packet dynamics can be monitored experimentally, and what substantial information can be gained from femtosecond spectroscopy. Intuitively we feel that the evolving wave packet moves into a different region of configurational space to pass the entire journey of a chemical reaction or any other photophysical event. Because the initial trajectories are synchronized by the pump electric field at zero time, a second femtosecond pulse, delayed with respect to the pump pulse, can probe the wave packet as it "visits" different regions in the configurational space corresponding to the different steps in the elementary reaction. Therefore, such real-time investigations permit study of the rearrangement of bonds during physical or chemical transformation, revealing the intermediate species as reactants are transformed into products. The atomic vibrations leading to the breaking of bonds, or the bonds' rearrangements, occur on the femtosecond time scale, so such real-time investigations require femtosecond temporal resolution.

The possibility of monitoring wave packet dynamics has been subject to intensive study in the past two decades, [3, 5, 9–10] and further details are in the books cited. We shall give some examples that are useful for educational purposes.

Femtosecond nonlinear spectroscopy can be carried out under various conditions. We can distinguish two main limits:

1. the impulsive limit,
2. the non-impulsive (hole-burning) limit.

In the impulsive limit a pump pulse is short compared with the characteristic period of the vibrational, torsional and rotational modes. In the impulsive limit, the nuclear motions are excited coherently by the pump pulse, which prepares the coherent state called the wave packet. As the ultrashort pulse intrinsically consists of a broad range of frequency components the excited wave packet contains information about the collective molecular state involved instead of a collection of separate eigenstates. For the electronic transition the impulsive limit comprises a coherent superposition of the Frank-Condon manifold of different vibrational modes. The shorter is the pump pulse the more vibrational states and normal vibrational modes are involved in the wave-packet. For very short pulses ($\sim$6 fs), corresponding to a spectral distribution of 3700 cm^{-1}, this collective nuclear coordinate depends, in principle, on all the nuclear degrees of freedom and offers no spectral resolution of vibrations within the wave packet! This does not sound helpful, but it is a fact. We cannot violate the Heisenberg Uncertainty Principle, $\Delta\omega\Delta\tau \geq 1$, so cannot have perfect time and frequency resolution at the same time. The better the time resolution, the worse is the frequency resolution. Consider the simplest configuration used in pump-probe spectroscopy, with the spectrometer and a CCD detector, as well as an independent photodiode to measure the delay-time, τ, between the pump and probe pulses. Such a configuration of a time and a frequency domain contains an apparent contradiction. One may think that "The temporal resolution, $\Delta\tau$, and the spectral resolution of the probe pulse, $\Delta\omega$, may be controlled independently, and so we can make both $\Delta\tau$ and $\Delta\omega$ as small as we desire". But this is wrong: if the temporal pulse is infinitely short, and has an infinitely broad spectrum, the resulting spectrogram would be completely independent of the frequency of the molecular system: all the frequency information is lost! So, using pulses which are too short is sometimes a bad idea. On the other hand, it makes no sense to perform ultrafast femtosecond experiments on a system with narrow spectral lines, since no more information is gained beyond that from *cw* (continuous wave) measurements in conventional spectrometers, which are much cheaper than the sophisticated femtosecond methods!

It is important to recognize that the temporal resolution of the laser pulses has to be adjusted to the molecular system that is being probed. Having time and frequency domain information simultaneously is a result of an interplay between the temporal pulse-duration and the dynamics of the molecular system. Even if we control the delay, τ with infinite accuracy, the actual delay between the pump and the probe events is not, in principle, the same. Quantum mechanical principles determine the actual delay in the system, and only when the system's dephasing is fast will it be equal to τ. However, the fast dephasing means a broad line in the frequency domain, and we return again to the beginning – the uncertainly principle.

We shall now concentrate on the impulsive limit with the temporal pulse duration chosen properly for the molecular system being probed. We will show that, in this case, we can obtain interesting information about both the time and frequency domain.

The idea of a typical impulsive limit experiment to observe the coherent wave packet dynamics is illustrated in Fig. 9.3.

One may see from Fig. 9.3a that the pump-probe signal has a strongly oscillatory character. The oscillations are slowly damped with time, owing to dephasing. The oscillations represent wave packet motions and arise from the quantum mechanical effects called, *quantum beats*. The quantum beats are described in chapter 8 of this book. In some cases they may obscure the pump-probe dynamics and provide trivial information on the frequencies of the vibrational oscillators—which can be obtained easily by conventional vibrational spectroscopy—IR and Raman (see the discussion above about the uncertainty principle). However, in many cases, when the interplay between the time-frequency domains is properly chosen, the quantum beats illustrate additional important features of the dynamics, such as the vibrational frequencies of the transition states, or the frequencies for intermediates which help to define the reaction coordinate [11]. If we observe the quantum beats in the pump-probe signal, the next step is shown in Fig. 9.3b. We perform the Fourier transform of the signal $\Delta T/T$ around an arbitrarily chosen delay-time, τ, for a given probe wavelength, λ_{probe}. The Fourier transform provides the amplitudes and the frequencies of the vibrations modulating the $\Delta T/T$ signal around the chosen delay, τ, i.e., we produce the spectrogram. By sliding the spectrogram for a given delay, τ, we can obtain the experimental vibrational spectrum of instantaneous frequencies that dominate in the wave packet at τ. By using this procedure for different values of τ we can monitor the time-evolution of the wave packet.

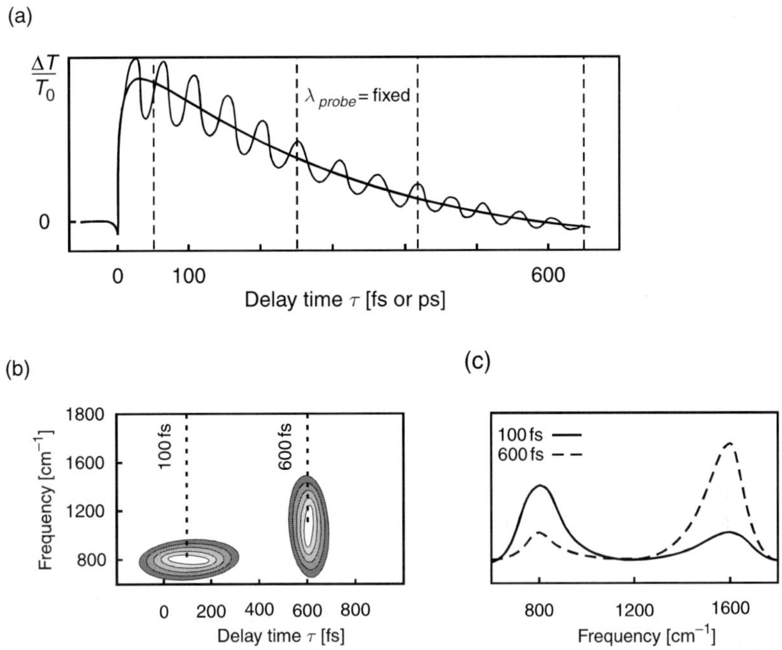

Fig. 9.3 (a) Time-dependence of the transmission change ($\Delta T/T$) at λ_{probe} after excitation at λ_{pump}. (b) Spectrogram at λ_{probe} as a function of the delay, τ, calculated from the Fourier transform of $\Delta T/T$ around a given delay time, τ. (c) Fourier power spectrum of instantaneous frequencies for a given delay time, τ.

9.2.1. Excited-State Vibrational Coherence

One of the unique aspects of femtosecond spectroscopy is that we can observe directly the real-time motion of the nuclei of molecules in their excited state by monitoring vibrational coherence. From that we can learn about the vibrational modes that are involved in the primary steps of reactions, or which make a major contribution to energy dissipation before returning to the ground state.

In Fig. 9.4 we present the results for early events in the photoexcitation of a semiconducting polymer, studied with sub-10 fs time resolution.

The experiment monitors the internal conversion, IC, ($S_2 \rightarrow S_1$) following photoexcitation by using sub-10 fs pulses in the visible range ($S_0 \rightarrow S_2$). Figure 9.4 presents a typical pump-probe signal. The impulsive collective vibrational coherence is evident from the transient spectrum as a periodic modulation of the signal. The vibrational coherence was assigned to the ground-state motions, giving a time-constant for the depletion of S_2, estimated from the $\Delta T/T$ kinetics at about 70 fs, while the oscillations are slowly damped with time, owing to dephasing, and persist longer than 1 ps. The authors calculated the Fourier power spectrum from Fig. 9.4 and found that it contains two main frequencies at 1510 cm^{-1} and 2115 cm^{-1}, assigned to collective vibrational coherence initiated by the short pump pulse.

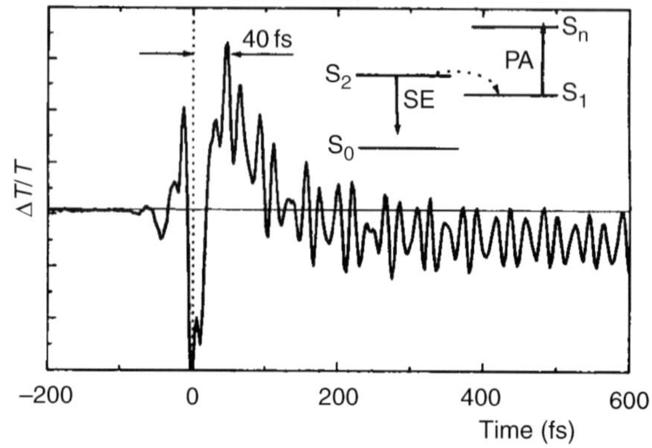

Fig. 9.4 Transmission difference as a function of pump-probe delay in poly-DCHD-HS (poly[1,6-bis(3,6-dihexadecyl-n-carbazolyl)-2,4-hexadiene) at 580 nm. The level scheme shows the internal conversion in S_2 following photoexcitation [12]. Reproduced from ref. [12]. Permission has been granted by the copyright holder.

9.2.2. Vibrational Coherence in "Reacting" Excited-State Molecules. Bacteriorhodopsin

Even more exciting than studying vibrational coherence in the excited state is the study of vibrational coherence in chemical reactions. One of the most frequently studied processes is a sequence of reactions in rhodopsin, or a photocycle of bacteriorhodopsin, in which photoisomerization is one of key events. A reason for studying ultrafast dynamics for these proteins is to understand these real-world processes. Rhodopsin is responsible for processes in vision, and bacteriorhodopsin is responsible for the transfer of a proton from the cytoplasm to the extracellular surface of the cell. The resulting proton gradient across the cell membrane triggers ATP synthesis, which is used for metabolism in the bacterium.

The absorption of a photon from the visible range (568 nm) initiates in bacteriorhodopsin a cyclic sequence of reactions that is completed on the millisecond time scale (Fig. 9.5) leading to the proton's moving from the cytoplasmic side to the extracellular surface, and generation of an electrochemical potential that is used by a bacterium to maintain its metabolism. As one can see from Fig. 9.5, the observed time constants of the BR (Bacteriorhodopsin) photocycle span a range of more than ten decades. The photocycle can be divided into two distinct parts. The first one comprises very fast molecular processes occurring on the femto- and picosecond time scales, in which BR-568 (all-*trans*) is excited to the formation of the K-intermediate that has the 13-*cis* configuration. The second part of the photocycle is much slower, and the mechanisms and processes involved have been reviewed recently [13–15]. The situation for the first part of the photocycle is not still clear. It is evident that the key event occurring within the first 3.5 ps is the all-*trans* → 13-*cis* isomerization. However, it is not clear whether this occurs immediately upon excitation, within the first 500 fs (with J-625 in the 13-*cis* configuration), or between the states J and K.

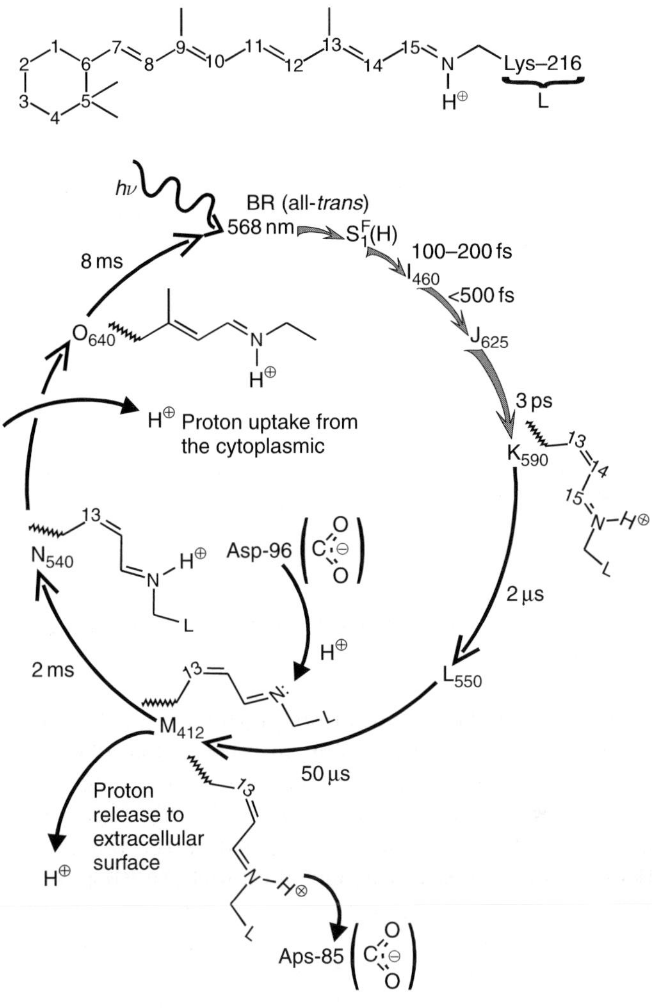

Fig. 9.5 Photocycle of bacteriorhodopsin. Reprinted with permission from [31]. Copyright [2004], American Institute of Physics.

Some key questions about the primary events of the BR photocycle remain unanswered. Understanding the primary mechanisms is a major challenge, and the subject of an active scientific debate [13–40]. There are still many discrepancies and inconsistencies between the experimental results and the theoretical models proposed so far. One of the main puzzles is a mismatch between the spontaneous fluorescence spectrum (730 nm) and the stimulated emission (860–950 nm), the origin of the fluorescent state I_{460}, and its role in the photocycle.

Compelling evidence that the all-*trans* → 13-*cis* isomerization is not the primary event from excitation was provided by comparing the ultrafast dynamics of the native BR with that of an artificially modified protein in which the retinal was replaced by a rigidified prosthetic group that is unable to twist about the active

$C_{13}{=}C_{14}$ band [16–18]. The results demonstrated that the initial ultrafast spectral changes observed in the native BR are virtually identical to those recorded for the modified analogs—including a rise of the absorption/emission bands (460/860 nm) in less than 30 fs [16–17]. This indicates that the initial spectral changes in excited BR must involve motions other than $C_{13}{=}C_{14}$ torsion [19–22]. This picture is consistent with results obtained from measurements of resonance Raman spectra [23–26] and computational studies [22, 27–30] which suggest that the initial motion out of the Franck-Condon region involves in-plane stretching modes (mainly C=C), with torsional motions coming into play at later times.

Recently, the following mechanism (two-state, many modes model) has been proposed to explain the femtosecond electronic dynamics of bacteriorhodopsin [31].

Light-adapted bacteriorhodopsin BR-568 (all-*trans*) is excited to the lowest lying excited state $S_1(B_u)$ with the pump pulse. After excitation to the $S_1(B_u)$ state the C=C stretching mode is the primary vibrational mode accepting the excitation energy. The vibrationally excited Franck-Condon state of the C=C stretching mode (H species) relaxes via intermolecular vibrational relaxation processes (path 1A in Fig. 9.6) to the vibrationally relaxed S_1 state (I species), or exchanges the energy, via an intramolecular energy-transfer process, to another vibrational mode of lower frequency (path 1B). The path 1A represents a non-reactive pathway of the BR photocycle with the return to the S_0 state in the all-*trans* configuration. The path 1B characterizes processes of intramolecular energy exchange between the C=C stretching mode and the vibrational mode of lower frequency. Figure 9.6 illustrates only the intramolecular energy exchange between the C=C stretching mode and the HOOP (hydrogen out of plane) vibrations. However, it can be expected that the process of energy transfer also involves other vibrational modes with frequencies between 1526 cm^{-1} (C=C stretch) and 800 cm^{-1} (HOOP). Figure 9.6 presents only those of the vibrational modes that are responsible for the most characteristic experimental features of the spectral time-evolution of the BR photocycle.

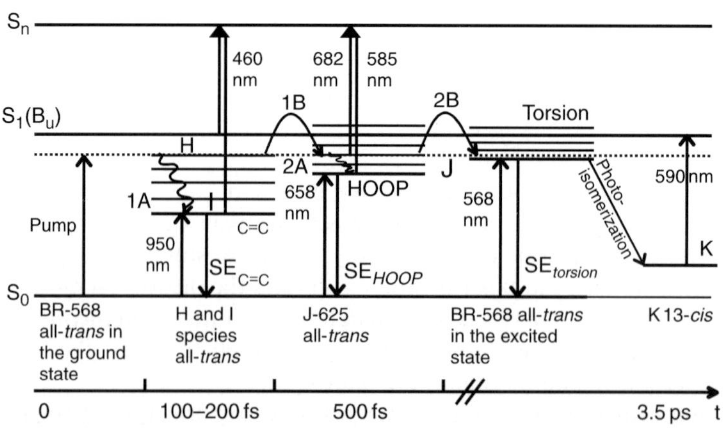

Fig. 9.6 Molecular mechanisms of primary events in BR and its retinal-modified analogs proposed in the two-state, many modes model. Reprinted with permission from [31]. Copyright [2004], American Institute of Physics.

The intramolecular energy exchange (1B) excites the HOOP mode to the vibrationally hot states that can dissipate the excess energy via path 2A (intermolecular energy relaxation), or 2B (intramolecular energy transfer to the vibrational modes of lower frequency).

The process of branching into reactive (photocycle) and non-reactive (return to S_0 and the regeneration of the original all-*trans* ground-state configuration) is continued until the intramolecular energy transfer reaches the torsional coordinate ((2B) in Fig. (9.6)). The coupling between the electronic transition and the torsional coordinate occurs at the end of the sequence of events consisting of a chain of intramolecular energy transfer steps. Once the intramolecular energy relaxation mechanism has brought the retinal chromophore to the $S_1(B_u)$-state coupled with the torsional coordinate, isomerization to the 13-*cis* configuration (K species) occurs within 3.5 ps.

This picture is consistent with the wave packet dynamics of bacteriorhodopsin obtained recently, [41] which is shown in Fig. 9.7. To give insight into the primary events of the bacteriorhodopsin photocycle Kobayashi *et al.* measured the time-resolved transmission change, $\Delta T/T$, of bacteriorhodopsin BR_{568} using sub-5 fs pump and probe pulses.

Figure 9.7 shows the transient signal $\Delta T/T$ at several wavelengths. The oscillatory character of the signal represents vibrational coherence. The Fourier transformation of $\Delta T/T$ around a given delay time, τ, gives the amplitudes and frequencies of the molecular vibrations involved in the early steps of the photocycle. The set of Fourier transforms for different delays produces a spectrogram given in Fig. 9.8.

Figure 9.9 shows the Fourier power spectrum of instantaneous frequencies that provides information about the dynamics of the wave packet shortly after excitation. One can see that the C=C and C–C stretching modes and HOOP modes at around 900 cm^{-1} are the primary vibrational modes detected upon excitation, in contrast to early suggestions [32] that the torsional motion around the C_{13}=C_{14} bond is the primary reaction coordinate.

9.2.3. H-Bond Dynamics

The examples of coherent nuclear wave packet dynamics given above provide information about the high-frequency intramolecular vibrations that are usually the primary modes accepting energy upon pump excitation. However, it has been shown recently that a coherent nuclear wave packet can also be monitored experimentally for intermolecular low-frequency modes [44]. The low-frequency intermolecular modes have been observed in H-bonded carboxylic acid dimers upon excitation of the O–D stretching mode. The low-frequency vibrations, such as large amplitude motions, contain essential information about the reaction coordinates of the reaction intermediates.

One of the most striking aspects of H-bonds is found in their dynamic behavior, particularly their vibrational dynamics. The IR hydrogen-stretching bands of hydrogen-bonded systems show a broad variety of IR band shapes, from vibrational substructure in the gas phase to extremely broad bands in the liquid phase, or a continuous IR absorption in $(BH\cdots B)^+$ or $(BH\cdots B)^-$ acid and base solutions. These IR band-shapes in the frequency domain correspond to the $S^{(1)}$ time-response function

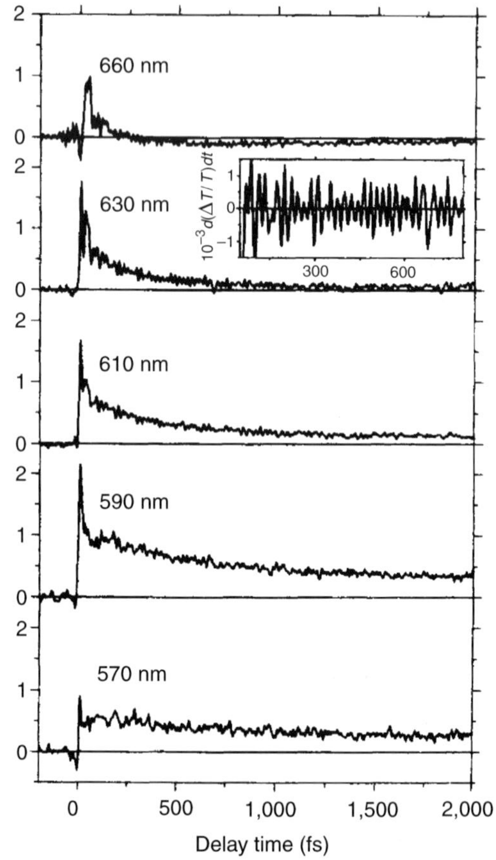

Fig. 9.7 Time-dependence of the transmittance change ($\Delta T/T$) following sub-5-fs pulse-excitation of a suspension of the light-adapted purple membrane of *Halobacterium Salinarium* containing BR_{568} at room temperature. Sub-5-fs pump and probe pulses (520–750 nm) were generated by non-collinear optical parametric amplification [42–43]. The positive $\Delta T/T$ observed at wavelengths shorter than 630 nm is due to the bleaching of BR_{568}. At wavelengths longer than 660 nm, a negative signal due to the absorption of I ($S_1 \rightarrow S_n$ transition) appeared at 250 fs. The inset shows an enlarged trace monitored at 630 nm. Reproduced from ref. [41]. Permission has been granted by the copyright holder.

(eq. 9.25) introduced in this chapter, and reflect intriguing dynamic and vibrational relaxation mechanisms.

Two main relaxation mechanisms have been proposed. The first group of models suggest that broadening of the band results from direct dipolar interaction of the dipole moment of the complex with the local fluctuating electric field produced by the solvent dipoles [45] or of the dipole moment of the proton donor and the dipole moment produced by the H-bond [46–47] in the complex regarded as a non-rigid supermolecule. The second group of models suggest that an indirect relaxation mechanism, caused by anharmonic coupling of the high-frequency $\nu_S(X-H)$ stretching-mode to the low-frequency hydrogen-bridge motion

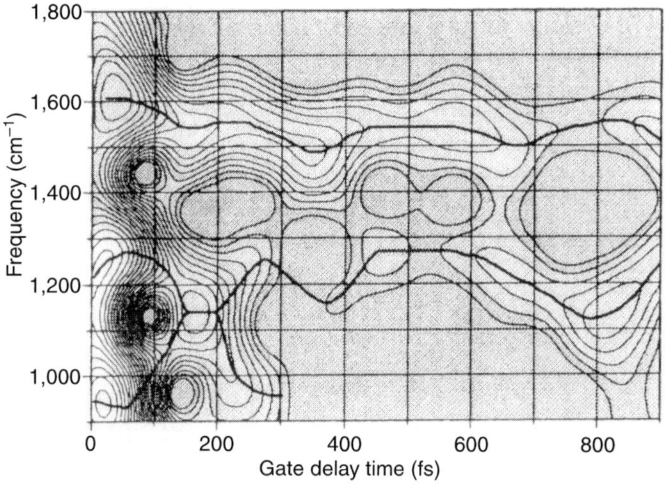

Fig. 9.8 Spectrogram calculated for the trace at 610 nm. The calculations used the Hanning window function $W(t - t_d) = 1/2 + 1/2\cos[\pi(t - t_d)/2\Delta t]$ with a 75-fs half-width at half-maximum (HWHM) of Δt and a peak at t_d (window delay time). Reproduced from ref. [41]. Permission has been granted by the copyright holder.

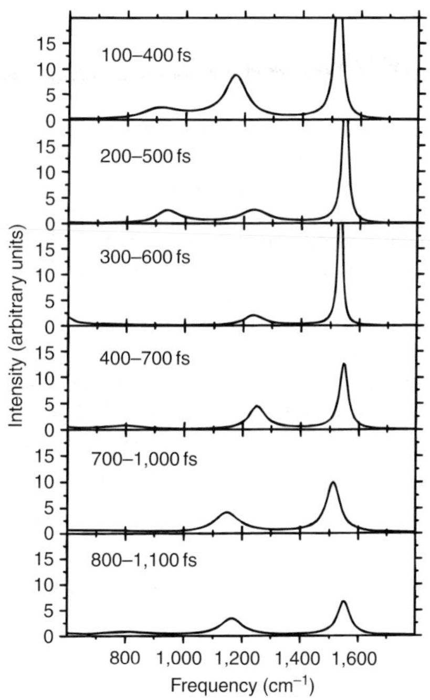

Fig. 9.9 Fourier power spectrum of instantaneous frequencies appearing in the trace of the transmittance change probed at 630 nm. The spectra were obtained by the maximum-entropy method for the 300-fs rectangular gated signal. Reproduced from ref. [41]. Permission has been granted by the copyright holder.

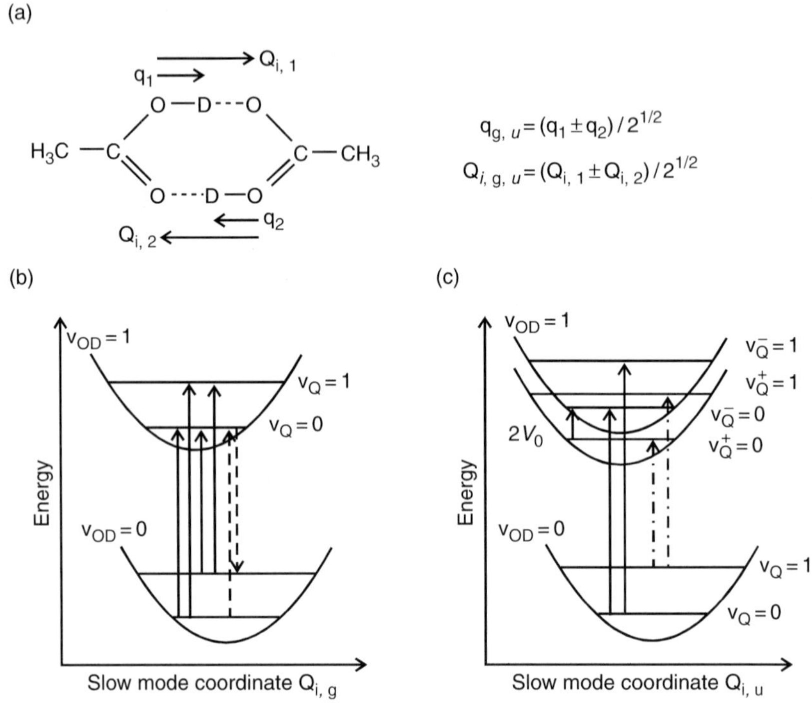

Fig. 9.10 (a) Scheme of the cyclic acetic acid dimer, including high ($q_{1,2}$) and low-frequency ($Q_{i,1,2}$) local-mode coordinates. (b, c) Schematic potential energy diagrams of low-frequency modes anharmonically coupled to the O–D stretching motion. The potential energy surfaces in the $v_{OD} = 0$ and $v_{OD} = 1$ state, and the two lowest vibrational levels of the low-frequency mode, are plotted *versus* the symmetrized low-frequency coordinates, (b) $Q_{i,g}$, and (c) $Q_{i,u}$. For $Q_{i,u}$ the resonant (Davydov) splitting, $2V_0$ in the $v_{OD} = 1$ state, results in two series of vibrational transition lines (solid and dash–dotted arrows) from the $v_Q = 0$ and $v_Q = 1$ levels. Dashed line in Fig. (b): resonantly enhanced Raman process. Reproduced from ref. [44]. Permission has been granted by the copyright holder.

$\nu_\sigma(\vec{X}{-}H\cdots\overleftarrow{Y})$, is responsible for the vibrational dynamics and band-broadening in H-bonded systems [48–57]. Therefore, the H-bond stretching mode $\nu_s(X{-}H)$ dynamics are controlled by the dynamics of the bridge motion $\nu_\sigma(\vec{X}{-}H\cdots\overleftarrow{Y})$. The idea of the anharmonic coupling is presented in Fig. 9.10. for the cyclic acetic acid dimer.

Figure 9.11 shows the time-dependent change of the O–D stretching absorbance, ΔA, as a function of the time delay between pump and probe pulses for different probe frequencies. The strong signal around zero delay is due to a coherent pump-probe coupling: however the oscillatory character at later delay-times exclusively reflects changes of vibrational dynamics.

The Fourier transforms of the oscillatory signals from Fig. 9.11 are presented in Fig. 9.12. The Fourier transform displays three frequency components – at 50, 145, and 170 cm^{-1} that can be compared with a steady-state low-frequency Raman spectrum of deuterated acetic acid. One can see that the vibrations predicted from

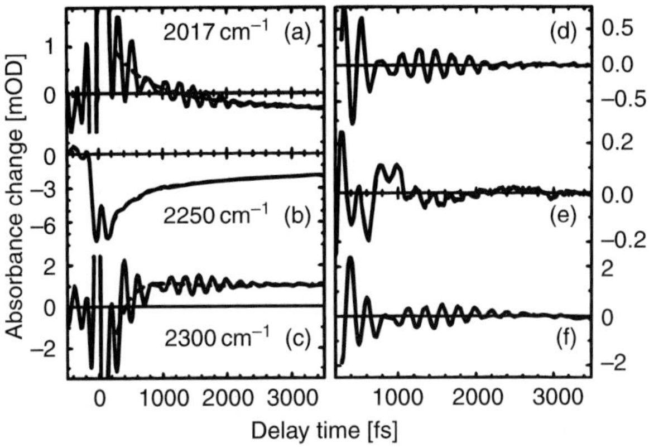

Fig. 9.11 (a)–(c) Time dependent change of the O–D stretching absorbance, ΔA, as a function of the time-delay between pump and probe for three different probe frequencies (solid lines). The strong signal around delay zero is due to a coherent pump-probe coupling and – in figs. a, c – exceeds the range plotted here. Dash-dotted lines: numerical fits of incoherent signals. (d)–(f) Oscillatory part of the pump-probe transients after subtraction of the incoherent signals. Reproduced from ref. [44]. Permission has been granted by the copyright holder.

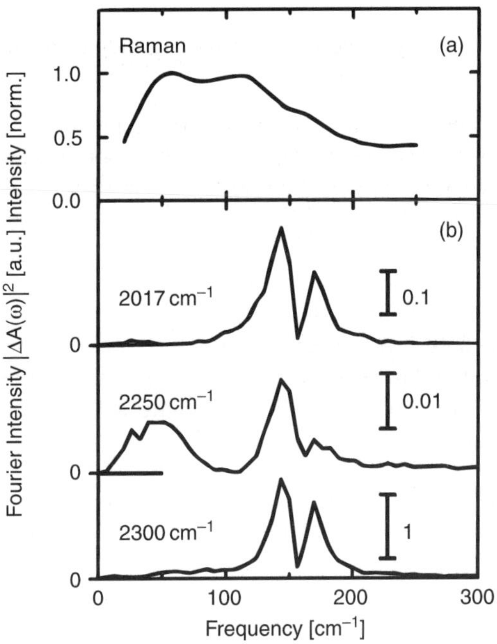

Fig. 9.12 (a) Steady-state low-frequency Raman spectrum of deuterated acetic acid [58]. (b) Fourier spectra of the oscillatory transients from Fig. 9.11 (d–f). The spectra, which are scaled relative to each other, display three components around 50-, at 145-, and at 170 cm^{-1}. Reproduced from ref. [44]. Permission has been granted by the copyright holder.

the oscillatory character of the time-dependent transient signals in Fig. 9.11 are similar to the vibrations from the Raman spectrum. Therefore, the IR pump-probe spectroscopy helped to identify the intermolecular vibrational motions that are coupled with the high-frequency O–D stretching mode providing an important basis for the theoretical understanding of both hydrogen-bonded structures and vibrational absorption bands.

9.3. PHOTOISOMERIZATION

In section 9.2.2. we illustrated the way in which light can convert into mechanical motion the photoinduced *trans–cis* isomerization about a double bond in bacteriorhodopsin. Photoisomerization processes have been studied by numerous groups, over many decades. With continual improvements in ultrashort-pulse lasers this subject is being continuously explored [59–63].

Shorter and shorter pulses have allowed access to new aspects of faster dynamics that could not be studied in the past. Early ultrafast measurements at the beginning of the 1980s monitored processes that could be observed with picosecond pulses. Molecular reorientations occur on the picosecond time scale, so most early studies involved the understanding of a solvent's role in condensed-phase chemistry. The reaction rates in the gas phase and condensed phase were compared.

9.3.1. Photoisomerization of *cis*- and *trans*-Stilbene

One of the most studied reactions is the photoisomerization of *cis*- and *trans*-1,2-diphenylethylene (*cis*- and *trans*-stilbene) (Fig. 9.13). This has been studying from the early days of ultrafast spectroscopy, largely because they are brightly colored, and absorb light in the visible region which is convenient experimentally and accessible.

The conventional picture of the photoisomerization of *cis*- and *trans*-stilbene has been the two-state, one-dimensional model, with torsional motion as the primary reaction coordinate (Fig. 9.13), with a rotation by 180° about the C=C bond. The models assume that there is a barrier which is the rate-limiting step in the *trans*- to *cis*-isomerization [64–66]. However, recent papers show that there is much reason to believe that a simple torsional reaction coordinate is not sufficient to describe the stilbene's dynamics [67].

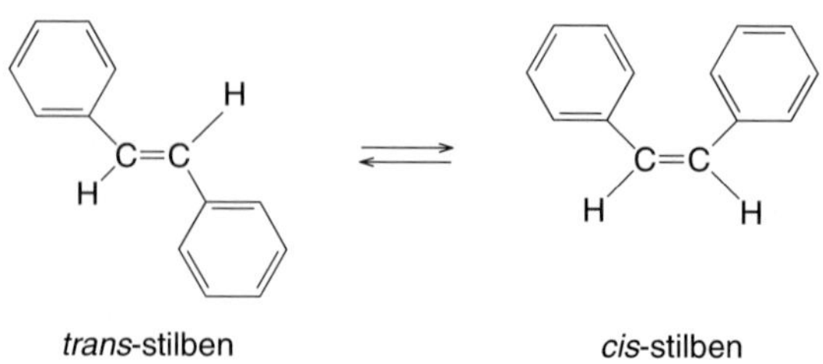

Fig. 9.13 Stilbene isomers.

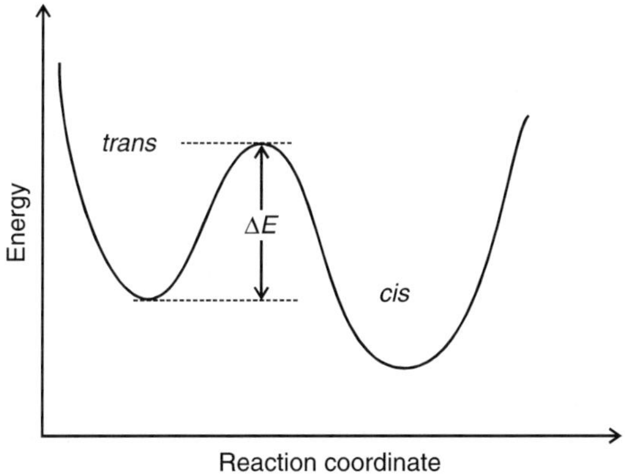

Fig. 9.14 Variation of potential energy of stilbene isomers with the reaction coordinate.

Some of the first time-resolved measurements of this reaction were reported by Hochstrasser's group [68], which determined the rate of isomerization in solution and in the gas phase. The methods of molecular-beam picosecond spectroscopy were used to study the stilbene's isomerization in the gas phase [69]. This method permits the excitation of gas molecules, with a laser beam, to specific vibrational states and to study the isomerization rate as a result of intramolecular vibrational energy relaxation. One of the most surprising results of these investigations was that the stilbene isomerization occurs more quickly in solution than in the gas phase. This result seems to run counter to accepted ways of thinking, which said that reactions involving large-amplitude motions, such as torsional rotation of bulky phenyl groups, should be slowed by solvent molecules.

Among various methods for monitoring isomerization dynamics is transient fluorescence decay from the excited state, S_1, of *trans* stilbene. In this method, a pumping pulse promotes a molecule from the ground electron state of the *trans* configuration to the excited state, S_1, followed by the fluorescence decay, $S_1 \rightarrow S_0$ (Fig. 9.15). Another method is the monitoring of the transient absorption by the pump-probe method (Fig. 9.16).

So far there is no definitive explanation of the fact that the rate of photoisomerisation of stilbene and its derivatives is larger in the condensed phase solution than in

Fig. 9.15 Electronic energy levels of stilbene isomers.

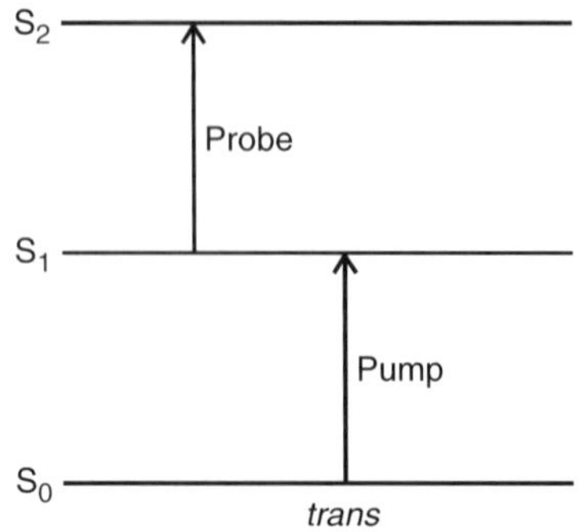

Fig. 9.16 Absorption of stilbene isomers measured by the pump-probe method.

the gas phase. Some researchers claim that this results from the fact that solvent reduces considerably the reaction barrier, ΔE (Fig. 9.14) which must be crossed in passing from the reactant to the product, which would accelerate the isomerization. Other researchers suggest that the potential-energy surface has a shape different from that presented in Fig. 9.14, and that photoisomerization is a result of energy conical intersection [67].

Femtosecond laser studies of the *cis*-stilbene photoisomerization reactions [60] have been performed to obtain the most detailed understanding of a polyatomic isomerization reaction in a condensed-phase environment. These experiments demonstrate that vibrationally hot product molecules are formed within a few hundred femtoseconds of the escape of the molecule from the *cis** region of the potential energy surface. The frictional effects on the *cis*- to *trans*-reaction coordinate were found to be important, and account for the anisotropy of the *trans* product molecules.

9.4. INTRAMOLECULAR CHARGE TRANSFER

The effect of solvent, and particularly its polarity, is essential to the reaction-rate in the excited electronic state. Upon light excitation a molecule is promoted to the excited state, which is accompanied by a considerable charge redistribution and dipole-moment change. The charge redistribution and significant modification of the dipole moment triggers isomerization in a molecule and, in turn, the solvent molecules' reorganization around the excited molecule. The archetypal example of such compounds is 4-N,N-dimethylaminobenzonitrile (DMABN) in which intramolecular charge-transfer effects were observed for the first time. The twisted intramolecular charge transfer (TICT) states occur as a result of rotation of the dimethylamino group around the N-phenyl bond in the excited state, that has a large dipole moment (Fig. 9.17). Electrostatic interactions between the molecular dipole moment and polar solvents stabilize the twisted form, and a new emission appears in the visible region, which is not observed in non-polar solvents. On excitation, the locally excited (LE) state of DMABN is produced which emits around 340 nm.

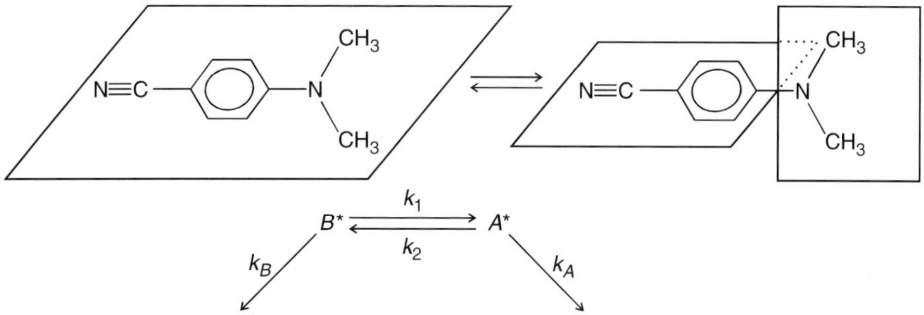

Fig. 9.17 Isomerization of 4-N,N-dimethylaminobenzonitrile (DMABN).

Twisting about the bond joining the dimethylamino group to the phenyl ring yields the TICT state which also emits, but at longer wavelength (400 nm or more, depending on the solvent). The double fluorescence observed in the UV and in the visible region was interpreted as corresponding to the planar structure and to the twisted form [70].

The isomerization reaction in DMABN has been studied by many groups using the methods of ultrafast laser spectroscopy [71–79]. It was shown that the viscosity plays a smaller role than solvent polarity in DMABN photoisomerization. The temperature effects were measured, keeping the same viscosity for different solvents of a specific series (e.g., alcohols). The results showed that the reaction rate increases with decreasing temperature. If the isomerization barrier is not temperature dependent, the reaction rate constant, k, should decrease, and not increase with temperature, owing to the Boltzmann factor, $\exp\left(-\frac{\Delta E}{kT}\right)$. The experimental results suggest that the barrier height depends on temperature, and decreases with temperature decrease, overcoming the usual Boltzmann temperature effect. This effect was explained by an increase of liquid polarity with decreasing temperatures.

9.5. MOLECULAR REORIENTATIONS

Laser spectroscopy with picosecond pulse durations is often used to study reorientations of molecules. Many laser methods are used for this purpose, including those based on the Kerr optical effect or induced dichroism [80–88].

If a cell containing a liquid is placed in an electrical field, the molecules tend to align with the field. This causes the liquid to become anisotropic and birefringent. The change in refractive index is given by

$$n_o - n_e = K \times E^2 \times \lambda \tag{9.38}$$

where λ is the wavelength of the light, E is the electric field strength, K is the Kerr constant, n_o and n_e are respectively the ordinary and extraordinary refractive indices (see Chapter 5). In a typical Kerr cell the liquid (typically nitrobenzene) is placed between two flat parallel plates, spaced several millimeters apart, to which a fairly high voltage (typically 10–20 kV) is applied. If a pair of polarizers is put around the cell, oriented at 90 degrees, the device acts as a shutter if no electric voltage is applied. However, when the voltage is provided to the cell, the extraordinary ray is produced, resulting in a phase difference for the ordinary and extraordinary beams,

$$\Phi = \frac{2\pi}{\lambda}(n_o - n_e)l, \tag{9.39}$$

that changes the polarization of the incident light making possible a light output. The phase difference depends on the magnitude of the applied electric field, E. If the field is such that the cell retards the extraordinary ray by a half-wavelength, the rotation of polarization will be 90 degrees; such a voltage is called the, "half-wavelength voltage". At the half-wave voltage, the following relationship applies

$$(n_o - n_e) \times l = \lambda/2. \tag{9.40}$$

The electric field provided by the voltage delivered to the plates can be replaced by an electric field of a laser pulse. So, if we apply an intense pumping pulse that induces birefringence in a liquid sample, the second pulse—polarized at 45° with respect to the pump pulse—probes the phase-difference when it passes through the liquid cell (eq. (9.39)). The liquid placed between the crossed polarizers transmits the probe light as long as the liquid maintains the birefringence induced by the first pumping pulse (Fig. 9.18). Thus, by measuring the intensity of the probe pulse which passed through the sample, as a function of the time-delay Δt between the probe and the pump pulse, we can measure the decay of the induced birefringence. In turn, the induced birefringence decay is a measure of molecular reorientations.

Generally, electronic polarization (and not only orientational polarization) also makes a contribution to the refractive index. However, the timescale of the electronic relaxation is much shorter than that of re-orientational times, and the birefringence due the induced electron polarization lasts only as long as the sample is irradiated with the pumping pulse.

Another method used to study reorientational processes is *induced dichroism*. The optical Kerr effect method can be useful for concentrated solutions, whereas the method of induced dichroism is used to study reorientational processes in dilute solutions. In concentrated solutions the reorientational dynamics are obscured by additional processes such as energy-transfer and energy-relaxation, T_1 processes. The induced

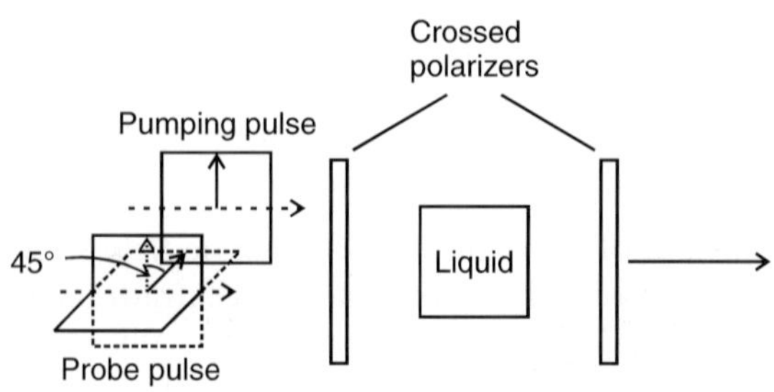

Fig. 9.18 Optical Kerr effect.

dichroism requires a laser-pulse frequency corresponding to the transition frequency, whereas in the Kerr method no specific requirements are imposed on the frequency.

9.6. INVESTIGATION OF INTERMEDIATES

Time-resolved spectroscopy is already well recognized as a powerful technique for studying the structure and dynamics of short-lived intermediate species. The photochemical and photophysical properties of short-lived intermediates are currently of interest owing to their ability to participate in many important photochemical reactions, such as hydrogen atom and electron and proton transfer. Many aspects of these reactions are important for understanding fundamental processes in chemical and biological systems [89].

9.6.1. Photoreduction

Photoreduction plays a key role in many chemical reactions and in biological systems, and is one of the reactions studied most frequently by time-resolved spectroscopic methods. The photoreduction involves the transfer of a hydrogen atom from the ground state of a molecule (BH) to a photo-excited molecule (A^*). The actual process occurs in two steps:

a) electron transfer,
b) proton transfer,

$$A^* + BH \xrightarrow{e^-} A^- \ldots (BH)^+ \xrightarrow{H^+} AH^\bullet + B^\bullet \tag{9.41}$$

where $A^- \ldots (BH)^+$ is an intermediate CT complex. As a result of photoreduction the radicals $AH^\bullet$ and $B^\bullet$ are generated. This mechanism is favored particularly in cases where one of the reactants has a low ionization potential and acts as an electron donor. Figure 9.19 shows an example of a photoreduction reaction between benzophenone (Bp) in the triplet-excited state, T_1, and N,N-dimethylaniline (DMA) in the singlet state, S_0.

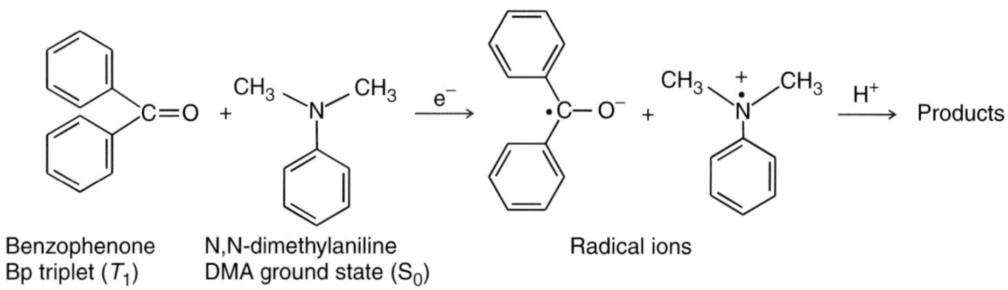

Fig. 9.19 Photoreduction reaction of benzophenone (Bp) [89].

In order to establish whether the reaction runs in two steps one can employ the transient absorption method:

1. the first laser-pulse excites the benzophenone molecule to the triplet state, $S_0 \rightarrow T_1$,
2. the second probe-pulse monitors a product absorption of the neutral radical, or a transient absorption of the charge-transfer complex, CT (scheme (9.41)). If the second step with the proton-transfer exists, the radical ion's spectrum $(Bp)^{\bullet -}$ should evolve to the neutral free-radical spectrum.

9.6.2. Carbenes

Another important group of compounds studied by time-resolved spectroscopic methods is the carbenes (Fig. 9.20) [89]. The large interest in this group of compounds results from their wide applications in microelectronics and semiconductor technology.

In Fig. 9.20, the group R denotes a chemical substituent, σ is a covalent bonding orbital, p denotes a p-orbital, and the arrows denote the spins of electrons occupying the orbitals, p or σ.

Picosecond and femtosecond lasers have allowed studies of many reactions involving carbenes, which have helped to establish the mechanisms of many reactions. One of the most interesting topics related to the photochemistry of carbenes is the dynamics of *spin relaxation*, observed, for example, in the reaction shown in Fig. 9.21, which leads to the generation of the triplet state in a carbene. The carbene precursors are diazo compounds, as shown in Fig. 9.21. Their excitation by a laser-light from the UV range stimulates a reaction in which nitrogen, N_2, and the carbene in a singlet state S_0 at energy E_S, are produced. Since for many carbenes the ground state is the triplet state T_0, the singlet state of the carbene dissipates its energy via intersystem crossing and evolves to the ground, triplet state, T_0, having energy E_T.

Picosecond spectroscopy methods have provided interesting results for the relationship between the intersystem-crossing rate, k_{ST}, and the energy gap, $\Delta E = E_S - E_T$. The gap's magnitude, ΔE, can be controlled by altering the solvent polarity, since the singlet state is better stabilized than the triplet state in polar solvents. The results on the spin relaxation lead to the unexpected conclusion that the rate constant, k_{ST}, decreases, and does not increase, with decreasing ΔE. Some authors have rationalized this effect by arguing that the smaller number of vibrational states of the triplet state, T_0, involved in the coupling to the excited singlet state, S_0, that leads to less efficient intersystem-crossing.

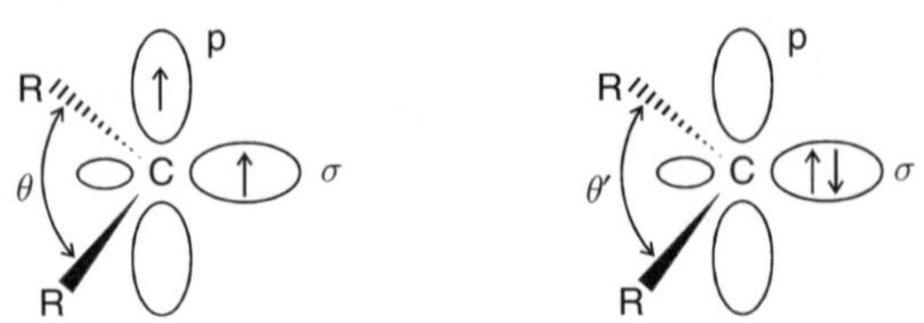

Fig. 9.20 Triplet (σ^1, p^1) and singlet (σ^2, p^0) electronic spin-states of carbenes.

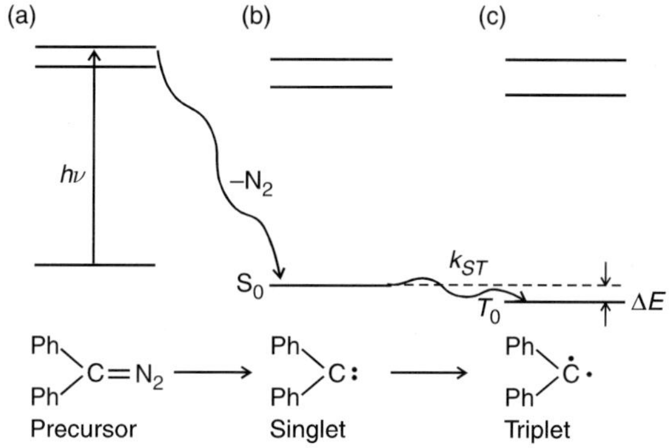

Fig. 9.21 Dynamics of spin-relaxation in carbenes [89].

9.6.3. Excited-State Proton Transfer

The proton-transfer reaction is one of the most elementary reactions. The reaction determines acid – base properties. Acids are substances that can donate H^+ ions to bases. An H^+ ion is just a proton since a hydrogen atom is a proton and one electron. So, an acid is a "proton donor", and a base is a "proton acceptor". The reaction between an acid and base is essentially *a proton transfer*. The acidity value of K_A, a measure of acidity, can change by 5–10 orders of magnitude when a molecule is excited from a ground state to an excited state, which indicates that it depends strongly on charge distribution. Because the process of proton transfer in the excited state (Fig. 9.22) is very fast, pico- and femtosecond laser spectroscopy provides a valuable method to study these processes [90]. The two wells in the ground-state potential-energy curve represent the primary (AH) and anionic (A^-) forms, and the two wells in the excited state curve represent the corresponding excited states $(AH)^*$ and $(A^-)^*$, respectively. It is clear from Fig. 9.22 that the AH form is more stable in the ground state. The first pump pulse promotes an AH molecule to the excited state, leading to the reorganization of bonding electrons which results in a proton transfer from $(AH)^*$ to B. Both excited forms, $(AH)^*$ and $(A^-)^*$, may exhibit emission—for example, the dual emission in the fluorescence spectrum of salicylic acid [91].

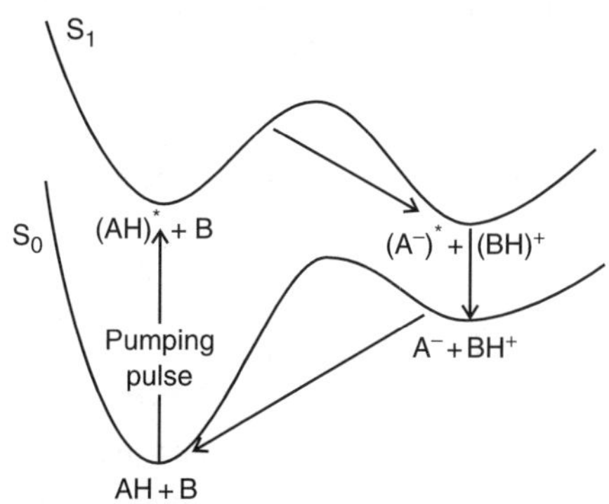

Fig. 9.22 Excited state proton transfer.

There are many time-resolved methods that can be employed to study excited state proton-transfer processes [92–99]:

a) fluorescence decay,
b) two-step transient absorption, $S_0 \rightarrow S_1 \rightarrow S_2$,
c) transient absorption from the ground state, S_0, by the pump-probe method.

Typical systems where the excited state proton transfer may occur are presented in Fig. 9.23.

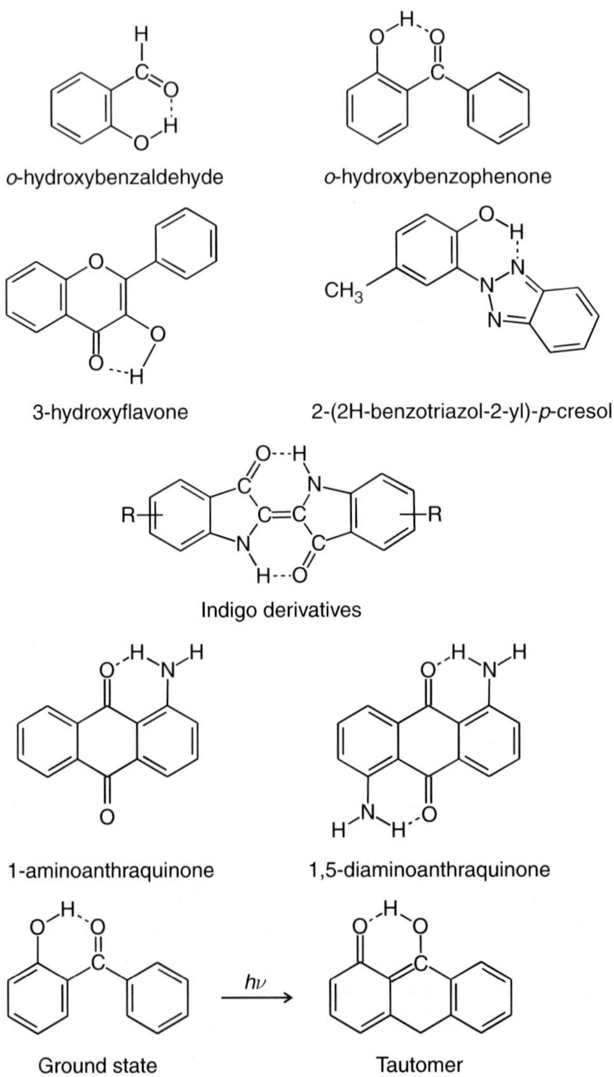

Fig. 9.23 Excited state proton transfer molecules.

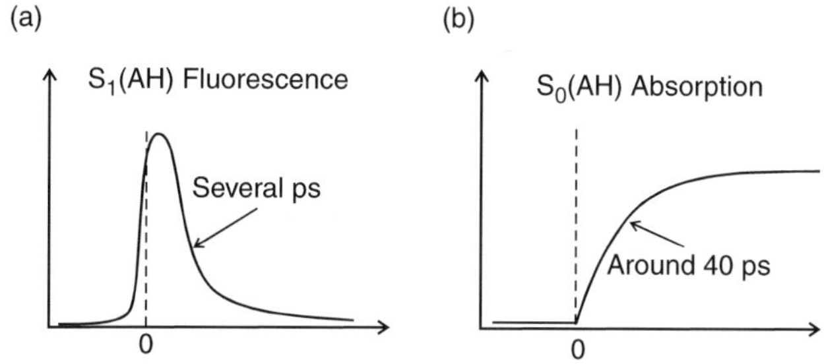

Fig. 9.24 Illustration of a signal typical of fluorescence decay and transient-absorption by the pump-probe method, in a proton transfer reaction in the excited state.

How can we obtain information about a proton-transfer from the pico- or femtosecond measurements? First, assume that we use two methods independently: fluorescence decay, and transient absorption by the pump-probe method (Fig. 9.24). Let us now suppose that from the fluorescence measurements we have received the information that the singlet-state population of the excited molecule $(AH)^*$ disappears within several picoseconds, while the time of return to the ground state, S_0, obtained from the transient absorption measurements is much longer, e.g., 40 ps. This finding indicates that an intermediate form (probably a protonated tautomer or a triplet-state) must exist. To eliminate the triplet-state one can use a triplet-state *quencher* (e.g., 1-methylnaphthalene).

9.7. ULTRAFAST COHERENT SPECTROSCOPY. VIBRATIONAL DYNAMICS

So far we have been interested in chemical reactions studied by the time-resolved spectroscopic methods, without a deeper insight into the mechanisms of energy dissipation. These mechanisms and reactions rates are closely related to the way the excess energy is transferred between the various degrees of freedom. The most important questions about chemical reaction dynamics are the following. How does the energy delivered to reactant molecules redistribute among the different degrees of freedom? How is the speed of energy redistribution related to the chemical bond breaking or forming of a new bond? The vibrational energy-transfer from a solute to the surrounding solvent molecules is a process of fundamental importance, and is one of the most important factors determining the rate of many reactions. These processes are called *vibrational relaxation.* Many references relate to this topic [100–131]. The vibrational-energy relaxation is characterized by a time, T_1, which defines the lifetime in the excited vibrational level. Vibrational relaxation comprises two distinct processes – *energy relaxation processes,* T_1 and the *phase-relaxation,* T_2. We have already mentioned these processes in section 8.3, where we discussed the CARS method, and in section 8.4 where the photon-echo method is discussed. In most cases, $T_1 > T_2$, which indicates that in a coherently excited medium the excess vibrational energy is maintained in a given vibrational mode for a longer time than the vibrational phase. As a result, the shape and the width of a vibrational line, obtained by linear or non-linear spectroscopy, are usually dominated by the phase relaxation T_2.

Originally, the dephasing times, T_2, were measured from isotropic spontaneous Raman line-shapes in the frequency domain [109–113]. The isotropic spontaneous Raman line-width is inversely related to the vibrational dephasing time. The relaxation processes are very fast, and occur on the picosecond and femtosecond timescale. Recent developments of picosecond and femtosecond lasers have allowed studies of vibrational dynamics directly in the time domain. The first time-resolved method employed to study vibrational coherence was CARS (coherent anti-Stokes Raman Spectroscopy) [100]. By analogy to NMR, the CARS method is sometimes called Raman free-induction decays (FID). The CARS technique is a 3rd-order non-linear spectroscopy, and was discussed in chapters 5 and 8. The CARS signal, measured as a function of the time delay between the pump and the probe pulses is exactly the Fourier transform of the Raman line-shape. If the FID signal is exponential, the corresponding Raman line-width is $\Delta(hwhh) = (2\pi c T_2)^{-1}$ where $\Delta(hwhh)$ is the half-width (i.e., width at half height) of the band. Therefore, both methods provide the same type of information about vibrational dynamics. However, both have an inherent limitation—that a difficulty in interpreting dephasing times arises when both homogeneous and inhomogeneous processes give rise to the same dephasing time, which is the most typical situation in molecular condensed-matter systems. The typical solute–solvent interactions consist of two distinct timescales: rapid force changes for short-range repulsive interactions resulting from collisions, and slow fluctuations for long-range attractive interactions. The frequency of an individual molecular oscillator is a function of time, owing to the solvent motions. For rapid fluctuations, the vibrational frequency also changes rapidly. If these fluctuations are faster than the time-resolution of the method (laser pulse-duration) the vibrational shifts are averaged over the statistical ensemble of molecules, and the system exhibits a single resonance vibrational frequency. In the frequency domain, the case with rapid fluctuations is called *homogeneous broadening*. In this case, the dephasing time, T_2, depends on both the distribution of frequency shifts and the rate of frequency fluctuations. In the other limiting case, if the solvent forces change slowly, each individual molecular oscillator retains its frequency for a long time. In this case, the dephasing time depends only on the width of the frequency distribution. Such a system shows a distribution of resonance frequencies, not a single resonance. In the frequency domain, the case with slow fluctuations is called *inhomogeneous broadening*.

In the case when both homogeneous and inhomogeneous processes give rise to vibrational dynamics and band-broadening, it is valuable to use higher-order spectroscopies, such as the 7th order time-domain Raman spectroscopy, called Raman echo, [114–120] or IR echo [121]. Vibrational echo experiments are the vibrational analog of the spin-echo in the pulsed NMR method, and were described in Chapter 8. Originally, vibrational-echoes experiments could be made only with free-electron lasers—which require large synchrotron facilities, but advances in tunable femtosecond IR-laser technology (OPA described in chapter 6) have now allowed the vibrational echoes to be performed on a table-top apparatus. Several groups have included 5th order non-linearities to study intermolecular, rather than intramolecular vibrations [122–129] and overtone dephasing [130]. Very recently, two-dimensional vibrational echo methods have been developed [131–132]. A combination of IR pump-probe methods and vibrational echo measurements can be used to completely determine vibrational dynamics, including T_1- and T_2-processes.

One may ask why vibrational dephasing is so important, and why so much effort has been made to develop new experimental methods to study vibrational dephasing? It is important to realize that vibrational dephasing is a window through which solvent–solute interactions can be monitored. The solvent–solute interactions which cause vibrational dephasing are the same forces which cause friction along the chemical reaction coordinate. Thus, a better understanding of the vibrational dynamics will result in better understanding of solvent effects on reaction dynamics.

9.7.1. Energy Relaxation T_1, and Phase Relaxation T_2

Although in the introduction the classical description was used to give a qualitative understanding of vibrational dynamics, a quantum description is needed for a quantitative description of real systems [100, 133].

The processes of phase-relaxation, T_2, occur in all spectral ranges and they can be monitored by the time-resolved coherent spectroscopic methods. The term, "coherent spectroscopy" denotes that interaction between the radiation of the laser beam and molecules in the sample generates non-linear polarization $\boldsymbol{P}$ in the medium, coherent in phase with the incident laser beam. The non-linear polarization is proportional to the macroscopic dipole-moment $N\langle\hat{\mu}\rangle$

$$\boldsymbol{P} = N\langle\hat{\mu}\rangle = N \cdot Tr\{\hat{\mu}\rho\} = N\mu(\rho_{ab} + \rho_{ba}) \tag{9.42}$$

where N is the number of molecules per volume unit, $\rho_{ab} = \rho^*_{ba}$ are non-diagonal elements of the density matrix ρ of the system, and $\langle\mu\rangle$ is a molecular dipole moment.

The density matrix, ρ, evolves in time according to the Liouville equation

$$\frac{\partial\rho}{\partial t} = \frac{\mathrm{i}}{\hbar}[\rho, H], \tag{9.43}$$

where the Hamiltonian, H,

$$H = H_0 + H_1 + H_2 \tag{9.44}$$

consists of the Hamiltonian, H_0, for a molecular system unperturbed by interaction with its surroundings, the electromagnetic-interaction Hamiltonian, H_1, describing the interaction between molecules, and H_2 which describes the interaction between the matter and electromagnetic field —$E(t) = E_0 \cos(\Omega t + kr)$ of the laser beam,

$$H_2 = -\hat{\mu}E(t) \tag{9.45}$$

where $\hat{\mu}$ is the dipole moment operator. In electron spectroscopy the dipole moment, μ_{nm}, of the electronic transition $n \rightarrow m$ is expressed by the dipole moment operator as, $\mu_{nm} = \int \Psi_n^* \hat{\mu} \Psi_m d\tau$. In IR spectroscopy the main contribution gives the transient dipole moment modulated by vibrations, $\hat{\mu} = \frac{\partial \mu}{\partial q} q$, where q is the normal coordinate of the vibrational mode. Raman scattering is determined by the induced transient dipole moment, μ_{ind}, modulated by vibrations, $\hat{\mu}_{ind} = \frac{\partial \alpha}{\partial q} qE$, where α is the polarizability coefficient.

If we consider a two-level system, E_a and E_b, the solution of the equation (9.43) is the following set of equations

$$\dot{\rho}_{aa} = \frac{i\mu E_0}{2\hbar}(\tilde{\rho}_{ba} e^{-ikr} - \tilde{\rho}_{ab} e^{ikr}) - \frac{\rho_{aa}}{T_1}, \tag{9.46}$$

$$\dot{\tilde{\rho}}_{ba} = \frac{i\mu E_0}{2\hbar}(\rho_{aa} - \rho_{bb}) e^{-ikr} + (i\Delta - \frac{1}{T_2})\tilde{\rho}_{ba}, \tag{9.47}$$

$$\rho_{bb} = 1 - \rho_{aa}, \tag{9.48}$$

where ω_0 is the resonance frequency expressed as $\omega_0 = (E_b - E_a)/\hbar$, $\Delta = \omega_0 - \Omega$ is a measure of detuning from the resonance frequency, and $\tilde{\rho}_{ba} = \rho_{ba} \exp(i\Omega t)$.

The physical sense of the relaxation times T_1 and T_2 can be explained by using the equations (9.46–9.47) when one assumes that the electric field that causes excitation to the upper level, b, is switched off, ($E_0 = 0$). Then we have

$$\rho_{aa}(t) = \rho_{aa}(0) \cdot \exp(-t/T_1). \tag{9.49}$$

and,

$$\boldsymbol{P}(t) = N\mu(\rho_{ba} + \rho_{ab}) = \boldsymbol{P}(0)\exp(-t/T_2). \tag{9.50}$$

Therefore, the equation (9.49) shows that the relaxation time, T_1, describes the return of the system perturbed by the electromagnetic field to the equilibrium characterized by the Boltzmann distribution

$$\frac{n_b}{n_a} = e^{-(E_b - E_a)/kT}, \tag{9.51}$$

where n_b and n_a are the numbers of molecules occupying E_b and E_a, respectively. The electromagnetic field disturbs the Boltzmann population distribution and the system needs some time to return to equilibrium. The time T_1 is a measure of the time needed to achieve equilibrium: it is a measure of the lifetime in the excited level, and depends on energy relaxation processes that lead to a decrease in the number of molecules occupying the excited energy state.

The equation (9.50) describes the time-evolution of the non-diagonal elements that contribute to the sample polarization $\boldsymbol{P}$. When a strong external electric field of the laser beam is applied, the transient dipole moments of individual oscillators have the same phase as the electric field. The contributions to the total polarization in the average add constructively, and the coherence is large. With time, when the external laser-field is switched off, the molecules no longer oscillate in the same phase, because each of them interacts in a distinct way with the surrounding molecules. As long as a strong external electric field forced them to oscillate coherently in the same phase, the effect of the intermolecular interactions was negligible, but after switching off the external field it becomes important again. The molecules get out of phase with each other, the individual dipole moments add destructively, there is a cancellation between the different contributions to the polarization, and the coherence decays. The coherence decay is characterized by T_2 in eq. 9.50, which is called the *time of phase-relaxation* or *dephasing time*. This is the time needed to randomize the phase of vibrations. It is important to understand what physical processes cause the decay of coherence which, in turn, results in decrease of the polarization vector $\boldsymbol{P}$.

In general, three mechanisms cause the phase coherence decay:

a) reorientational processes, T_R,
b) pure dephasing, T_2^*, resulting from interaction with surrounding molecules,
c) energy relaxation processes, T_1.

The effective dephasing time, T_2, is related to the processes (a)–(c) in the following way:

$$\frac{1}{T_2} = \frac{1}{T_R} + \frac{1}{T_2^*} + \frac{1}{2T_1}. \tag{9.52}$$

The reorientational effects can be removed by using the pump and the probe beams of suitable polarizations ("magic angle"). Moreover, the energy relaxation processes are usually slower than the dephasing processes ($T_1 > T_2$). Therefore the relationship, $1/T_2 \cong 1/T_2^*$ is often a good approximation.

The time-resolved spectroscopic methods allow us to determine the energy-relaxation time, T_1, and the dephasing time, T_2. The energy-relaxation time, T_1, is usually obtained using the IR pump-probe transient absorption. To obtain the dephasing time, T_2, one can use various IR or Raman techniques including coherent Raman scattering and vibrational echo spectroscopy. Femtosecond three-pulse four-wave mixing (FWM) techniques can be used to sort and measure the processes that contribute to coherence relaxation. The relaxation times for vibronic coherence can be monitored using photon echo (PE) and reverse-transient grating (RTG) measurements at different temperatures to isolate inhomogeneous and homogeneous components. Different pulse sequences can be used to select ground or excited state vibrational coherences [104]. Femtosecond Raman experiments using variable polarization configurations can be used to measure directly the isotropic and anisotropic polarizability dynamics of intramolecular vibrational modes [105]. Two-dimensional infrared vibrational echo spectroscopy [107] is able to make important contributions to the determination both of vibrational dynamics, and structures evolving rapidly in time.

As we mentioned at the beginning of chapter 9.7, the main difference between the various methods originates from the number of pulses employed for excitation:

1. one excitation pulse (CARS),
2. two excitation pulses (photon echo),
3. three excitation pulses.

The method (1) based on one excitation pulse permits the determination of the dephasing time, T_2, but it contains unseparated dephasing originating from fast (homogeneous) and slow (inhomogeneous) processes. Method (2), with two excitation pulses, permits determination of the dephasing time, T_2, for the homogeneous dephasing processes (which we denote as T_2^{h}), eliminating the part of dephasing related to inhomogeneous processes. Method (3) permits determination of both the T_2^{h} and T_1. A schematic picture of the polarization decay $\boldsymbol{P}(t)$ in the experiments (1)–(3) is presented in Fig. 9.25.

In method (1) (Fig. 9.25 a) the system is excited with one pulse, denoted in Fig. 9.25 with a symbol, *a*, followed by monitoring the polarization in the direction $\boldsymbol{k}_a$ determined by the direction of the exciting beam (or two simultaneous beams in the CARS method described in Chapter 8). The polarization signal decays in time, and can be probed by the second laser pulse, which is delayed with respect to the pumping pulse. If the system is homogeneous, the polarization decay is exponential $\boldsymbol{P}(t) = \boldsymbol{P}(0)\exp(-t/T_2)$ (dashed curve in Fig. 9.25 a). If the system undergoes inhomogeneous dephasing with a certain static resonance-frequencies distribution, $f(\omega_0)$, the polarization decay is initially exponential; at longer times, interferences between different resonances induce a faster decay (dotted curve in Fig. 9.25 a). The most typical situation is that both mechanisms, homogeneous and inhomogeneous dephasing, participate in spectral-line broadening in the frequency domain, and it is not possible to separate them by one-pulse excitation methods, in contrast to the two-pulses excitation (photon echo) or three-pulses excitation methods.

As we have said, the time-resolved CARS method is one of the most frequently used one-pulse excitation methods for the dephasing time, T_2, measurements. The

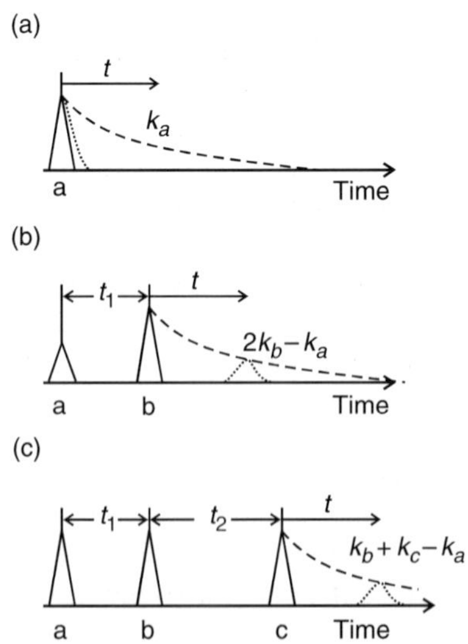

Fig. 9.25 The polarization decay in the following techniques: (a) One-pulse excitation, (b) Two-pulses excitation, (c) Three-pulses excitation; --- homogeneous dephasing, ... inhomogeneous dephasing [100].

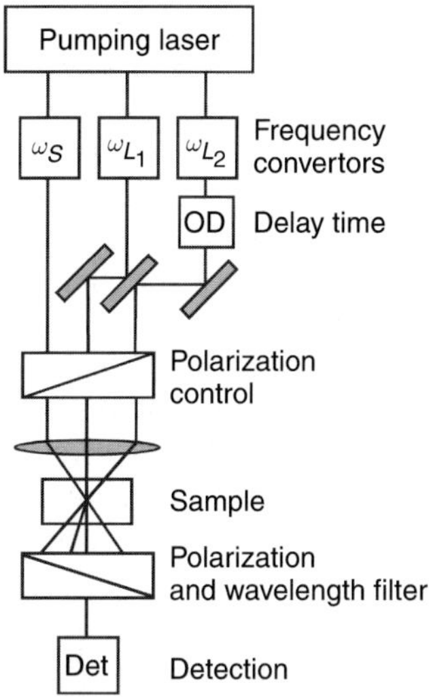

Fig. 9.26 Scheme of CARS experimental setup.

principles of CARS were discussed in section 5.5. A typical time-resolved CARS experimental setup is presented schematically in Fig. 9.26.

Two laser beams of visible light, at frequencies ω_{L_1}, ω_S, chosen to match a specific vibrational-mode frequency, $\omega_{vib}(\omega_{L_1} - \omega_S = \omega_{vib})$, illuminate the sample simultaneously, inducing a coherent phase for all molecular oscillators in the sample. Then the probing beam at frequency ω_{L_2}, delayed with respect to the pump beams ω_{L_1}, ω_S, is used to monitor the anti-Stokes Raman scattering. The intensity of the anti-Stokes Raman scattering is directly proportional to the square of the vibrational coherence, which decays with time. The dephasing time is measured by monitoring the intensity of the anti-Stokes Raman signal as a function of the delay-time between the excitation (pump) and the scattering (probe) pulses. To recall, the coherent anti-Stokes scattering is emitted as a highly directional beam in a direction determined by the appropriate phase-matching condition described in Chapter 5.

At the present level of development of laser technology, adjusting the frequencies of the pumping beams to achieve resonance with the frequency of a specific vibrational mode does not make any experimental problems. There is a wide selection of methods that can be used:

1. tunable dye lasers or solid-state tunable lasers,
2. nonlinear effects in crystals such as second-harmonic, third-harmonic, or white continuum generation,
3. nonlinear effects in optical parametric amplifiers.

A typical result obtained from the CARS experiment is presented in Fig. 8.12. One can see that for short times the CARS signal differs considerably from the exponential decay (dashed line in Fig. 8.12). This is a consequence of the fact that the signal is strongly disturbed by non-resonant phenomena of the third order ($\boldsymbol{P} = \chi^{(3)}\varepsilon_0\mathbf{EEE}$) that do not obey the condition $\omega_{L_1} - \omega_S = \omega_{vib}$. For longer times, the CARS signal decays exponentially ($\exp(-t/T_2)$). Therefore, from the dependence of the CARS signal as a function of the time delay between the pumping and the probing one can determine the vibrational dephasing time, T_2. Typical vibrational dephasing times in liquids are of the order of 0.1–10 ps. However, the CARS signal frequently shows significant deviation from simple exponential relaxation dynamics. Figure 9.27 illustrates such a case for binary liquid mixtures.

Time-resolved CARS spectroscopy can also deliver valuable information about T_2 and T_1 in glasses and in organic and inorganic molecular crystals at low temperatures. The typical dephasing times, T_2, in solids are much longer than in liquids, [110, 114] and spontaneous Raman scattering methods in the frequency domain with a typical spectral resolution of 0.1 cm^{-1}, may not be able to determine vibrational

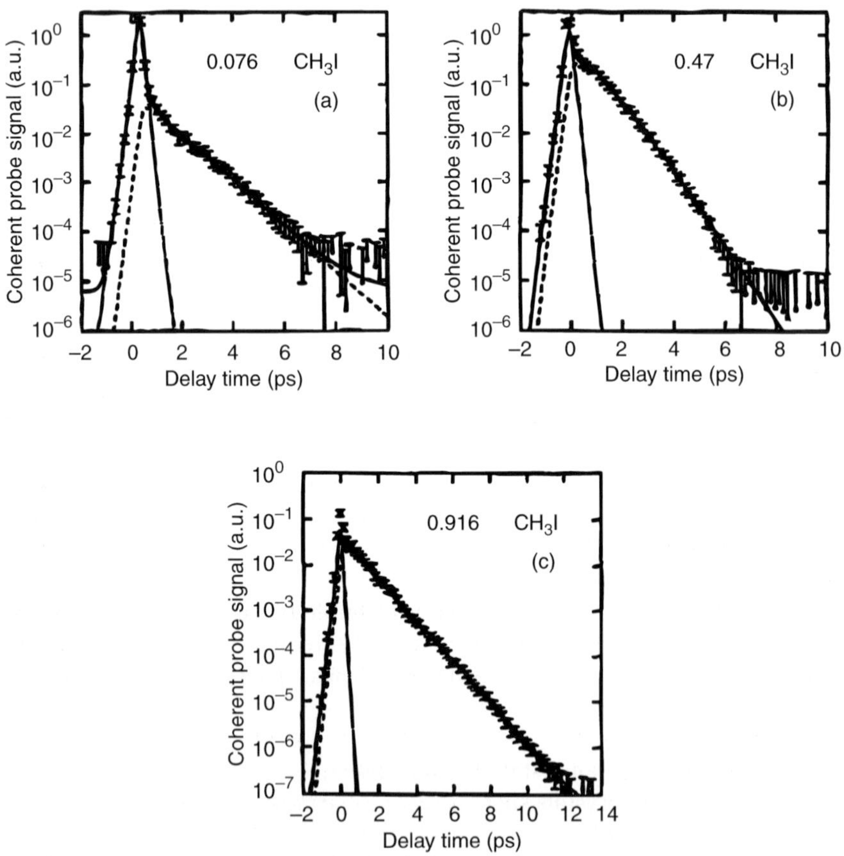

Fig. 9.27 Coherent probe-scattering signal of the symmetric stretching vibration, υ_1, of CH_3I in admixture with $CDCl_3$, *vs* delay time for the magic angle $\theta = 49.1°$ and three mole fractions of CH_3I: $x = 0.076$ (a), 0.47 (b), and 0.916 (c); experimental points (298 K); calculated curves of isotropic scattering component (dashed), non-resonant part (dotted), and total CARS signal (solid line) [134].

dynamics at low temperatures. The homogeneous linewidth is dominated by the vibrational lifetime at low temperatures and by pure dephasing in the liquid. The contribution from orientational relaxation is significant at some temperatures, but never dominant. Restricted orientational relaxation at temperatures below ~120 K causes the homogeneous line-shape to deviate from Lorentzian, while at higher temperatures the line shape is Lorentzian [103, 106]. Sometimes the CARS signals and vibrational echoes are modulated by quantum beats, [100, 107] as discussed in Chapter 8.

9.8. DYNAMICS OF AN EXCESS ELECTRON. SOLVATED ELECTRON

The ionization of dielectric molecular liquids yields positive and negative charges. Upon ionization, the negative charges – electrons are completely delocalized, but owing to interactions with the solvent they become trapped. The localization of an excess electron in a solvent cavity gives rise to bound eigenstates which are modulated by the coupling to solvent fluctuations. Therefore, an excess electron is an unique probe of solvation dynamics. The apparent simplicity of an excess electron makes it an ideal solute for studying solvation in polar and non-polar solvents. The electron solvated in water was discovered over 40 years ago [135–152]. The dynamics of the location, trapping, and solvation of an excess electron at room temperatures is very rapid (fractions of picoseconds, and femtoseconds). Direct monitoring of these processes became possible in the 1990s with the development of the femtosecond laser technology. The nature of the stabilized electron depends on the solvent. In some liquids, such as water, ammonia, or saturated alkenes, the excess electron exists as a "solvated electron" trapped in a cavity formed by several solvent molecules. In other liquids, the excess electron "sticks" to the solvent molecule, forming a radical anion.

Under high energy photons of a laser beam, or $\gamma-$ radiation, an electron can be ejected from the valence shell of the molecule and promoted to the conduction band. To make this possible, the energy of photons has to be at least equal to the ionization energy (e.g., ~10 eV in water). However, generation of an excess electron can be observed at considerably lower energies of incident photons, owing to multiphoton processes. The excess electron that has just been promoted in the conduction band begins to localize in a cavity, owing to interactions with solvent molecules. It happens that the electron becomes trapped by a single molecule, forming a so-called radical anion. After a sufficient number of nanoseconds the electron recombines with the geminate radical cations which were produced during the ionization. The processes of localization, trapping, and solvation, depend on mechanisms of electron relaxation, which are very fast, as well as reorientations of solvent molecules around the electron, which are much slower and occur on the picosecond time-scale. Table 9.1 illustrates the most typical events that may happen in water upon photon-excitation and the fate of the excess electrons produced in the high-energy ionization channel, local ionization channel, and low energy dissociative channel [153].

Table 9.1

Ionization and dissociation channels for water

High energy ionization channel	$H_2O + h\nu \rightarrow H_2O^+ + e^-$	local ionization
	$H_2O + h\nu \rightarrow H_2O^+ + e_{qf}$	
	$H_2O^+ + 2H_2O \rightarrow H_5O_2^+ + OH^\bullet$	
	$H_2O^+ + H_2O \rightarrow H_3O^+ + OH^\bullet$	
	$H_5O_2^+ + e_{qf}^- \rightarrow H_5O_2^\bullet$	
Local ionization channel	$H_5O_2^\bullet \rightarrow 2H_2O + H^\bullet$	local capture unimolecular decay
	$H_5O_2^\bullet + OH^\bullet \rightarrow 3H_2O$	geminate recombination
	$H_3O^+ + e_{qf}^- \rightarrow H_3O^\bullet$	local capture ultrafast unimolecular
	$H_3O^\bullet \rightarrow H_2O + H^\bullet$	decay geminate recombination
	$H_3O^\bullet + OH^\bullet \rightarrow 2H_2O$	
Low energy dissociation channel	$3H_2O + h\nu \rightarrow H_5O_2^\bullet + OH^\bullet$	dissociative photolysis
	$H_5O_2^\bullet \rightarrow 2H_2O + H^\bullet$	unimolecular decay
	$H_5O_2^\bullet + OH^\bullet \rightarrow 3H_2O$	geminate recombination
	$H_2O + h\nu \rightarrow H_3O^\bullet + OH^\bullet$	dissociative photolysis
	$H_3O^\bullet \rightarrow H_2O + H^\bullet$	ultrafast unimolecular decay
	$H_3O^\bullet + OH^\bullet \rightarrow 2H_2O$	geminate recombination

Figure 9.28 illustrates the characteristic energy levels of water, and an excess electron solvated in water (called a hydrated electron). The hydrated electron's energy in the ground state, *1s*, is 3.2 eV. This value obtained from the ionization-potential measurements by photophysical methods differs considerably from the energy of 1.7 eV estimated from the thermodynamics of the reaction:

$$H^\bullet + OH^- \leftrightarrow (e^-)_{aq} + H_2O, \tag{9.53}$$

where $(e^-)_{aq}$ denotes the hydrated electron. The thermodynamic value does not seem to be reasonable because it would indicate that half of the absorption band of the hydrated electron 1s $\rightarrow$ 2p (1.7 eV) would lie in the vacuum region (Fig. 9.28). This discrepancy suggests that the model of a hydrated electron as a negative charge surrounded by a solvation shell of water molecules adopted to the thermodynamic considerations may not be valid. Thus, questions arise about the structure of the solvated electron.

Figure 9.29 presents the most popular models of the hydrated electron. Figure 9.29a symbolizes a group of continuous models, in which the response of solvent molecules to the excess electron localization is described by a dielectric continuum polarization. The polarization of the medium in the immediate vicinity of the electron leads to a low-energy well where the electron is trapped. The models in which the solvation shell of the dielectric continuum was replaced with the discrete solvation shell of several solvent dipole oriented molecules is given in Fig. 9.29b. These models, also called models of the trapped electron, are very popular, although they suffer from many deficiencies, including the above-mentioned inconsistency between the electron stabilization energy obtained from the photophysical methods and the energy calculated from the thermodynamic reaction (9.53). In another group of models (Fig. 9.29c) an excess electron is treated as a conventional anion (center F)

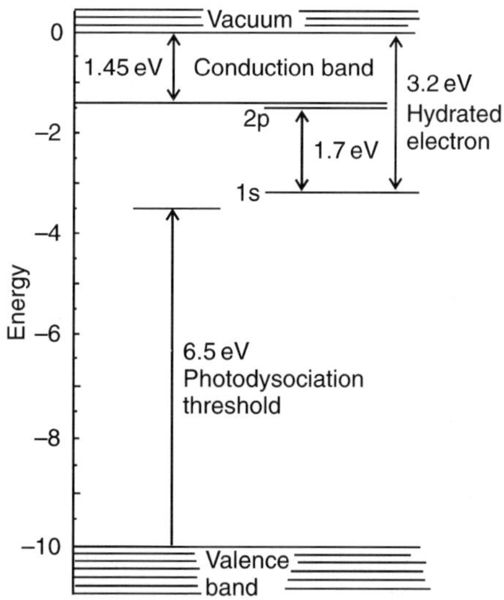

Fig. 9.28 Energy levels of water and hydrated electron.

when the anionic gap in the imperfect crystal lattice becomes replaced by an excess electron. This type of solvation is expected to exist in ion crystals, e.g., in frozen solutions of alkali metal halides. In the first three models presented in Fig. 9.29 a–c it is assumed that an excess electron does not react with a solvent molecule. The excess electron is trapped by surrounding molecules as a result of long and short range interactions with a solvent, and it does not occupy the molecular orbitals of the solvent molecules. The other extreme case is when a radical anion is formed with an excess electron occupying a molecular orbital of the solvent molecule. Molecular beam spectroscopy suggests that the smallest cluster able to bind an excess electron is a dimer, $(H_2O)_2^{\bullet -}$. The binding energy is estimated to be very small –17 meV. The

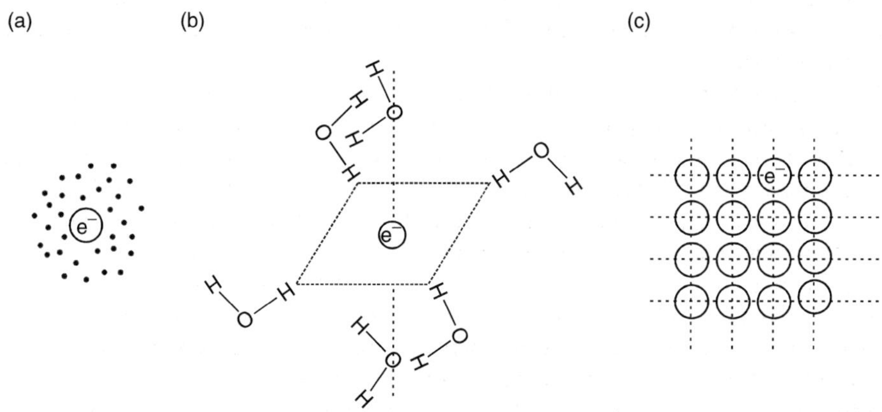

Fig. 9.29 Models of solvated electron.

creation of a radical anion $H_2O^{\bullet-}$ on a single molecule of water is impossible because the interaction of an electron with a water molecule leads to reaction

$$e^{-}_{aq} + H_2O \rightarrow H^{\bullet} + OH^{-}. \tag{9.54}$$

9.9. EXCESS ELECTRON SPECTROSCOPY

To understand the structure of the solvated electron-trap at the molecular level one should find a physical probe to investigate the processes of localization, trapping, and stabilization. The best probe for an excess electron is its electronic optical spectrum. The optical spectrum of an excess electron consists of two characteristic bands—a band in the near infrared, and a band in the visible range. In the pump-probe experiments the first pump pulse generates an excess electron, and the second, probing pulse, monitors the transient absorption in the near infrared (~1000 nm) and in the visible range (720 nm for a hydrated electron). [149] The first results obtained with this method show that the band in the near-infrared region appears just after the pumping pulse—faster than 110 fs, the experimental time resolution in the early pump-probe experiments, followed by a decrease in absorption in the near IR, and an increase of the band in the visible range. After a fraction of a picosecond the near-infrared band disappears, in contrast to the band in the visible range—which is observed for a considerably longer time, until all excess electrons undergo recombination. In the liquid phase, the recombination occurs in nanoseconds. In frozen matrices at the low temperature of 77 K the recombination processes are several orders of magnitude slower.

Some modifications of the pump-probe experiments, known as photoexcitation methods, [154] have been applied to study the dynamics of excess electron-solvation. In this method the sequence of three laser pulses has been applied: a) the first pulse generates an excess electron, b) after some nanoseconds, when the electron is already equilibrated, the second pulse is applied, which pumps the electron from the 1s-state to the 2p-state, c) the third pulse, which is a probe pulse, monitors the changes of the transient absorption. The sequence of the three pulses has an advantage over the previously discussed sequence of two laser-pulses because the probing laser-pulse monitors the clearly defined energy states, 1s and 2p, instead of the non-equilibrium states of a delocalized electron that were monitored in the early pump-probe experiments.

Figures 9.30 and 9.31 show the typical transient absorption ($\Delta A = -\log(T/T_0)$) of the sample, where T is the transmission when the pump pulse precedes the probing pulse, T_0 is the transmission when no pump pulse is used. The transients obtained at probing-wavelengths in the near infrared region show a fast *increase* followed by decay. In contrast, the transients in the visible range show a fast *reduction* in optical density (bleaching) followed by a recovery. The absorption in the near-IR and the recovery in the visible region exhibit a fast (~30–80 fs) component [155–156] followed by a ~1 ps and 190–300 fs decays [150, 155–156].

The solvated electron's transient absorption in the near-IR and the visible has been interpreted in terms of the three models presented in Table 9.2.

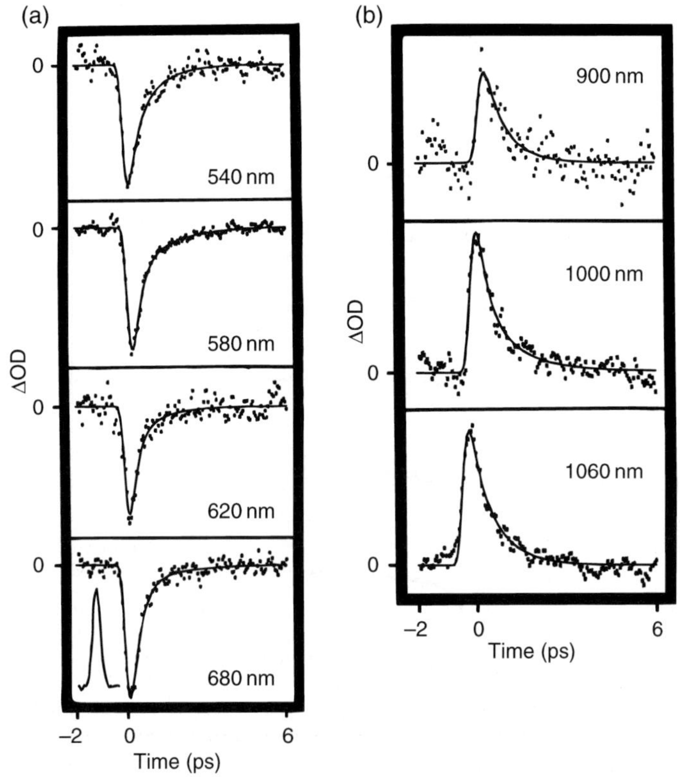

Fig. 9.30 Transient absorption spectra of solvated electron in water *versus* delay-time at room temperature, (a) 540–680 nm visible probing range, (b) 900–1060 nm, near-IR probing range. Reprinted with permission from [154]. Copyright [1993], American Institute of Physics.

Although the first two models in Table 9.2 assume a similar origin of the near-infrared and visible transient absorption, they differ significantly in the details of the electron dynamics. The first model [155] has interpreted the fast dynamics $\leq$300 fs as p-state solvation followed by a slower (300 fs or 1000 fs) [155–158] non-adiabatic transition. In the second model, [150] an ultrafast ($\sim$50 fs) or ($\sim$190 fs) component is interpreted as a 2p $\rightarrow$ 1s transition followed by a slow, $\sim$1 ps, cooling in the ground state [151–150]. The femtosecond transient absorption experiments of photo-injected electrons seem to support the slower, non-adiabatic transition proposed in the first model, rather than the second model [159–160].

The first model has been supported by early semi-classical molecular dynamics (MD) simulations [158, 161]. These, although they have contributed significantly to the interpretation of the transient solvation experiments, suffered from many serious inherent deficiencies. The first MD simulations used rigid water molecules interacting with an excess electron in an adiabatic approximation [161], so the coupling of the electron to the solvent's molecular vibrations was completely neglected. MD simulations with SPC flexible water [158] were closer to the real-world systems, but were still unable to include quantum effects in the solvent molecular

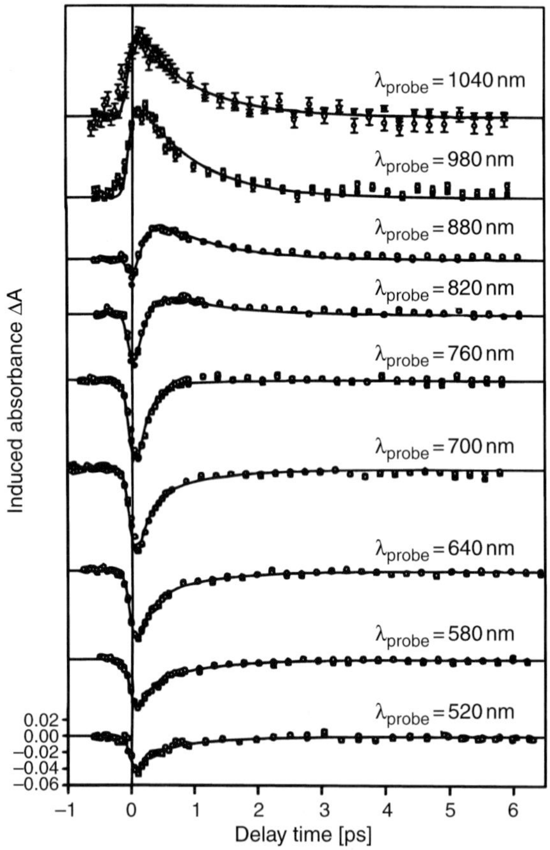

Fig. 9.31 Transient absorption changes *versus* delay-time for solvated electron in water at room temperature in the wavelength range 520–1040 nm with parallel (filled circles) and perpendicular probe polarization (hollow circles) [150].

vibrations (Franck-Condon states' relaxation). Further improvements in the MD simulation by including both phase-decoherence effects (dephasing) and Franck-Condon states' relaxation [139,162] has led to a more realistic physical picture. The improved MD simulations have supported the third model from Table 9.2 [147–148].

The third model, proposed for the first time by Abramczyk, [163] assumes that both the excess-electron dynamics and the equilibrated cavity structure are determined by the coupling between the electronic states of the excess electron and the intramolecular and intermolecular vibrational modes (also called torsional or librational motions) of the solvent. It has been shown that the introduction of the vibronic coupling leads to a decrease of the adiabatic energy of the electronic transition [163]. The higher is the frequency of the vibrational mode the lower is the energy of the electronic transition. After excitation to the 2p-state, the high-frequency stretching mode of water, ν_S(OH) (3650 cm^{-1} in an isolated water molecule, which is red-shifted in the liquid phase) is the primary accepting vibrational

Table 9.2

Origin of the near-infrared and visible absorption of a hydrated electron

	References	Near infrared absorption	Visible absorption
model 1 [155]	C. Sliva, P.K. Walhout, K. Yokoyama, P.F. Barbara, Phys. Rev. Lett. 80 (1998) 1086	Electron in the first excited state (p-like) 2p → c transition	Electron in the first excited state (p-like) 1s → 2p transition
model 2 [150]	M. Assel, R. Laenen, A. Laubereau, Chem. Phys. Lett. 317 (2000) 13		
model 3 [147–148]	H. Abramczyk, J. Kroh, J. Phys. Chem. 95 (1991) 6155–6159; J. Phys. Chem. 96 (1992) 3653–3658	1s → 2p transition. Coupling of an excess electron with the stretching mode of water, ν_s(OH)	1s → 2p transition. Coupling of an excess electron with the torsional (librational) motions of water, ν_t(OH) at 710 cm^{-1}

mode. This coupling is responsible for the near-infrared absorption. The dynamics in this spectral region is determined by vibrational relaxation of the vibrationally excited Franck-Condon states to the vibrationally relaxed 2p state (fast component, ~50–190 fs) and intramolecular energy transfer to another vibrational mode of lower frequency (the bending mode of water) that occurs on the time-scale of the slow component, ~1 ps in the pump-probe experiment. According to the vibrational coupling model [148] the second, slow intermediate observed in the near-infrared can be interpreted as the 1s → 2p transition coupled to the bending mode of water. The process of vibrational relaxation and intramolecular energy-transfer is continued until complete equilibration is reached. The equilibrated solvated electron is coupled to librational motions of water. Thus, the dynamics, band-shape, and bandwidth, of the visible band of the hydrated electron are determined by vibrational dephasing of the librational motions of water due to the coupling with the bath.

Figure 9.32 shows the absorption bands of the solvated electron coupled with: 1) the torsional (librational) mode of water, ν_t(OH) at 710 cm^{-1}, 2) the bending mode of water at 1595 cm^{-1}, 3) the stretching mode of water, ν_S(OH) at 3651 cm^{-1}. One can see an excellent agreement between the experimental band of the hydrated electron in the visible region, and the theoretical band calculated from the model 3 for the coupling with the low-frequency torsional mode of water, ν_t(OH), at 710 cm^{-1}. The coupling with the high-frequency bending and stretching modes leads to the red-shift, and is responsible for the near-infrared bands.

The model of vibrational coupling has been tested in a series of papers for an excess electron in water [147], alcohols [147–148], ammonia and amines [164], ethers [165], hydrocarbons [166].

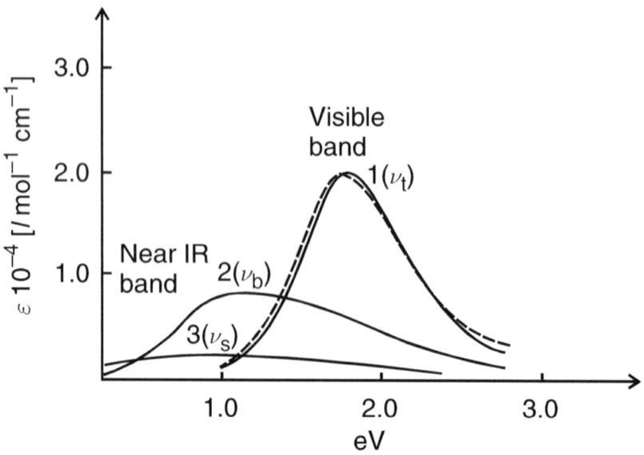

Fig. 9.32 Near-IR absorption and visible absorption spectra of solvated electron in water [148], coupling with: (1) the torsional (librational) mode of water ν_t(OH) at 710 cm^{-1}, (2) the bending mode of water ν_b(OH) at 1595 cm^{-1}, (3) the stretching mode of water ν_s(OH) at 3651 cm^{-1}.

Recently, the vibrational coupling (model 3 from Table 9.2) has been supported directly by resonance Raman spectroscopy [137–138]. Resonance Raman spectroscopy is an ideal technique for probing the coupling of an electronic transition to Franck-Condon active vibrations, and revealing the vibrational dynamics along the structural coordinates after photoexcitation. If there is a coupling with an excess electron the vibrational modes should show Raman enhancement upon resonant excitation of a solute molecule. The experiment showed strong resonance enhancement of the water's Raman librational bands, the intramolecular bending and the stretching modes, which indicates that the conclusions are exactly the same as those obtained from the vibrational coupling model [148].

REFERENCES

1. T.S. Rose, M.J. Rosker, A.H. Zewail, *J. Chem. Phys.* 88 (1988) 6672
2. G.R. Fleming, *Chemical Applications of Ultrafast Spectroscopy*, Oxford University Press (1985), QD96.L3F57
3. *Femtochemistry and Femtobiology: Ultrafast Reaction Dynamics at Atomic-Scale Resolution* (Nobel Symposium 101), Villy Sundström, Sture Forsén, Ulf Eberg, Helen Hjalmarsson, September 9–12, 1996, Björkborn, Karlskoga, Sweden, Proceedings: *Femtochemistry* and Femtobiology*:* Ultrafast Reaction Dynamics at Atomic-Scale Resolution*, ed. V. Sundström, Imperial College Press, London (1996)*
4. *Femtosecond Chemistry: A Handbook in Two Volumes*, ed. J. Manz, Wiley (1995), QD502.F46, vols. 1, 2
5. *World Scientific Series in 20th Century Chemistry*, vol. 3, *Femtochemistry: Ultrafast Dynamics of the Chemical Bond*, ed. A.H. Zewail, vols. I–II, Imperial College Press (1994)
6. S. Mukamel, *Principles of Nonlinear Optical Spectroscopy*, Oxford University Press, New York (1995)
7. *Topics in Applied Physics*, vol. 18, Ultrashort Laser Pulses, ed. S.L. Shapiro, Springer-Verlag, Berlin Heidelberg (1977)
8. *Topics in Applied Physics*, vol. 60, Ultrashort Laser Pulses and Applications, ed. W. Kaiser, Springer-Verlag, Berlin Heidelberg (1988)

9. *Ultrafast Infrared and Raman Spectroscopy*, ed. M.D. Fayer, Marcel Dekker, New York (2001)
10. *Ultrafast Phenomena XII*, Springer Series in Chemical Physics 66, eds. T. Elsaesser, S. Mukamel, M.M. Murnane, N.F. Scherer, Springer Verlag, Heidelberg (2001)
11. J.C. Polanyi, A.H. Zewail, Accounts of Chemical Research (Holy-Grail Special Issue), 28 (1995) 119.
12. G. Cerullo, M. Zavelani-Rossi, G. Lanzani, S. De Silvestri, p. 581, Fig. 1, in *Ultrafast Phenomena XII*, eds. T. Elsaesser, S. Mukamel, M.M. Murnae, N.F. Scherer, Springer Verlag, Heidelberg (2000)
13. T.G. Ebrey, in *Thermodynamics of Membranes, Receptors and Channels*, CRC Press, New York (1993), p. 353
14. J.K. Lanyi, Biochim. Biophys. Acta 1183 (2) (1993 Dec 7) 241, Review
15. J.K. Lanyi, J. Phys. Chem. B 104 (2000) 11441
16. T. Ye, N. Friedman, Y. Gat, G.H. Atkinson, M. Sheves, M. Ottolenghi, S. Ruhman, Phys. Chem. B 103 (1999) 5122
17. T. Ye, E. Gershgoren, N. Friedman, M. Ottolenghi, M. Sheves, S. Ruhman, Chem. Phys. Lett. 314 (1989) 429
18. L. Ujj, Y. Zhou, M. Sheves, M. Ottolenghi, S. Ruhman, G.H. Atkinson, J. Am. Chem. Soc. 122 (2000) 96
19. R. Akiyama, A. Yoshimori, T. Kakitani, Y. Imamoto, Y. Schichida, J. Hatano, J. Phys. Chem. A 101 (1997) 417
20. G.H. Atkinson, Y. Zhou, L. Ujj, A. Aharoni, M. Sheves, M. Ottolenghi, J. Phys. Chem. A 106 (2002) 3325
21. L. Ujj, Y. Zhou, M. Sheves, M. Ottolenghi, S. Ruhman, G.H. Atkinson, J. Am. Chem. Soc. 122 (2000) 96
22. M. Garavelli, F. Negri, M. Olivucci, J. Phys. Chem. Soc. 121 (1999) 1023
23. L. Song, M.A. El-Sayed, J. Am. Chem. Soc. 120 (1998) 8889
24. G. Eyring, B. Curry, A. Brock, J. Lugtenburg, R. Mathies, Biochem. 21 (1982) 384
25. S.W. Lin, M. Groesbeek, I. van der Hoef, P. Werdegem, J. Lugtenburg, R.A. Mathies, J. Phys. Chem. B 102 (1998) 2787
26. L. Zhu, J. Kim, R.A. Mathies, J. Raman Spect. 30 (1999) 777
27. A. Warshel, N. Barboy, J. Am. Soc. Chem. 104 (1982) 1469
28. A. Warshel, Nature 260 (1976) 679
29. R.G. Luque, M. Garavelli, F. Bernardi, M. Merchan, M.A. Robb, M.O. Livucci, Proc. Natn. Acad. Sci. USA 97 (2000) 9379
30. A. Warshel, W.W. Parson, Quarterly Rev. Biophys. 34 (2001) 563
31. H. Abramczyk, J. Chem. Phys. 120 (2004) 11120
32. R.A. Mathies, C.H. Brito Cruz, W.T. Pollard, CH.V. Shank, Science 240 (1988) 777
33. J.W. Petrich, J. Brenton, J.L. Martin, A. Antonetti, Chem. Phys. Lett. 137 (1987) 369
34. J. Dobler, W. Znith, W. Kaiser, D. Oesterhelt, Chem. Phys. Lett. 144 (1988) 215
35. J. Herbst, K. Heyne, R. Diller, Science 297 (2002) 822
36. G.H. Atkinson, L. Ujj, Y. Zhou, J. Phys. Chem. A 104 (2000) 4130
37. M. Garavelli, P. Celani, F. Bernardi, M.A. Robb, M. Olivuci, J. Am. Chem. Soc. 119 (1997) 6891
38. M. Du, G.R. Fleming, Biophysical Chemistry 48 (1993) 101
39. K.C. Hasson, F. Gai, P.A. Anfinrud, Proc. Natl. Acad. Sci. USA 93 (1996) 15124
40. F. Gai, K.C. Hasson, J. Cooper McDonald, P.A. Anfinrund, Science 79 (1998) 1886
41. T. Kobayashi, T. Saito, H. Ohtani, Nature 414 (2001) 531
42. A. Shirakawa, I. Sakane, T. Kobayashi, *Pulse-front-Matched Optical Parametric Amplification for sub-10-fs Pulse Generation Tunable in the Visible and Near Infrared*, Opt. Lett. 23 (1998) 1292
43. A. Shirakawa, I. Sakane, M. Takasaka, T. Kobayashi, *Sub-5-fs Visible Pulse Generation by Pulse-front-Matched Non-collinear Optical Parametric Amplification*, Appl. Phys. Lett. 74 (1999) 2268–2270
44. K. Heyne, N. Huse, E.T.J. Nibbering, T. Elsaesser, Chem. Phys. Lett. 369 (2003) 591

45. G. Zundel, in: *The Hydrogen Bond*, vol. 2, eds. P. Schuster, G. Zundel, C. Sandorfy, North-Holland, Amsterdam (1976), p. 683
46. H. Abramczyk, Chem. Phys. 94 (1985) 91
47. H. Abramczyk, Chem. Phys. 116 (1987) 249
48. S. Bratos, J. Chem. Phys. 63 (1975) 3499
49. S. Bratos, H. Ratajczak, J. Chem. Phys. 76 (1982) 77
50. G.N. Robertson, J. Yarwood, Chem. Phys. 32 (1978) 267
51. V.P. Sakum, Chem. Phys. 99 (1985) 457
52. A. Witkowski, J. Chem. Phys. 47 (1967) 3645
53. Y. Maréchal, A. Witkowski, J. Chem. Phys. 48 (1968) 3697
54. Y. Maréchal, Chem. Phys. 79 (1983) 69
55. H. Abramczyk, *IR *s(XH) Absorption Band shape of H-Bonded Complex. I. Theory*, Chem. Phys. 144 (1990) 305
56. H. Abramczyk, *IR *s(XH) Absorption Band shape of H-Bonded Complex. II. Numerical Calculations of the Absorption Profiles in Weak and Strong H-bond complexes*, Chem. Phys. 144 (1990) 319
57. O. Henri-Rousseau, P. Blaise, in Theoretical Treatments of Hydrogen Bonding, ed. Dusan Hadzi, Wiley, 1997, p. 165
58. O. Fauskov Nielsen, P.A. Lund, J. Chem. Phys. 78 (1983) 652
59. *Ultrafast Dynamics of Chemical Systems*, ed. John D. Simon, Kluwer, Dordrecht (1994)
60. R.J. Sension, S.T. Repinec, A.Z. Szarka, R.M. Hochstrasser, *Femtosecond Laser Studies of the cis-Stilbene Photoisomerization Reactions*, J. Chem. Phys. 98 (8) (1993) 6291–6315
61. D.L. Phillips, Jon-Marc Rodier, A.B. Myers, *The cis-trans Stilbene Photoreaction Probed Through Two-color UV Picosecond Pump-probe Raman Spectroscopy*, in *Time-Resolved Vibrational Spectroscopy VI*, eds. A. Lau, F. Siebert, W. Werncke, Springer-Verlag, Berlin (1994), 95
62. J.K. Rice, A.P. Baronavski, *Ultrafast Studies of Solvent Effects in the Isomerization of cis-Stilbene*, J. Phys. Chem. 96 (8) (1992) 3359–3366
63. Hiro-o Hamaguchi, *Picosecond Solvation Dynamics of S_1 trans-Stilbene by Nanosecond Transient Raman Spectroscopy with "Optical Depletion Timing"*, J. Chem. Phys. 89 (4) (1988) 2587
64. G. Orlandi, W. Siebrand, Chem. Phys. Lett. 30 (1975) 352
65. P. Tavan, K. Schulten, Chem. Phys. Lett. 56 (1978) 200
66. V. Molina, M. Merchan, B.O. Ross, Spectrosc. Acta A 55 (1999) 433
67. J. Quenneville, T.J. Martinez, J. Phys. Chem A. 107 (2003) 829
68. R.M. Hochstrasser, Pure & Appl. Chem. 52 (1980) 2683
69. J.A. Syage, P.M. Felker, A.H. Zewail, *J. Chem. Phys.* 81 (1984) 4706
70. K. Rotkiewicz, K.H. Grellmann, Z.R. Grabowski, Chem. Phys. Lett. 19 (1973) 315
71. Y. Wang, K.B. Eisental, J. Chem. Phys. 107 (1982) 77
72. E. Kosower, J. Am. Chem. Soc. 107 (1985) 1114
73. J. Hicks, M.T. Vandersall, E.V. Sitzmann, K.B. Eisental, Chem. Phys. Lett. 135 (1987) 413
74. C. Ma, W.M. Kwok, P. Matousek, A.W. Parker, D. Phillips, W.T. Toner, M. Towrie, *Excited States of 4-Aminobenzonitrile (ABN) and 4-Dimethylaminobenzonitrile (DMABN): Time-resolved Resonance Raman, Transient Absorption, Fluorescence, and ab Initio Calculations*, J. Phys. Chem. A 106 (14) (2002) 3294
75. C. Ma, W.M. Kwok, P. Matousek, A.W. Parker, D. Phillips, W.T. Toner, M. Towrie, *Resonance Raman study of Ring Deuterated 4-Dimethylaminobenzonitrile (DMABN-d_4): the Ground, ICT and Triplet States*, Journal of Photochemistry and Photobiology A: Chemistry 142 (2–3) (2001) 177
76. C. Ma, W.M. Kwok, P. Matousek, A.W. Parker, D. Phillips, W.T. Toner, M. Towrie, *Time-Resolved Study of the Triplet State of 4-dimethylaminobenzonitrile (DMABN)*, J. Phys. Chem. A 105 (19) (2001) 4648
77. S.A. Trushin, T. Yatsuhashi, W. Fuß, W.E. Schmid, *Coherent Oscillations in the Charge-transfer System 4-dimethylamino-benzonitrile*, Chem. Phys. Lett. 376 (3–4) (2003) 282

78. W. Rettig, S. Lutze, *Mechanistic Considerations for the Dual Fluorescence of Dimethylaminobenzonitrile: a Fluorescence Anisotropy Study*, Chem. Phys. Lett. 341 (3–4) (2001) 263
79. G. Saielli, D. Braun, A. Polimeno, P.L. Nordio, Chem. Phys. Lett. 257 (1996) 381
80. Richard A. Farrer, John T. Fourkas, *Orientational Dynamics of Liquids Confined in Nanoporous Sol-Gel Glasses Studied by Optical Kerr Effect Spectroscopy*, Acc. Chem. Res. 36 (8) (2003) 605
81. Alessandra Scodinu, John T. Fourkas, *Intermolecular Dynamics and Structure of Carbon Disulfide in Isoviscous Alkane Solutions: An Optical Kerr Effect Study*, J. Phys. Chem. B 107 (1) (2003) 44
82. Geoffrey L.D. Ritchie, Ewan W. Blanch, *Temperature Dependence of the Electrooptical Kerr Effect: Anisotropic Electric Dipole Polarizabilities of* NH_3, CH_3NH_2, $(CH_3)_2NH$, *and* $(CH_3)_3N$, J. Phys. Chem. A 107 (12) (2003) 2093–2099
83. Jonathan Segal, Dan Huppert, Boris Fainberg, *Resonance heterodyne optical Kerr effect spectroscopy of population wave packets with intense chirped pulses*, Journal of Luminescence, 102–103 (2003), 669, Proceedings of the 2002 International Conference on Luminescence and Optical Spectroscopy of Condensed Matter
84. Kenji Kiyohara, Yoshifumi Kimura, Yoshihiro Takebayashi, Noboru Hirota, Koji Ohta, *Density Dependence of Optical Kerr Effect of Linear Molecules*, J. Chem. Phys. 117 (2002) 9867
85. C.J. Fecko, J.D. Eaves, A. Tokmakoff, *Isotropic and Anisotropic Raman Scattering from Molecular Liquids Measured by Spatially Masked Optical Kerr Effect Spectroscopy*, J. Chem. Phys. 117 (3) (2002) 1139
86. Byung-Ryool Hyun, Sergei V. Dzyuba, Richard A. Bartsch, Edward L. Quitevis, *Intermolecular Dynamics of Room-Temperature Ionic Liquids: Femtosecond Optical Kerr Effect Measurements on 1-Alkyl-3-Methylimidazolium bis-(Trifluoromethyl)sulfonyl)imides*, J. Phys. Chem. A 106 (33) (2002) 7579
87. Niko P. Ernsting, George M. Photiadis, Horst Hennig, Thomas Laurent, *Rotational Friction Kernel in Water from the Femtosecond Time-Resolved Optical Kerr Effect of Acetonitrile/Water Mixtures*, J. Phys. Chem. A 106 (40) (2002) 9159
88. S. Yoshioka, T. Yagi, S. Nakashima, T. Okada, *Temperature Dependence of Low-frequency Responses of Liquid 1-Alkenes Studied by Femtosecond Optical Kerr Effect Spectroscopy*, J. Mol. Liquids 96–97 (2002) 275
89. K.B. Eisental, *Ultrafast Chemical Reactions in the Liquid State*, in *Topics in Applied Physics*, vol. 60, Ultrashort Laser Pulses and Applications, ed. W. Kaiser, Springer-Verlag, Berlin Heidelberg (1988)
90. *Ultrafast Hydrogen Bonding Dynamics and Proton Transfer Processes in the Condensed Phase*, eds. T Elsaesser, H.J. Bakker, Understanding Chemical Reactivity, vol. 23, Kluwer Academic Publishers, Dordrecht (2002)
91. A. Weller, Z. Elektrochem. 60 (1956) 1144
92. M. Ziolek, J. Kubicki, A. Maciejewski, R. Naskrecki, A. Grabowska, *Excited State Proton Transfer and Photochromism of an Aromatic Schiff Base. Pico- and Femtosecond Kinetics of the N,N′-bis-(Salicylidene)-p-Phenylenediamine (BSP)*, Chem. Phys. Lett. 369 (1–2) (2003) 80
93. H. Wang, H. Zhang, O.K. Abou-Zied, C. Yu, F.E. Romesberg, M. Glasbeek, *Femtosecond Fluorescence Upconversion Studies of Excited-State Proton-Transfer Dynamics in 2-(2′-Hydroxyphenyl)benzoxazole (HBO) in Liquid Solution and DNA*, Chem. Phys. Lett. 367 (5–6) (2003) 599
94. Satoru Shiobara, So Tajima and Seiji Tobita, *Substituent Effects on Ultrafast Excited-state Proton Transfer of Protonated Aniline Derivatives in Aqueous Solution*, Chem. Phys. Lett. 380 (5–6) (2003) 673
95. Sergei Yu. Arzhantsev, Satoshi Takeuchi, Tahei Tahara, Ultrafast Excited-state Proton Transfer Dynamics of 1,8-Dihydroxyanthraquinone (Chrysazin) Studied by Femtosecond Time-Resolved Fluorescence Spectroscopy, Chem. Phys. Lett. 330 (1–2) (2000) 83–90
96. J.L. Herek, S. Pedersen, L. Bañares, A.H. Zewail, J. Chem. Phys. 97 (12) (1992) 9046

97. Satoshi Takeuchi, Tahei Tahara, *Excitation-wavelength Dependence of the Femtosecond Fluorescence Dynamics of 7-Azaindole Dimer: Further Evidence for the Concerted Double Proton Transfer in Solution*, Chem. Phys. Lett. 347 (1–3) (2001) 108
98. Wei-Ping Hu, Ru-Min You, Shih-Yao Yen, Fa-Tsai Hung, Po-Hung Chou, Pi-Tai Chou, *Dynamics of Ground-state Reverse Proton Transfer in the 7-Azaindole/carboxylic Acid Systems*, Chem. Phys. Lett. 370 (1–2) (2003) 139
99. A.D. Roshal, J.A. Organero, A. Douhal, *Tuning the Mechanism of Proton-transfer in a Hydroxyflavone Derivative*, Chem. Phys. Lett. 379 (1–2) (2003) 53
100. W. Zinth, W. Kaiser, *Topics in Applied Physics*, vol. 60, Ultrashort Laser Pulses and Applications, ed. W. Kaiser, Springer-Verlag, Berlin Heidelberg (1988), p. 235
101. S. Schlücker, M. Heid, R.K. Singh, B.P. Asthana, J. Popp, W. Kiefer, *Vibrational Dynamics in Hydrogen-bonded (Pyridine + Water) Complexes Studied by Spectrally Resolved Femtosecond CARS*, Z. Phys. Chem. 216 (2002) 267
102. P. Vöhringer, D.C. Arnett, R.A. Westervelt, M.J. Feldstein, N.F. Scherer, J. Chem. Phys. 102 (10) (1995) 4027
103. A. Vierheilig, T. Chen, P. Waltner, W. Kiefer, A. Materny, A.H. Zewail, *Femtosecond Dynamics of Ground-state Vibrational Motion and Energy Flow: Polymers of Diacetylene*, Chem. Phys. Lett. 312 (5–6) (1999) 349
104. Igor Pastirk, Vadim V. Lozovoy, Marcos Dantus, *Femtosecond Photon Echo and Virtual Echo Measurements of the Vibronic and Vibrational Coherence Relaxation times of Iodine Vapor*, Chem. Phys. Lett. 333 (1–2) (2001) 76
105. M. Khalil, O. Golonzka, N. Demirdöven, C.J. Fecko, A. Tokmakoff, *Polarization-selective Femtosecond Raman spectroscopy of Isotropic and Anisotropic Vibrational Dynamics in Liquids*, Chem. Phys. Lett. 321 (3–4) (2000) 231
106. A. Tokmakoff, M.D. Fayer, *Homogeneous Vibrational Dynamics and Inhomogeneous Broadening in Glass-forming Liquids: Infrared Photon Echo Experiments from Room Temperature to 10 K*, J. Chem. Phys. 103 (8) (1995) 2810
107. K.D. Rector, A.S. Kwok, C. Ferrante, A. Tokmakoff, C.W. Rella, M.D. Fayer, *Vibrational Anharmonicity and Multilevel Vibrational Dephasing from Vibrational Echo Beats*, J. Chem. Phys. 106 (24) (1997) 10027
108. O. Golonzka, M. Khalil, N. Demirdöven, A. Tokmakoff, *Coupling and Orientation between Anharmonic Vibrations Characterized with Two-dimensional Infrared Vibrational Echo Spectroscopy*, J. Chem. Phys. 115 (23) (2001) 10814
109. M. Kolodziejski, G. Waliszewska, H. Abramczyk, *I. Vibrational Relaxation in Liquid Phenylacetylene*, J. Phys. Chem. 102 (1998) 1918
110. H. Abramczyk, B. Brozek, *Is There a Correlation Between Phase Transition of Phenylacetylene in Liquids and Frozen Matrices and Vibrational Dynamics*? Chem. Phys. 250 (1999) 145
111. M. Kolodziejski, G. Waliszewska, H. Abramczyk, Vibrational Dephasing in Bromocyclohexane: How to Separate Contributions from Different Mechanisms, Chem. Phys. 213 (1996) 341–356
112. H. Abramczyk, K. Paradowska-Moszkowska, *Crystallization and Glass Formation Processes in Methylcyclohexane. Vibrational Dynamics as a Possible Molecular Indicator of the Liquid-glass Transition*, J. Chem. Phys. 24 (2001) 11221
113. D.W. Oxtoby, Adv. Chem. Phys. 40 (1979) 1
114. D. Vanden Bout, L. Muller, M. Berg, Phys. Rev. Lett. 67 (1991), 3700
115. L.J. Muller, D. Vanden Bout, M. Berg, J. Chem. Phys. 99 (1993) 810
116. D. Vanden Bout, J.E. Freitas, M. Berg, Chem. Phys. Lett. 229 (1994) 87
117. D. Vanden Bout, M. Berg, J. Raman Spectrosc. 26 (1995) 503
118. R. Inaba, K. Tominaga, M. Tasumi, K.A. Nelson, K. Yoshihara, Chem. Phys. Lett. 211 (1993) 183
119. K. Tominaga, R. Inaba, T.J. Kang, Y. Naitoh, K.A. Nelson, M. Tasumi, K. Yoshihara, in Proceedings of the XIV International Conference on Raman Spectroscopy; Wiley, New York (1994)

120. K. Yoshihara, R. Inaba, H. Okamoto, M. Tasumi, K. Tominaga, K.A. Nelson, in *Femtosecond Reaction Dynamics*, ed. D. Wiersma, North-Holland, Amsterdam (1994)
121. A. Tokmakoff, M.D. Fayer, Acc. Chem. Res. 28 (1995) 437
122. Y. Tanimura, S. Mukamel, J. Chem. Phys. 99 (1993) 9496
123. V. Khidekel, S. Mukamel, Chem. Phys. Lett. 240 (1995) 304
124. S. Palese, J.T. Buontempo, L. Schilling, W.T. Lotshaw, Y. Tanimura, S. Mukamel, R.J.D. Miller, J. Phys. Chem. 98 (1994) 12466
125. K. Tominaga, G.P. Keogh, Y. Naitoh, K. Yoshihara, J. Raman Spectrosc. 26 (1995) 495
126. K. Tominaga, K. Yoshihara, Phys. Rev. Lett. 74 (1995) 3061
127. K. Tominaga, K. Yoshihara, J. Chem. Phys. 104 (1996) 4419
128. K. Tominaga, K. Yoshihara, J. Chem. Phys. 104 (1996) 1159
129. T. Steffan, K. Duppen, Phys. Rev. Lett. 76 (1996) 1224
130. K. Tominaga, K. Yoshihara, Phys. Rev. Lett. 76 (1996) 987
131. K.A. Merchant, D.E. Thompson, M.D. Fayer, Phys. Rev. Lett. 86 (2001) 3899–3902
132. T. Elsaesser, S. Mukamel, M.M. Murnane, *Ultrafast Phenomena XII*, eds. N.F. Scherer, Springer, Chemical Physics (2001) p. 499
133. S. Mukamel, *Principles of Nonlinear Optical Spectroscopy*, Oxford University Press, Oxford (1995)
134. F. Lindenberger, R. Stuckl, B.P. Asthana, A. Laubereau, J. Phys. Chem. A 103 (1999) 5655
135. E.J. Hart, J.W. Boag, *has inspired many experimental, theoretical and computer simulation studies*, J. Am. Chem. Soc. 84 (1962) 4090
136. M. Mostafavi, M. Lin, H. He, Y. Muroya, Y. Muroya, *Temperature-dependent Absorption Spectra of the Solvated Electron in Ethylene Glycol at 100 atm. Studied by Pulse Radiolysis from 296 to 598 K*, Chem. Phys. Lett. 384 (1–3) (2004) 52
137. Michael J. Tauber, Richard A. Mathies, *Structure of the Aqueous Solvated Electron from Resonance Raman Spectroscopy: Lessons from Isotopic Mixtures*, J. Am. Chem. Soc. 125 (5) (2003) 1394
138. M.J. Tauber, R.A. Mathies, *Resonance Raman Spectra and Vibronic Analysis of the Aqueous Solvated Electron*, Chem. Phys. Lett. 354 (5–6) (2002) 518
139. Kim F. Wong, Peter J. Rossky, *Dissipative Mixed Quantum-classical Simulation of the Aqueous Solvated Electron System*, J. Chem. Phys. 116 (19) (2002) 8418
140. Kim F. Wong, Peter J. Rossky, *Solvent-induced Electronic Decoherence: Configuration Dependent Dissipative Evolution for Solvated Electron Systems*, J. Chem. Phys. 116 (19) (2002) 8429
141. Andrzej L. Sobolewski, Wolfgang Domcke, *Hydrated Hydronium: a Cluster Model of the Solvated Electron?*, Phys. Chem. Chem. Phys. 4 (2002) 4
142. Michael J. Tauber, Richard A. Mathies, *Fluorescence and Resonance Raman Spectra of the Aqueous Solvated Electron*, J. Phys. Chem. A 105 (49) (2001) 10952
143. T. Scheidt, R. Laenen, *Ionization of Methanol: Monitoring the Trapping of Electrons on the fs Time Scale*, Chem. Phys. Lett. 371 (3–4) (2003) 445
144. J.C. Alfano, P.K. Walhout, Y. Kimura, P.F. Barbara, J. Chem. Phys. 98 (1993) 5996
145. H. Lu, F.H. Long, R.M. Bowman, K.B. Eisenthal, J. Phys. Chem. 93 (1989) 27
146. H. Abramczyk, *Absorption spectrum of the solvated electron. I. Theory*, J. Phys. Chem. 95 (1991) 6149
147. H. Abramczyk, J. Kroh, *Absorption Spectrum of the Solvated Electron. II. Numerical Calculations of the Profiles of the Electron in Water and Methanol at 300 K*, J. Phys. Chem. 95 (1991) 6155
148. H. Abramczyk, J. Kroh, J. Phys. Chem. 96 (1992) 3653
149. Y. Gauduel, S. Pommeret, A. Migus, A. Antonetti, *Hydrogen/Deuterium Isotope Effects on Femtosecond Electron Reactivity in aqueous Media*, J. Phys. Chem. 95 (2) (1991) 533
150. M. Assel, R. Laenen, A. Laubereau, Chem. Phys. Lett. 317 (2000) 13
151. M. Assel, R. Laenen, A. Laubereau, J. Phys. Chem. A 102 (1998) 2256
152. P.J. Reid, C. Silva, P.K. Walhout. P.F. Barbara, Chem. Phys. Lett. 228 (1994) 658

153. F.F. Muquet, G.W. Robinson, *Ultrafast Reaction Dynamics and Solvent Effects: International Research Workshop*, eds. Y. Gauduel, P.J. Rossky, AIP Conference Proceedings 298, France, August 1993, p. 158
154. J.C. Alfano, P.K. Walhout, Y. Kimura, P.F. Barbara, J. Chem. Phys. 98 (1993) 5996
155. C. Sliva, P.K. Walhout, K. Yokoyama, P.F. Barbara, Phys. Rev. Lett. 80 (1998) 1086
156. K. Yokoyama, C. Silva, D.H. Son, P.K. Walhout, P.F. Barbara, J. Phys. Chem. A 102 (1998) 6957
157. Y. Kimura, J.C. Alfano, P.K. Walhout, P.F. Barbara, J. Phys. Chem. 98 (1994) 3450–3458
158. B.J. Schwartz, P.J. Rossky, J. Chem. Phys. 101 (1994) 6902
159. F.H. Long, H. Lu, K.B. Eisenthal, Phys. Rev. Lett. 64 (1990) 1469
160. X. Shi, F.H. Long, H. Lu, K.B. Eisenthal, J. Phys. Chem. 100 (1996) 11903
161. J. Schnitker, P. Rossky, J. Chem. Phys. 86 (1987) 3462
162. E.R. Bittner , B.J. Schwartz, P.J. Rossky, J. Mol. Structure (Theochem) 389 (1997) 203
163. H. Abramczyk, *Absorption Spectrum of the Solvated Electron. I. Theory*, J. Phys. Chem. 95 (1991) 6149
164. H. Abramczyk, J. Kroh, *Absorption Spectra of the Solvated Electron in Ammonia and Amines*, J. Phys. Chem. 95 (1991) 5749
165. H. Abramczyk, J. Kroh, *Absorption Spectra of the Solvated Electron in Ethers*, Chem. Phys. 157 (1991) 373
166. H. Abramczyk, B. Werner, J. Kroh, *Absorption Spectra of the Solvated Electron in Hydrocarbons*, J. Phys. Chem. 96 (1992) 9674

– 10 –

Lasers in Medicine

10.1. INTRODUCTION

In the last decade the application of lasers in medicine has increased dramatically. With the development of laser technology, lasers from various spectral ranges, characterized by different energies, peak powers, intensities and pulse durations, are being tested in medicine.

An understanding of how the new and traditional sources of laser light interact with human tissue is an essential prerequisite for choosing suitable methods for treatment and diagnostics. The development and commercialization of femtosecond lasers has opened up new applications in biomedicine. Particularly in surgery, femtosecond pulses allow for much more precise cutting than do nanosecond lasers [1]. Ultrashort lasers have also opened up exciting areas of optical tomography. The development of a reliable optical biopsy has been an elusive goal for some years. Researchers are developing a technique in which two-dimensional images of biological tissue can be recorded using an ultrafast electronic-gated imaging camera system. This imaging typically uses 120 fs pulses at a 1 kHz repetition rate from ultrafast laser amplifiers. The biggest advantage of ultrafast lasers in surgical applications is limiting biological tissue damage. The pulse interacts with the tissue faster than thermal energy can diffuse to surrounding tissues. It simply means less, if any, burning and destruction of neighboring tissue. In addition, cuts on the dimensions of microns may result in new applications to repair nerve damage.

This section does not intend to provide a comprehensive, integrated survey of the current status and latest innovations in the biomedical uses of lasers. Those who are interested in detailed reviews of the most important applications of lasers in medicine, surgery, and dentistry are referred to [2–3]. Here, we only present the principle of interaction between laser radiation and biological tissue. Before going into details one can state that the radiation–biological tissue interaction is determined mainly by the laser irradiance [W/cm^2], which depends on the pulse energy, pulse duration, and the spectral range of the laser light. The interaction depends also on thermal properties of tissue – such as heat conduction, heat capacity and the coefficients of reflection, scattering and absorption [4].

The main components of biological tissue that contribute to the absorption are melanin, hemoglobin, water and proteins.

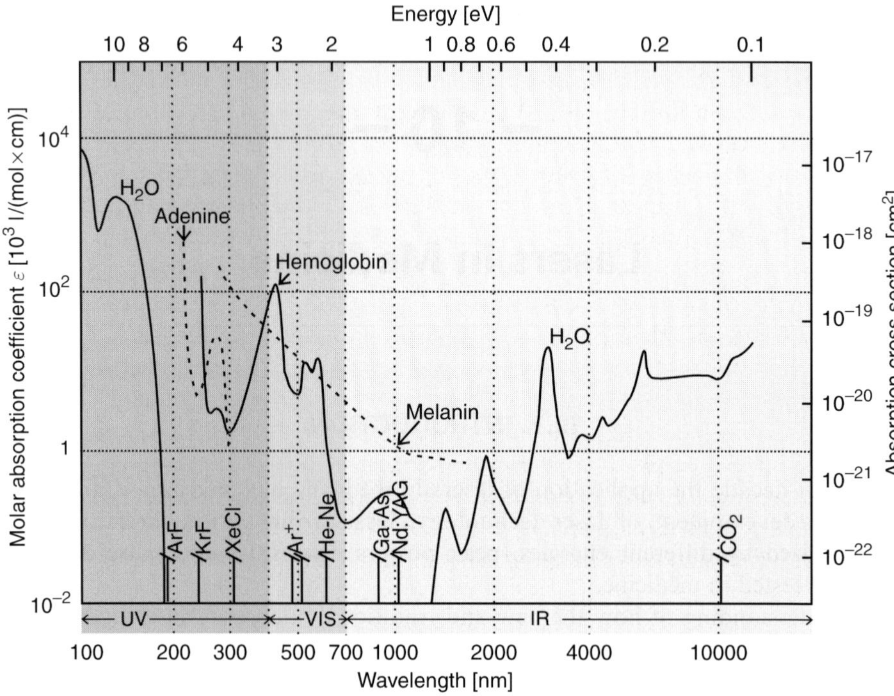

Fig. 10.1 Absorption spectra of main absorbers in biological tissue. Modified from [5].

Figure 10.1 presents the absorption spectra of the main absorbers in biological tissue. One can see that absorption in the IR region (2000–10000 nm) originates from water, which is the main constituent of most tissues. Proteins absorb in the UV region (mainly 200–300 nm). Pigments such as hemoglobin in blood and melanin, the basic chromophore of skin, absorb in the visible range.

The absorption properties of the main biological absorbers determine the depth of penetration of a laser beam. The penetration depth in biological tissue for typical lasers is shown in Table 10.1. The CO_2 laser (10.6 μm) does not penetrate deeply because of water absorption in this region, in contrast to the Nd:YAG laser (1064 nm) for which absorption of water, pigments and proteins is low. This property is therefore obviously important for medical applications. For example, the Nd:YAG laser can penetrate deeper and a cut made with the Nd:YAG laser will

Table 10.1

Comparison of penetration depths in biological tissue for different types of lasers

Laser type	Wavelength [μm]	Penetration depth [mm]
CO_2	10.60	0.10
Nd:YAG	1.064	6.00
Ar^+	0.4880, 0.5145	2.00
Excimer	0.193–0.351	0.01

not bleed due to tissue coagulation, in contrast to the CO_2 laser which is a better "scalpel" for precise thermal cutting of tissue due to vaporization by focusing on the tissue along a short optical path.

Now, we will consider the optical and thermal responses of laser-irradiated tissue. Although the variety of interaction mechanisms between the laser radiation and biological tissue may occur, there are five main categories of interaction:

1. photochemical interactions,
2. thermal interactions,
3. photoablation,
4. plasma-induced ablation,
5. photodisruption.

Figure 10.2 illustrates the main laser–tissue interactions. One can see that all types of interactions can be placed between two diagonals illustrating the energy fluences of 1 J/cm^2 and 1000 J/cm^2. This indicates that lasers used for medical applications must have fluences ranging from 1 J/cm^2 to 1000 J/cm^2. The fluences are controlled both by the energy (controlled by exposure time) and degree of focusing the laser beam on the tissue.

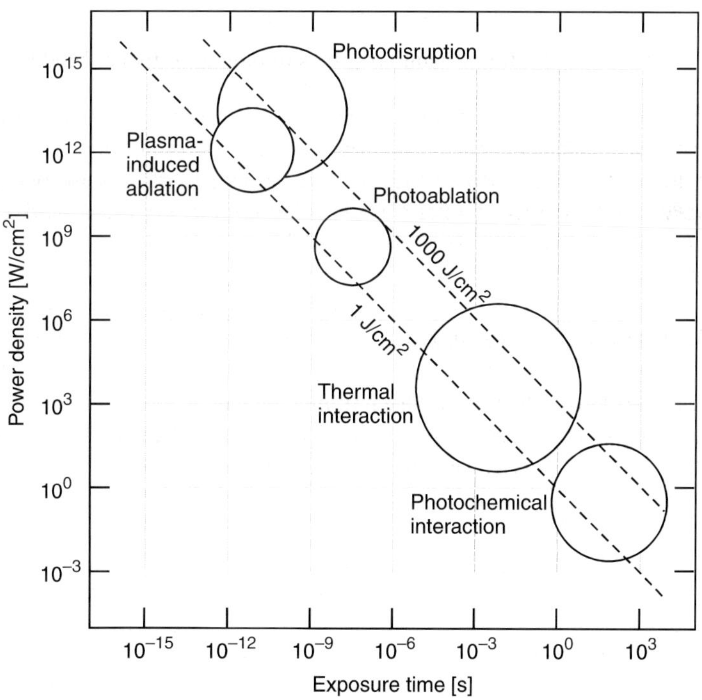

Fig. 10.2 Double logarithmic plot of the power density as a function of exposure time. The circles show the laser parameters required from a given type of interaction with biological tissue. Reproduced from [2]. Permission has been granted by the copyright holder.

With *cw* lasers or exposure time >1 s, only *photochemical interaction* can be induced. Powers of only a few mW can be used for these purposes.

For *thermal interactions* shorter exposure times (1 min–1 μs) and higher energies must be used. Thermal effects can be induced both by *cw* or pulsed lasers of 15–25 W power.

Photoablation occurs at exposure time between 1 μs and 1 ns. In practice, nanosecond pulses of 10^6–10^9 W/cm^2 irradiance should be employed.

Plasma-induced ablation and *photodisruption* occur for pulses shorter than nanoseconds. In practice, pico- and femtosecond lasers with an irradiance of 10^{12} W/cm^2 should be used. Note that although both phenomena occur at a similar time exposure and irradiance, they differ according to the energy densities that are significantly lower for plasma-induced ablation. The plasma-induced ablation is solely based on tissue ionization whereas photodisruption is primarily mechanical disruption.

10.2. PHOTOCHEMICAL INTERACTIONS

Photochemical interactions do not need a high power density. Lasers of 1 W/cm^2 power density and long exposure times ranging from seconds to *cw* light are sufficient. For this category of interactions, a laser induces chemical effects by initiating chemical reactions in tissue. In Chapter 9 we have shown many reactions induced by lasers. For example, vision processes in rhodopsin or proton pumping in bacteriorhodopsin are initiated by a laser beam from the visible range. Photochemical interactions are used in photodynamic therapy (PDT) as described in section 10.2.1.

Another application of very low irradiance of a laser beam is *biostimulation*. Although the term biostimulation has not been well defined scientifically, there are many contradictory reports on wound healing, hair growth, pain relief, and stimulated collagen synthesis effects obtained with IR or near IR lasers of extremely low laser power (1–5 mW) [6–15].

10.2.1. Photodynamic Therapy

Photodynamic therapy utilizes the laser light effect on various chemical substances (e.g., some porphyrins) in an oxygen-rich environment. Light induces a sequence of reactions that produce toxic substances such as singlet oxygen or free radicals. These substances are very reactive and can damage proteins, lipids, nucleic acids as well as other cell components. In contrast to chlorophyll in plants, which is a metal complex comprising a porphyrin ring with Mg^{+2} in the center, where light is used in photosynthesis for sugar production, other porphyrins use light for producing toxic components. The first attempts at using light in medical therapy started at the beginning of the twentieth century when Tappeiner and Jesionek in Germany proposed employing light to induce reactions in some dyes. In 1942 Auler and Bauer found that some porphyrins are accumulated at much larger concentrations and for a much longer time in cancer cells comared to healthy cells. It was suggested that this effect can be used for killing cancer cells. If porphyrins are transferred to toxic states in any way, e.g., by light, cancer cells would be damaged first. The method is very selective because it does not destroy healthy cells. The synthesis of hematoporphyrin derivative (HPD) and the discovery of its preferential accumulation in cancer cells by Schwartz in 1950 can be considered the beginning of modern *photodynamic therapy (PDT)*.

In the PDT method, a chemical substance known as a *sensitizer* is injected intravenously. During the next several hours the sensitizer is distributed to all of the organism's soft tissues, both healthy and diseased. At first, the substance concentration is the same in healthy and diseased cells, but after about 48–72 hours the sensitizer leaves the healthy cells in contrast to cancer cells, where it remains accumulated for 7–10 days. After about 3 days post injection, the concentration of the sensitizers (e.g., hematoporphyrin derivatives HPD) is about 30 times higher in diseased cells than in healthy ones. It should be emphasized that the sensitizer remains inactive and nontoxic in an organism as long as it is not irradiated. Side effects of this phase of therapy are negligible. About 3 days after the sensitizer injection, a patient is irradiated by a laser light. The laser light induces a sequence of reactions with the excited singlet state of oxygen ${}^1O_2{}^*$ as a final product. The singlet oxygen ${}^1O_2{}^*$ is very reactive, which makes it extremely toxic as it reacts with components of biological cells and destroys them. To protect healthy cells carotene is injected. Carotene reacts with ${}^1O_2{}^*$ causing oxygen transfer to the harmless triplet oxygen state ${}^3O_2$. The advantage of photodynamic therapy in cancer treatment over commonly used radio– and chemotherapy is selective destruction of diseased cells while saving healthy cells to a large extent. The side effects of many PDT therapies include skin sensitivity, therefore patients have to avoid intense light for a few weeks. The aim of modern PDT therapies is to eliminate these burdensome side effects.

10.2.2. Sensitizers

In most clinical applications hematoporphyrin derivatives (HPD) as well as dihematoporphyrin ethers (DHE) are used. The commercial name for DHE is sodium porfimer.

Figure 10.3 shows the porphine structure as well as the numbering used in this class of compounds.

Porphyrins consist of four aromatic pyrrole rings bonded by methene bridges =C— like porphine, but hydrogen is replaced by:

– A	acetic acid	($-CH_2COOH$)
– P	propionic acid	($-CH_2CH_2COOH$)
– M	methyl group	($-CH_3$)
– V	vinyl group	($-CH{=}CH_2$)

Porphyrins are divided onto:

– uroporphyrins	(containing only A and P)
– coproporphyrins	(containing only M and P)
– protoporphyrins	(containing M, P and V)

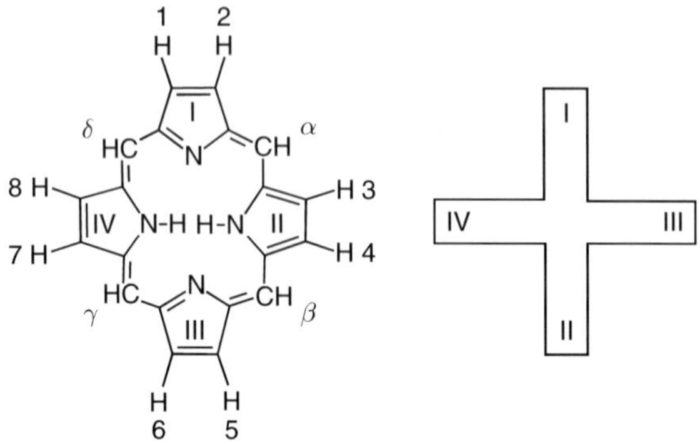

Fig. 10.3 Porphine structure.

The names of porphyrins contain also a number, e.g., uroporphyrin I. The number I defines a regular substituent repetition, e.g., AP AP AP AP, beginning with the pyrrole ring I. For porphyrins numbered with III, the order in ring IV is reversed: AP AP AP PA [16–20].

Some porphyrins such as dihematoporphyrin (Fig. 10.4) have already found application in photodynamic therapy or they have reached the III phase of clinical tests. HPD and DHE belong to the first generation of sensitizers. Their main side effect is skin photosensitivity. To reduce the side effects and increase efficacy, investigations have been made to synthesize second- and third-generation sensitizers, which absorb at longer wavelengths (>650 nm). This group consists of porphyrin, purpurin, benzoporphyrin, phthalocyanine, and naphthalocyanine derivatives. For phthalocyanine or naphthalocyanine, which absorb at 670 nm and 770 nm, the photosensitivity side effect disappears. The second- and third-generation sensitizers are in a relatively early stage of clinical tests in treatments of primary skin cancer, skin tumor metastases and chronic stable skin psoriasis.

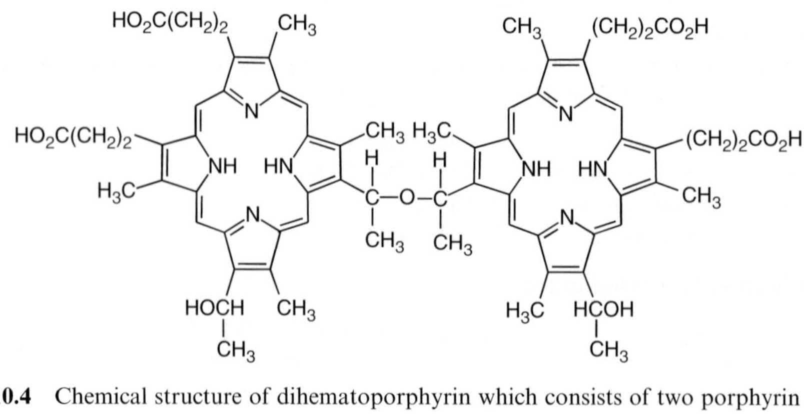

Fig. 10.4 Chemical structure of dihematoporphyrin which consists of two porphyrin rings.

10.2.3. Photochemistry of Sensitizers

There are two main mechanisms of photochemical reactions in sensitizers – I and II type photooxidation. In type I photooxidation, the sensitizer reacts directly with another chemical entity by hydrogen or electron transfer to yield transient radicals, which react further with oxygen. In type II photooxidation, the sensitizer triplet interacts with oxygen, most commonly by energy transfer, to produce an electronically excited singlet state of oxygen, which can react further with a chemical entity susceptible to oxidation.

The sensitizer in a singlet state, ^1S, absorbs a photon of energy $h\nu$ and is promoted to the singlet excited state, $^1S^*$. The excited singlet state $^1S^*$ emits the energy as fluorescence or in a radiationless way, returning to the ^1S state or crossing to the excited triplet state $^3S^*$ as a result of intersystem crossing (ISC) with breaking of the selection rule (spin change). The return from the triplet state to the ground singlet state ^1S may occur via emission of phosphorescence (Fig. 10.5).

The triplet state $^3S^*$ can also vanish as a result of proton transfer or electron transfer between the sensitizer and another chemical entity (RH) (for example substances that are the components of a human cell)

$$^3S^* + RH \rightarrow SH^\bullet + R^\bullet \quad (10.1)$$

$$^3S^* + RH \rightarrow S^{\bullet-} + RH^{\bullet+} \quad (10.2)$$

This is the type I photooxidation reaction. The reactions (10.1) or (10.2) induce further reactions with the oxygen triplet state $^3O_2$ contained in a cell environment

$$SH^\bullet + {}^3O_2 \rightarrow {}^1S + HO_2{}^\bullet \quad (10.3)$$

$$S^{\bullet-} + {}^3O_2 \rightarrow {}^1S + O_2{}^{\bullet-} \quad (10.4)$$

The other type of energy dissipation from the triplet state $^3S^*$ of the sensitizer is type II photooxidation. In type II photooxidation the triplet state of a sensitizer $^3S^*$ interacts directly with the oxygen triplet state $^3O_2$, leading to generation of the singlet excited oxygen state $^1O_2{}^*$

$$^3S^* + {}^3O_2 \rightarrow {}^1S + {}^1O_2{}^* \quad (10.5)$$

Oxygen in the singlet excited state is very reactive. This leads to oxidation of cell components such as proteins, lipids, and nucleic acids and eventually to necrosis of the cell.

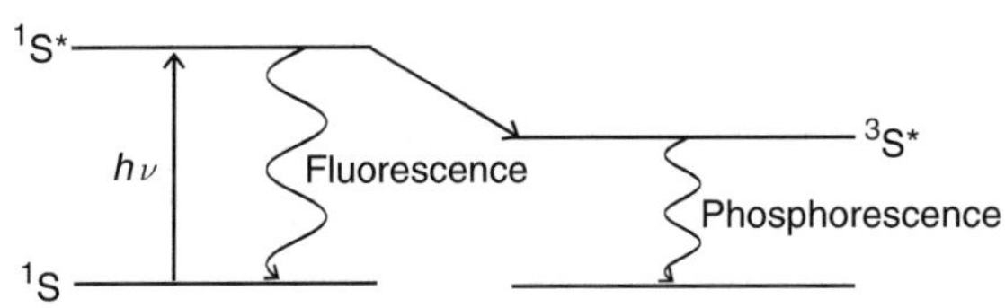

Fig. 10.5 Jablonski's diagram for a sensitizer S.

10.3. THERMAL INTERACTION

Thermal interactions are induced in a tissue by the increase in local temperature caused by a laser beam. In contrast to photochemical interactions, thermal interaction may occur without only specific reaction path and is highly non-selective and non-specific. Depending on the temperature achieved, the thermal effect on the tissue can be classified as:

1. reversible hyperthermia ($T > 31\,°C$) – some functions of the tissue can be perturbed but the effect is reversible,
2. irreversible hyperthermia ($T > 42\,°C$) – some fundamental functions of the tissue can be destroyed irreversibly,
3. coagulation ($T > 60\,°C$) – the tissue becomes necrotic,
4. vaporization ($T \geq 100\,°C$),
5. carbonization ($T > 150\,°C$),
6. pyrolysis ($T > 300\,°C$).

In some cases all of these thermal effects can be observed as a result of interaction with the laser (Fig. 10.6). In most applications one effect usually dominates, depending

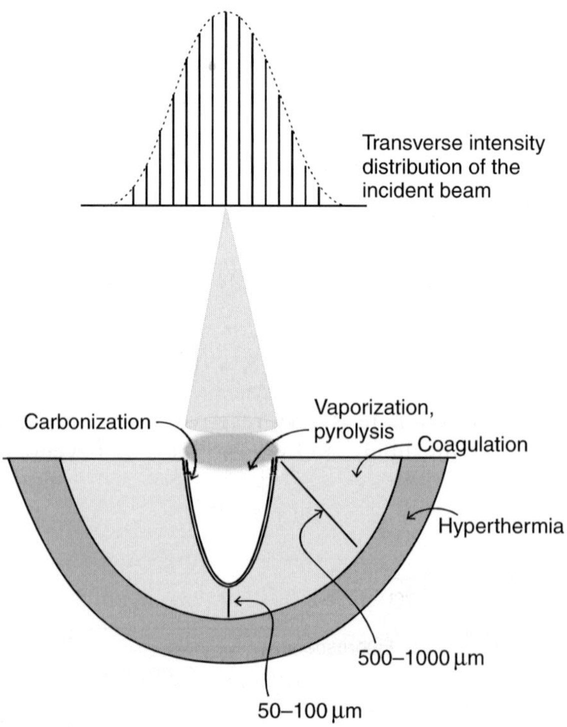

Fig. 10.6 The thermal effects on tissue.

on the goal of the surgery. For example, an Nd:YAG laser beam traveling long path in the tissue (see Table 10.1) is used for coagulation, whereas CO_2 lasers are more suitable for vaporization.

10.4. PHOTOABLATION

The mechanism of photoablation is illustrated in Fig. 10.7. A molecule is promoted to the repulsive excited state (a) (or to the Franck-Condon vibrationally hot state) followed by dissociation (b). The chemical bond is broken, leading to the destruction of biological tissue. As electronic transitions occur usually in the UV range, the photoablation process is usually limited to UV lasers. Therefore, excimer lasers (ArF, KrF, XeCl, XeF) are mainly employed but higher harmonics of other lasers can also be applied.

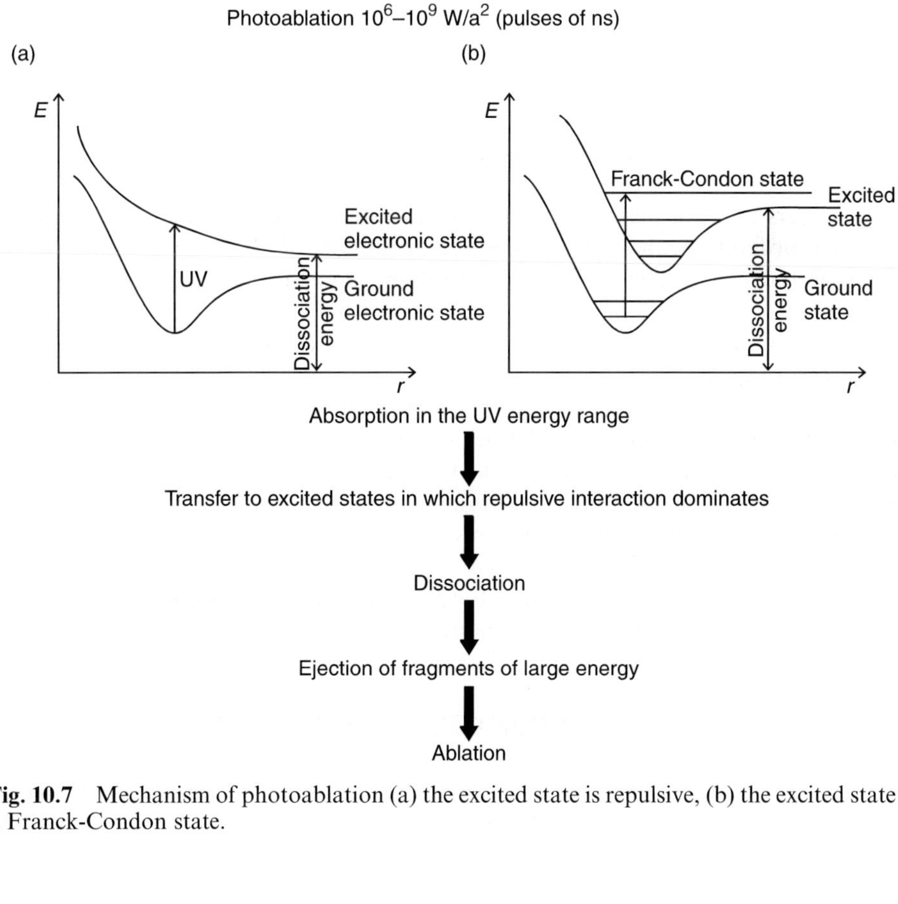

Fig. 10.7 Mechanism of photoablation (a) the excited state is repulsive, (b) the excited state is a Franck-Condon state.

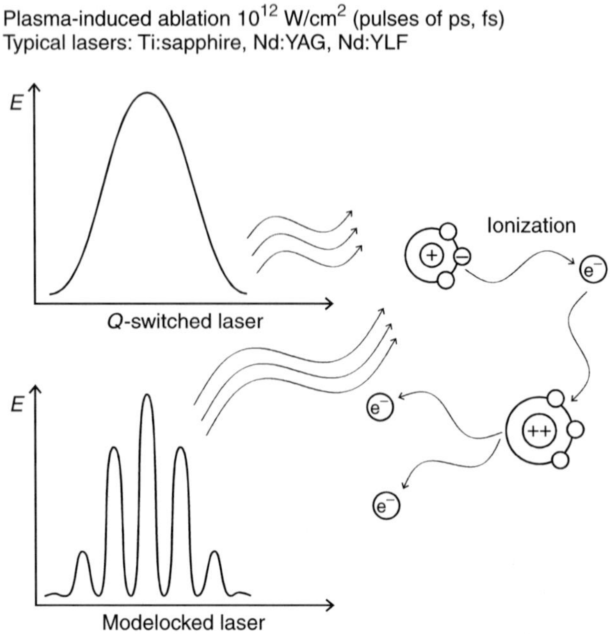

Fig. 10.8 Illustration of plasma-induced ablation.

10.5. PLASMA-INDUCED ABLATION

Typical lasers used for plasma-induced ablation are Nd:YAG, Nd:YLF, Ti:sapphire with pico- or femtosecond pulses generating irradiance at about $10^{12}\,W/cm^2$. Such a high power density leads to fields of $10^7\,V/cm$, comparable with the energies of electrons revolving in atoms. Therefore, the Q-switched or modelocked lasers can ionize molecules in biological tissue. The mechanism of plasma-induced ablation is illustrated in Fig. 10.8.

An ultrashort pulse from a Q-switched or modelocked laser ionizes biological tissue and generates a very large density of free electrons in a very short period of time with typical values of $10^{18}\,cm^{-3}$ due to an avalanche effect. Free electrons from ionization accelerate to high energies and collide with molecules, leading to further ionization. Light electrons and heavy ions move at different velocities, leading to the effect similar to that in the acoustic wave with areas of compression and dilation. The detailed mechanism of disruption for plasma-induced ablation based on fundamental relations of classical electrodynamics and plasma physics is developed in [2].

10.6. APPLICATION OF LASERS IN MEDICINE

Lasers have many varied applications in dentisty, cardiovascular medicine, dermatology, gastroenterology, gynecology, neurosurgery, ophthalmology and otolaryngology [2–3].

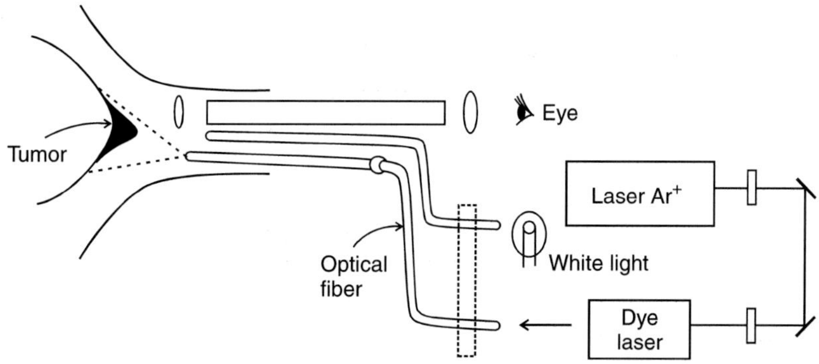

Fig. 10.9 Experimental setup for laser-induced photodynamic thermotherapy. The fiber is placed in the tissue with a transparent catheter.

In medicine it is necessary to evaluate the advantages and disadvantages of laser applications, their costs and benefits. First, one needs to ask whether the laser technology can do more than conventional approaches or whether they can do it cheaper. Second, in order to extend future applications, developments must be based on understanding of the effects of various parameters on laser–tissue interactions. In recent years laser-based applications have increased and become more attractive with the development of fiberoptics allowing access to biological tissue (Fig. 10.9). In the past, laser applications were limited to the surface of the skin.

In cardiovascular diseases, lasers are mainly used for laser angioplasty, laser thrombolysis, photochemotherapy, laser treatment of arrhythmias and transmyocardial revascularization. The thermal interaction, photoablation and photochemical interactions are used in these treatments. For example, laser angioplasty uses thermal effects to vaporize the plaque material, in contrast to balloon angioplasty where the plaque material is fractured, compressed or displaced. With the development of flexible catheters (Fig. 10.10) the level of complications due to perforation or occlusion has been reduced. Argon, Nd:NAG, and CO_2 lasers are used in thermal

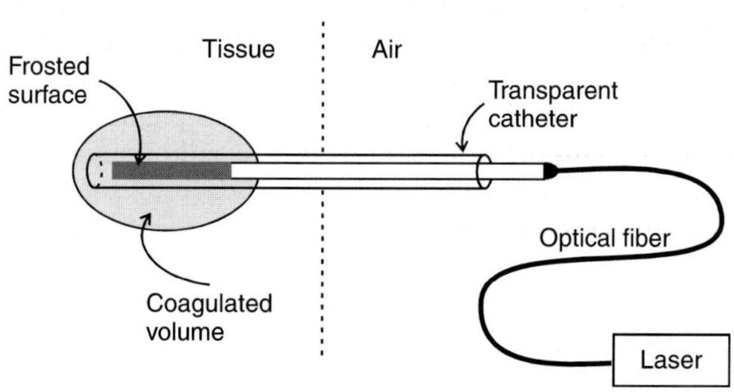

Fig. 10.10 Experimental setup for laser-induced interstitial thermotherapy. The fiber is placed in the tissue with a transparent catheter.

angioplasty. However, artheriosclerotic lesions have also been treated by photoablation with minimal thermal damage and this can be achieved using a variety of pulsed lasers at UV, VIS and IR wavelengths [21].

If lasers can cut metals and other hard objects, why not use them to drill teeth? One of the most obvious applications of lasers is removal of dental enamel, dentin, bone or cementum, instead of using an uncomfortable drill. The major problem with hard tissue in dental laser applications using Er:YAG ($\lambda = 2.94\,\mu m$), Er:YSGG ($\lambda = 2.79\,\mu m$), holmium, CO_2 and excimer lasers is the rise in temperature that causes damage to the pulp of the tooth. The best results appear to be obtained when the tissue is simultaneously cooled with a water spray. The hard tissue procedures have not been cleared yet by the FDA (Food and Drug Administration) in contrast to laser soft tissue dental applications with the CO_2, Nd:YAG, Ho:YAG and argon lasers. Another accepted procedure includes the use of lasers for the polymerization of light-activated restorative materials. There are numerous other potential applications, including selective removal of malignant tissue with the aid of porphyrin compounds and laser welding of dental bridges.

Lasers have become an essential tool in clinical dermatology practice because of the numerous benefits they can provide. A CO_2 laser is commonly used to ablate or vaporize superficially thin layers of soft tissue or to perform excisional surgery. In the ablation regime a small, focused beam of 0.1–0.2 mm in diameter and high power of 15–25 W is used to produce an irradiance of 50000–100000 W/cm^2. These conditions permit surgeons to perform bloodless incisions in most soft tissue. For example, blood vessels up to 0.5 mm in diameter are immediately sealed by the laser [22].

CO_2 lasers working in excisional mode (ablation) are used to remove vascular tumors, in nail surgery, to remove stromal-independent tumors such as melanoma or squamous cell carcinoma as well as in cosmetic surgery to treat keloids, rhinophyma, and blepharoplasty. CO_2 lasers working in vaporization mode (200–600 W/cm^2, 1–4 W, 1–3 mm diameter) [22–23] are employed to treat benign tumors, cosmetic conditions (tattoos, acne scars, rhinophyna, epidermal nevus), cysts, vascular and infectious lesions.

Nd:YAG lasers have also been commonly used by dermatologists. Originally, this laser was primarily used to photocoagulate large caliber blood vessels [24] commonly found in hemagniomas and nodular port wine stains. However, deep penetration and poor absorption in this spectral region (1064 nm) lead to a wide zone of thermal destruction which limits the use of Nd:YAG in treating many dermatologic disorders [22].

The most common imperfections of the skin, such as pigmented lesions (portwine stains, hemangiomas, lentigines) and tattoos, are usually treated with visible lasers including dye, argon, diode and ruby lasers.

Laser are applied in gastoenterology to treat gastrointestinal hemorrhage from peptic ulcers (Nd:YAG) [25–26], lithotripsy to fragment common duct stones in humans (tunable dye, *Q*-switched Nd:NAG, pulsed Nd:YAG) [27–28] and many other applications [29].

Ophthalmologists have been at the forefront of developing medical uses for new laser technology since the discovery of the first laser [30]. Photocoagulation was the earliest therapeutic laser procedure. This thermal effect is used for photocoagulation of the macula and peripheral retina, with argon, diode, Nd:YAG, krypton, and dye lasers. The other laser procedure is iridectomy and thermal [31] or ablative methods with pulsed Nd:YAG [32]. Recently, the most common use of lasers in ophthalmology is laser modification of corneal curvature. Excimer lasers have received FDA approval to correct myopia or hyperopia. These corrections can be done in LASIK surgery (*laser-assisted-in situ-keratomileusis*). LASIK surgery involves a suction ring that holds the eye steady while the platform for the microkeratome, a cutting instrument, is put in place. The microkeratome glides across the surface of the cornea, cutting through the outer layers. The instrument leaves an uncut part of the outer layer of the cornea to act as a hinge. The microkeratome is removed, the attached corneal flap is lifted out of the way, exposing the underlying layers of cornea to the laser beam, which corrects the curvature of the surface by ablation [33].

REFERENCES

1. Tibor Juhasz, Ron Kurtz, Ferenc Raksi, Carlos Suarez, Christopher Horvath, Greg Spooner, *The Femtosecond Blade: Applications in Corneal Surgery*, Opt. Photon. News 13 (1) (January, 2002) 24–29
2. M.H. Niemz, *Laser-Tissue Interactions, Fundamentals and Applications*, Springer-Verlag, Berlin Heidelberg New York (1996)
3. *Laser Surgery and Medicine. Principles and Practice*, Carmen A. Puliafito ed., Wiley-Liss (1996)
4. *Optical-Thermal Response of Laser-Irradiated Tissue*, A.J. Welch, M.J.C. van Gemert eds., Plenum Press, New York (1995)
5. J.-L. Boulnois, *Photophysical Processes in Recent Medical Laser Developments: A review*, Lasers in Med. Sci. 1 (1986) 47
6. E. Mester, G. Gyenes, J.G. Tota, *Experimentelle Untersuchungen über die Wirkung der Laserstrahlen auf die Wundheilung*, Z. Exp. Chirur. 2 (1969) 94–101
7. E. Mester, T. Spiry, B. Szende, J.G. Tota, *Effect of Laser Rays on Wound Healing*, Am. J. Surg. 122 (1971) 532–535
8. E. Mester, B. Szende, P. Gärtner, *Die Wirkung der Laserstrahlen auf den Haarwuchs der Maus*, Radiobiol. Radiother. 9 (1969) 621–626
9. R. Brunner, M. Landthaler, D. Haina, W. Waidelich, O. Braun-Falco, *Experimentelle Untersuchungen zum Einfluss von Laserlicht niedriger Leistungsdichte auf die Epidermisregeneration*, in: *Optoelektonik in der Medizin*, W. Waidelich ed., Springer-Verlag, Berlin Heidelberg New York (1984)
10. R.F. Lyons, R.P. Abergel, R.A. White, R.M. Dwyer, J.C. Castel, J. Uitto, *Biostimulation of wound healing in vivo by a helium-neon laser*, Ann. Plast. Surg. 18 (1987) 47–50
11. D.J. Castro, R.P. Abergel, C. Meeker, R.M. Dwyer, M.A. Lesavoy, J. Uitto, *Effects of the Nd:YAG laser on DNA synthesis and collagen production in human skin fibroblast cultures*, Ann. Plast. Surg. 11 (1983) 214–222
12. T. Kubasova, L. Kovacs, Z. Somosy, P. Unk, A. Kokai, *Biological effect of He-Ne laser: investigations on functional and micromorphological alterations of cell membranes, in vitro*, Lasers Surg. Med. 4 (1984) 381–388
13. M. Boulton, I. Marshall, *He-Ne laser stimulation of human fibroblast proliferation and attachment in vitro*, Lasers Life Sci. 2 (1986) 125–134
14. D.J. Carrillo, J. Calatayud, F.J. Manso, E. Barberia, J.M. Martinez, M. Donado, *A randomized double-blind clinical trial on effectiveness of helium-neon laser in the prevention in pain, swelling and trismus after removal of impacted third molars*, Int. Dent. J. 40 (1990) 31–36

15. S. Taube, J. Piironen, P. Ylippavalniemi, *Helium-neon laser therapy in the prevention of post-operative swelling and pain after wisdom tooth extraction*, Proc. Finn. Dent. Soc. 86 (1990) 23–27
16. *The Porphyrins*, vol. 1, Structure and Synthesis. Part A, D. Dolphin ed., Academic Press, New York (1978)
17. I. Rosenthal, E. Ben-Hur, *Phthalocyanines. Properties and Applications*; C.C. Leznoff, A.B. Lever eds., VCH Publishers Inc., New York (1989) 393
18. R. Bonnett, G. Martinez, Tetrahedron 57 (2001) 9513
19. S.K. Wu, H.C. Zhang, G.Z. Cui, D.N. Xu, H.J. Xu, Acta Chim. Sinica 43 (1985) 10
20. R. Bonnett, Rev. Contemp. Pharmacother. 10 (1999) 1
21. *Laser Surgery and Medicine. Principles and Practice*, Carmen A. Puliafito ed., Wiley-Liss (1996) p.1
22. *Laser Surgery and Medicine. Principles and Practice*, Carmen A. Puliafito ed., Wiley-Liss (1996) p.61
23. J.L. Ratz, P.L. Bailin, *The case for use the carbon dioxide laser in the treatment of port-wine stains*, Arch. Dermatol. 123 (1987) 74–75
24. M. Landthaler, D. Haina, R. Brunner, W. Waidelich, O. Braun-Falco, *Neodymium-YAG laser therapy for vascular lesions*, J. Am. Acad. Dermatol. 14 1(986) 107–17
25. C.P. Swain, S.G. Bown, D.W. Storey, J.S. Kirkham, T.C. Northfield, P.R. Salmon, *Controlled trial of argon laser photocoagulation in bleeding peptic ulcers*, Lancet 2 (1981) 1313–1316
26. C.P. Swain, J.S. Kirkham, P.R. Salmon, S.G. Bown, T.C. Northfield, *Controlled trial of Nd-YAG laser photocoagulation in bleeding peptic ulcers*, Lancet 1 (1986) 1113–1117
27. N.S. Nishioka, P.C. Levins, S.C. Murray, J.A. Parrish, R.R. Anderson, *Fragmentation of biliary calculi with tunable dye lasers*, Gastroenterology 93 (1987) 250–255
28. C. Ell, G. Lux, J. Hochberger, D. Muller, L. Demling, *Laser-lithotripsy of common bile duct stones*, Gut 29 (1988) 746–751
29. N.S. Nishioka, *Laser Surgery and Medicine. Principles and Practice*, Carmen A. Puliafito ed., Wiley-Liss (1996) chapter 4, p. 83
30. J.M. Krauss, C.A. Puliafito, *Laser Surgery and Medicine. Principles and Practice*, Carmen A. Puliafito ed., Wiley-Liss (1996) chapter 8, p. 249
31. M.M. Rodrigues, B. Streeten, G.L. Spaeth, L.W. Schwartz, *Argon laser iridotomy on primary angle closure or pupillary block glaucoma*, Arch Ophthalmol. 96 (1978) 2222–2230
32. L. Schwartz, *Laser iridectomy*, in *Laser therapy of the anterior segment: A practical approach*, L. Schwartz, G. Spaeth, G. Brown, eds., N.J. Thorofare, Charles B. Slack (1984) pp 29–58
33. I.G. Pallkaris, M.E. Papatzanaki, E.Z. Stathi, O. Frenschock, A. Georgiadis, *Laser in situ keratomileusis*, Lasers Surg. Med. 10 (1990) 463–468

– 11 –

Potential Hazards Associated with Inappropriate Use of Lasers

Lasers are used in many fields such as medicine, transport, the military, industry, trade, and printing. The small group of researchers using lasers in scientific laboratories in the early days has expanded considerably and every user needs to be conscious of the potential hazards associated with the inappropriate use of lasers. Of course, universally accessible commercial devices such as printers, barcode readers or audio-visual players use small power lasers, and their construction ensures safety of usage.

Laser safety standards were established by national work safety organizations. In most European countries, requirements are based on standards provided by the ANSI (American National Standards Institute, Z136.1, Safe Use of Lasers **1993**). The main criterion of safety classification is based on the potential hazard associated with the use of lasers. The most important parameters characterizing the potential hazard are AEL (*Accessible Emission Limit*) and MPE (*Maximum Permissible Exposure*). Based on these parameters, lasers have been classified into four classes; the higher the class, the more the potential hazard. The values of AEL and MPE were summarized for radiation of different wavelengths. The laser radiation is characterized by using units of radiant energy (J), radiant power (W), radiant exposure ($J\,m^{-2}$), irradiance ($W\,m^{-2}$), radiant energy density ($J\,m^{-3}$), radiant intensity ($W\,sr^{-1}$), and radiance ($W\,m^{-2}\,sr^{-1}$) and can be found in regulatory materials of national standards agencies.

Class 1: denotes entirely safe laser systems that do not pose a hazard, maximum level of acceptable MPE exposition must not be exceeded under normal operating conditions.

Class 2A: denotes low power lasers or laser systems that do not pose a hazard if the beam is viewed directly for a period not exceeding 1000 s.

Class 2: denotes low power devices emitting VIS radiation (400–700 nm) that do not pose a hazard because of the normal human aversion (blinking, eye movement) but present some potential hazard if viewed directly for an extended period of time.

Class 3A: denotes lasers having a cautionary label. They do not pose a hazard if viewed for only momentary periods but may present a hazard if viewed using collecting optics. The lasers have a DANGER label and are capable of exceeding permissible exposure levels in 0.25 s.

Class 3B: denotes lasers that can pose a hazard if viewed directly. They do not produce a hazardous diffuse reflection.

Class 4: denotes lasers that pose a hazard not only from direct viewing and direct reflection but may also produce hazardous diffuse reflection.

Lasers pose hazards that can be classified as:

- radiation hazards,
- chemical hazards,
- electrical hazards,
- other hazards associated with a specificity of a laser system.

The following safety procedures for each laser classification are proposed.

Class 1: warning sign should be located at an access panel to alert a user that more hazardous laser radiation is contained therein.

Class 2: a) cautionary label, b) indicator light to indicate laser emission, c) do not permit a person to stare at the laser, d) do not point the laser at a person's eye.

Class 3: a) never point the laser at a person's eye, b) use safety eyewear, c) NOTICE label posted outside the area of laser work, d) labeling a laser with warning sign DANGER, e) a warning light or buzzer may be used to indicate when the laser is operating, f) permit only experienced personnel to operate the laser, g) a key switch should be used if untrained person may gain access to the laser, h) terminate the primary and secondary beams at the end of their paths, i) attempt to keep laser beam path above or below eye level, j) mount the laser on a firm support mounting.

Class 4: a) install appropriate DANGER signs, b) interlock entrances to assure that the laser cannot emit when the door is open, c) wear eye protection, d) if the laser produces skin or a fire hazard, put a suitable shield between the laser beam and personnel, e) enclose the entire laser beam if possible, f) a key switch must be used to permit only authorized personnel to operate the laser, g) terminate the path with an absorbing, fire resistant target.

11.1. RADIATION HAZARDS

Lasers emit radiation from the ultraviolet to infrared. The radiation emitted from lasers is non-ionizing radiation in contrast to more dangerous ionizing radiation emitted by X-ray and gamma ray sources. However, the continuous development of laser technology, particularly the extension toward new spectral ranges into the deep UV and soft X, as well as higher intensities and repetition rates are bringing new problems related to the safety of users. To date, radioprotection has not been so important, but the development of high power lasers (>10 TW) with strong emission that may lead to multi-phonon ionization raises questions about new measures of radiation protection. Radiation from lasers is most dangerous to eyes and skin.

11.1.1. Eye Hazards

Let us discuss briefly the most typical potential hazard for eyes. The laser beam of UV light from the 180–315 nm region (actinic or middle UV) is absorbed by the cornea and may cause inflammation (*photokeratitis*) (Fig. 11.1). Light from this range is emitted by the excimer lasers such as ArF (193 nm), KrF (248 nm), XeCl (308 nm) and the fourth harmonic of the solid-state laser Nd:YAG laser (266 nm).

The light from the 315–400 nm range (near UV-A) is absorbed by the lens and can induce some forms of *cataract* (Fig. 11.2). Light from this range is emitted by the excimer XeF laser, nitrogen laser (337 nm) as well as the third harmonic of the Nd:YAG laser (354 nm).

The laser light from the 400–1400 nm range (visible and near IR) is not absorbed by external layers of the eye and passes by them non-attenuated. It is dangerous because the light passing through the lens is focused on the retina where it creates a point image of 10–20 μm in size (Fig. 11.3). Focusing a laser beam on a small area of retina causes the radiant exposure ($J\,m^{-2}$) to increase about 100,000 times in comparison with the incident exposure. Radiation from the 400–550 nm range (the second harmonic of the Nd:YAG (532 nm) laser, argon laser, krypton laser, dye lasers) is particularly dangerous. The point where the retina merges with the optic nerve is called the *optic disk*. If the focused laser beam strikes this area it will most likely cause blindness in the eye.

The light from the mid- and far infrared (IR-B and IR-C) range of 1400 nm–1 mm is absorbed in the front surface of the eye, in the cornea. For some wavelengths from the range of 1.4–3 μm, the radiation penetrates the tissue of the eye more deeply and can cause cataracts, similar to the effect generated by the UV range of 315–400 nm. The

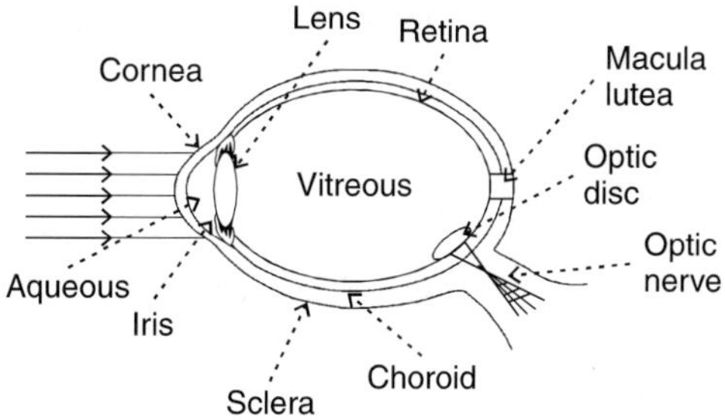

Fig. 11.1 180–315 nm radiation absorption in an eye.

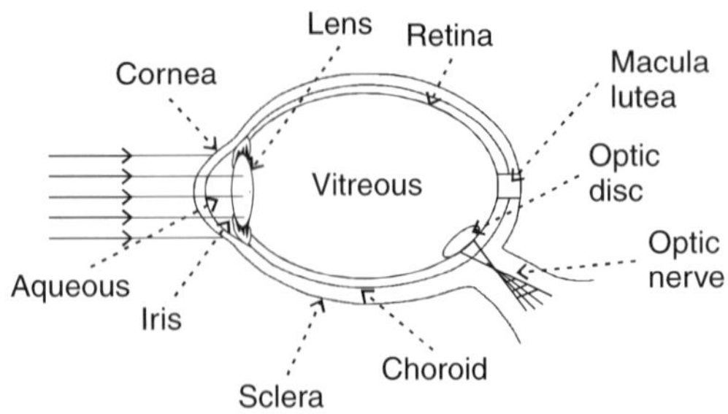

Fig. 11.2 Absorption of 315–400 nm radiation in the eye.

emission of CO (5–6.5 μm), CO_2 (10.4 μm), Ho:YAG (2120 nm), Er:YAG (2940 nm) occurs in this area. Table 11.1 summarizes basic biological effects of light in an eye.

One should avoid viewing laser beams directly, reflected from different surfaces as well as diffuse or scattered reflection (for the 4th class). We must wear appropriate protective eyewear when the beam is exposed to prevent all of the hazards mentioned above. It is obvious the protective filters in the eyewear should be carefully matched to the spectral range and laser parameters. The parameter that is used to specify eye protective filters is the *optical density* $OD(\lambda)$ for a given radiation wavelength λ, defined as

$$OD(\lambda) = \log_{10}\frac{H_0}{\mathrm{MPE}}, \tag{11.1}$$

where H_0 is the most likely irradiance or radiant exposure (expressed in W/cm^2 for *cw* lasers and J/cm^2 for pulsed lasers). MPE denotes the maximum permissible exposure that characterizes the radiation level, which humans can be exposed to without any harmful results. The MPE values depend on the wavelength of radiation, the pulse duration, exposure time as well as the kind of tissue subjected to exposure. The MPE values have been classified by national organizations for commercially available *cw* lasers and pulsed lasers. Most of the European countries adopted the ANSI standards. Some MPE values for the Nd:YAG laser and argon lasers are summarized in Table 11.2.

Let us calculate $OD(\lambda)$ for the pulsed Nd:YAG laser emitting at 1064 nm. Let the pulse energy be 80 mJ, and the time duration 15 ns. According to Table 11.2, the

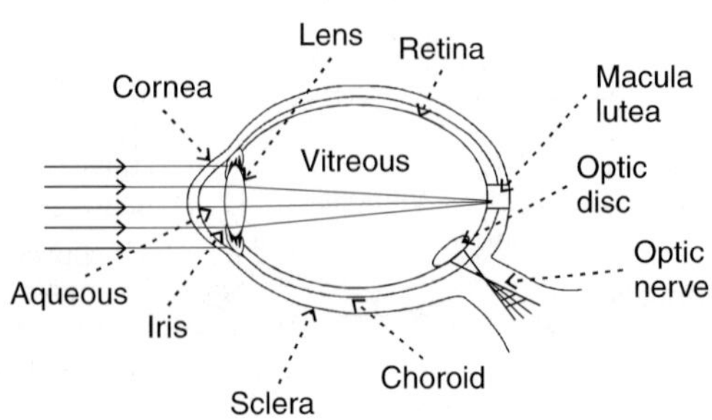

Fig. 11.3 Absorption of 400–1400 nm radiation in the eye.

Table 11.1

Basic biological effects of light in an eye

Range	Eye effects
Ultraviolet C (200–280 nm)	Photokeratitis
Ultraviolet B (280–315 nm)	Photokeratitis
Ultraviolet A (315–400 nm)	Photochemical UV cataract
Visible (400–780 nm)	Photochemical and Thermal Retinal Injury
Infrared A (780–1400 nm)	Cataract Retinal burns
Infrared B (1400–3000 nm)	Corneal burn Aqueous flare IR cataract
Infrared C (3000–10000 nm)	Corneal burn

MPE is $5 \cdot 10^{-6}$ J/cm^2. Let the diameter of the output laser beam be $d = 2$ mm, and the beam divergence of $\Phi = 1.0$ mrad (Fig. 11.4). In the worst case the maximum radiant exposure is

$$H = \frac{80\,\text{mJ}}{\pi(0.1\,\text{cm})^2} = 2.55\,\text{J/cm}^2. \tag{11.2}$$

In practice, the beam divergence reduces the H value. The standard values of MPE were established for larger diameter of the output beam –7 mm (maximum pupil size) giving H_0 value

$$H_0 = \frac{80\,\text{mJ}}{\pi(0.35\,\text{cm})^2} = 0.21\,\text{J/cm}^2. \tag{11.3}$$

Therefore, the filter will be characterized by the optical density value $OD(\lambda)$

$$OD(\lambda) = \log_{10}\frac{0.21}{5 \cdot 10^{-6}} = 4.62 \qquad \text{for} \qquad \lambda = 1064\,\text{nm}. \tag{11.4}$$

Table 11.2

MPE values for selected lasers

Laser type	Wavelength [nm]	MPE value for eye
Nd:YAG Laser	1064	5 μJ/cm^2 for pulsed lasers from 1 ns to 50 μs, 9 t$^{3/4}$ mJ/cm^2 from 50 μs to 1000 s, 1.6 mW/cm^2 for longer time
Ion Argon Laser	488/514	0.5 μJ/cm^2 from 1 ns to 18 μs, 1.8 t$^{3/4}$ mJ/cm^2 from 18 μs to 10 s, 10 mJ/cm^2 from 10 s to 10 000 s, 1 μW/cm^2 for longer pulses

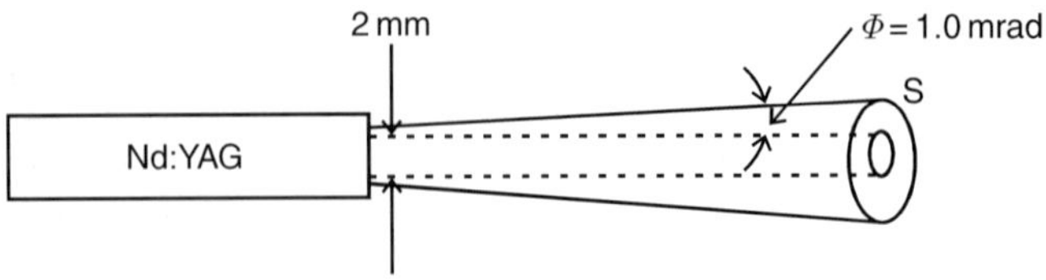

Fig. 11.4 Illustration of parameters needed to estimate a proper optical density OD.

This indicates that the filter in the eyewear required for this specific application should have the optical density 5. Analogous estimations for the argon laser emitting continuous radiation at $\lambda = 514\,\text{nm}$ at a power of 1 W give the following values: $\text{MPE} = 1 \cdot 10^{-6}\,\text{W/cm}^2$ and

$$H_0 = \frac{1\,\text{W}}{\pi(0.35\,\text{cm})^2} = 2.59\,\text{W/cm}^2 \qquad \text{for} \qquad \lambda = 514\,\text{nm}. \tag{11.5}$$

Therefore,

$$OD(\lambda) = \log_{10}\frac{2.59}{1 \cdot 10^{-6}} = 6.4. \tag{11.6}$$

11.1.2. Skin Hazard

In most laser applications, hazard to the skin is smaller than that to the eyes. However, extending the range of laser emission to the ultra-violet (and recently even to the soft X region) poses a hazard to the skin as well. The 230–380 nm range is the most dangerous. Particularly dangerous is the UV-B (280–315 nm) range. Radiation from this range may accelerate skin aging, skin burns or even skin cancer. In the 310–600 nm range (near UV/VIS), undesired photochemical reactions on the skin may occur whereas radiation from the 700–1000 nm range (IR) may cause burning of the skin. Table 11.3 summarizes the most typical biological effects of light on skin.

Table 11.3

Basic biological effects of light in skin

Range	Skin effects
Ultraviolet C (200–280 nm)	Erythema (sunburn)
Ultraviolet B (280–315 nm)	Accelerated skin aging Increased pigmentation
Ultraviolet A (315–400 nm)	Pigment darkening
Visible (400–780 nm)	Photosensitive reactions Skin burn
Infrared A (780–1400 nm)	Skin burn
Infrared B (1400–3000 nm)	Skin burn
Infrared C (3000–10000 nm)	Skin burn

11.2. OTHER HAZARDS

Some lasers can cause a chemical hazard. As described in Chapter 4 the excimer, dye and chemical lasers contain toxic substances as an active medium that can be dangerous to health. Gas lasers such as CO and CO_2 lasers contain toxic gases, which under faulty operation can leak from the laser pipe. Moreover, lasers induce many reactions, which can emit dangerous gas products.

Electrical hazard is another potential danger. In many lasers high voltage is used (>1 kV), but some *Q*-switched pulsed lasers may use 20 kV. Electrical hazard particularly concerns high power lasers in which inappropriate construction or faulty electrical systems during exploitation may lead even to death. General guidelines for electrical hazard are the following:

1. do not touch high-voltage devices when turned on,
2. do not change capacitors and any other elements with voltage switched on,
3. avoid wearing rings, metallic watch bands and other metallic objects,
4. when possible use only one hand in working on a circuit or control device,
5. never handle the electric equipment when hands, feet or body are wet, or when standing on a wet floor,
6. learn rescue procedure for helping victims,
7. do not engage in laser alignment, operating or adjusting when fatigued, under medication or when your mental activity would incline you toward risk.

One should also mention the X radiation hazard emitted as a result of inappropriate protection of such components as lamps working with a voltage higher than 15 kV. Recent developments extending toward new spectral ranges into the deep UV and soft X as well as higher intensities and repetition rates are bringing new problems related to protection against X-radiation. Flash lamps in optical pumping pose a dual hazard. First, UV radiation is emitted if a quartz tube is used. Moreover, the quartz tube can explode on occasion, so they should be provided with covers. Another associated hazard is noise. The primary source of noise comes from the capacitor bank discharges.

– 12 –

Detectors

In the previous chapters we focused mainly on sources of laser light, methods of generating ultra-short pulses, spectral range extension, amplification, and time-resolved laser spectroscopy. It is obvious that in the final stage of every experiment the optical signal has to reach a detector, which transforms the light signal into an analog or digital signal. So far we have not concentrated on the characteristics of the detectors, but it is obvious that proper detection determines the final success of the experiment or application. A detector has to be chosen that is able to record the proper spectral range, to convert the optical signal intensity linearly, to record the intensity with a suitable accuracy, and to generate low noise in comparison with the level of the signal. In some kinds of measurements, for example fluorescence measurements monitored in real-time, one should also consider the response time of the detector.

Nowadays, we can find different types of detectors in every home. They are in such devices as electronic toys, alarm devices, and TV remote controls. Although invisible to the telephone user, detectors play a crucial role in fiberoptic communication. These examples illustrate two main streams of everyday applications of detectors: fiberoptic communication and remote sensing. Detectors used in optical fibers record signals from the near-infrared range and they do not require very high detectivity, since they usually record a high intensity laser beam, but they have to be characterized by very fast responses. The detector most often used in fiberoptic communication is the InGaAs photodiode with a capability of sending information at the rate of 2.5 Gbits per second, which is 200,000 times faster than that sent by traditional telephone copper cables.

For the second application, remote sensing, those detectors that record in the near-infrared range are also used, but the requirements concerning the response time are lower than those in fiberoptic communications. This is not surprising as remote sensing serves to transmit small amount of information, usually simple commands such as: switch on, switch off, change the channel. More sophisticated methods of detection must be applied in another important field of remote sensing – the control of technological processes.

The third important group of applications comprises video cameras with charge coupled devices (CCDs).

The fourth field of applications, such as scientific investigations, medical diagnostics, atmosphere pollution monitoring or military applications, usually needs detectors of very high quality. These applications stimulate the further development of detection techniques, with military applications still being the main motor of progress. As an example, consider the far-infrared detectors recording radiation at wavelengths longer than 5 μm, which were constructed for antimissile defense program. With the end of the cold war, military technologies are finding an increasing number of civil applications.

In this chapter we describe the basic types of detectors as well as the basic parameters that should be taken into account in order to choose the most appropriate detector for a specific application. For more detailed description the reader is referred to [1–2].

12.1. DETECTORS TYPES AND DETECTORS CHARACTERIZING PARAMETERS

We can distinguish five basic types of detectors:

a) *Photoemissive detectors,*
b) *Thermal detectors,*
c) *Semiconductor detectors,*
d) *Photodiode arrays PDA*, known as *diode lines,*
e) *Charge coupled devices*, known as *CCD cameras.*

The first group is represented by photomultipliers designed to detect radiation from the UV-VIS–near-IR range.

The second group is represented by calibrated calorimeters with a *thermocouple* or *thermistor*, *bolometers*, *Golay detectors* as well as *pyroelectric detectors*. The thermal detectors utilize heat produced by incident radiation, therefore their detectivity is not dependent on the incident radiation wavelength and they can be used in any spectral range. In the past bolometers and Golay detectors were used nearly exclusively in the infrared range because of the lack of other detectors of high detectivity in this spectral range. The situation has now changed and there are many detectors offering detection in the infrared.

The third group is represented by *photoconductive* and *photovoltaic semiconductor detectors*, which were discussed briefly in Chapter 4.8.1. They are used for UV-VIS–IR detection.

All detectors that belong to groups (a)–(c) are *single channel detectors*. In other words, at a given time period only a single point of the spectrum is measured. If we place a single channel detector behind the exit slit of a monochromator, the detector can measure only one spectral component at a given time. To obtain the entire spectrum scanning across the slit, measurements are needed at each point. The detector has to record radiation many times for different positions of diffraction gratings of the monochromator corresponding to different wavelengths at the exit slit.

Detectors that belong to the groups (d)–(e) are *multichannel detectors*. The diode line PDA permits multiple wavelengths to be detected simultaneously. The CCD cameras also permit simultaneous recording of components corresponding to different wavelengths, but additionally they can record the spatial intensity distribution along the slit. Thus, the CCD camera permits a new type of spectroscopic investigations (*imaging spectroscopy*), giving information not only about spectral distribution but also about the spatial distribution of molecules inducing absorption, emission or scattering. The obvious advantage of diode lines (PDA) and CCD cameras is the much shorter time needed to record the whole spectrum. Sometimes the price paid for this advantage is lower spectral resolution than that of single channel detectors if the number of pixels of the CCD is not sufficient for a given application. Fig. 12.1 illustrates the principle of differences between a single and multichannel detection.

In order to choose the most appropriate detector for a given application, one needs to know the fundamental parameters that characterize detectors. The basic parameters specifying detectors are the following:

a) responsivity, R,
b) noise equivalent power, NEP,
c) detectivity, D,
d) normalized detectivity, D^*,
e) quantum efficiency, QE,
f) dynamic range,
g) response linearity,
h) detector response time τ.

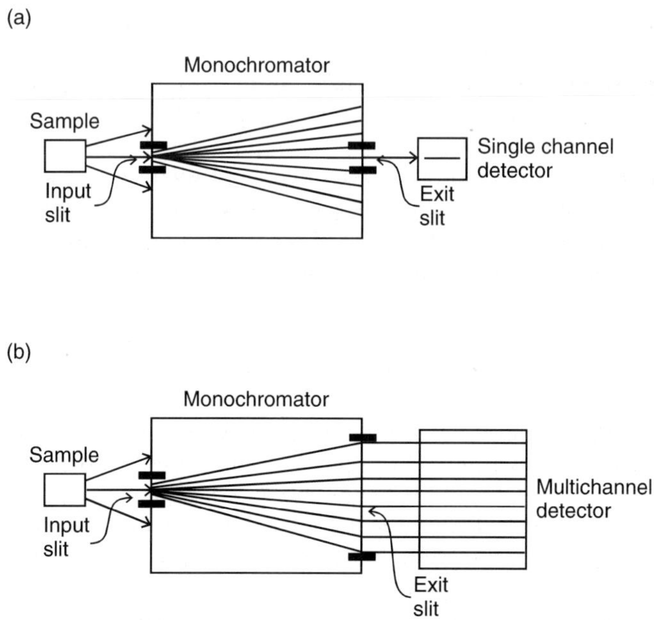

Fig. 12.1 Illustration of differences in detection in (a) a single channel, (b) multichannel detection.

Responsivity, R, is defined as the ratio of the output voltage (or the output current) S to the incident light power per 1 cm^2 (P) of a detector

$$R = \frac{S}{PA} \quad ([\text{V/W}] \text{ or } [\text{A/W}]), \tag{12.1}$$

where A is the sensitive area of the detector (cm^2) illuminated by the incident radiation. The larger the responsivity R of the detector, the higher is the sensitivity to detect the incident radiation.

Noise equivalent power (NEP) defines the lowest value of the incident light power, P, per 1 cm^2 of a detector for which the detected signal, S, is equal to the noise, N, that is $S/N = 1$. For values higher than the NEP, the signal begins to be distinguished from the noise. Since the ratio S/N is proportional to the square root of the bandwidth $\Delta\nu$, the NEP is expressed by the following

$$\text{NEP} = \frac{PA}{(S/N)\sqrt{\Delta\nu}} \quad ([\text{W/Hz}^{1/2}]). \tag{12.2}$$

The smaller the NEP, the smaller is the value of light intensity that can be recorded as a signal that does not disappear in the background noise.

Detectivity, D, is expressed as

$$D = \frac{1}{\text{NEP}}. \tag{12.3}$$

The larger the value of D, the lower are the powers of the incident radiation that can be measured by a detector.

Normalized detectivity, D*, permits a comparison of different detectors as it defines the detectivity per 1 cm^2 of a detector surface, that is

$$D^* = \frac{(S/N)\sqrt{\Delta\nu A}}{PA} = \frac{(S/N)\sqrt{\Delta\nu}}{P\sqrt{A}} = \frac{\sqrt{A}}{\text{NEP}} \quad ([\text{cm} \cdot \text{Hz}^{1/2}/\text{W}]). \tag{12.4}$$

Quantum efficiency, QE, characterizes only those quantum detectors that fit all groups mentioned at the beginning of the section, except thermal detectors. The quantum efficiency is defined as the ratio of the number of photoelectrons generated in a detector to the number of incident radiation photons. If, for example, the quantum efficiency is 20% this indicates that every fifth photon will generate a photoelectron.

Dynamic range of a detector is determined by the ratio of the smallest distinguishable signal, which can be measured by a detector, to the largest signal before saturation. The most intense signal is equal to the saturation level of a CCD (which is determined by its potential well capacity, which is discussed below) and the weakest signal is limited by the read out noise, which comes from the electronic process of reading the signal from the detector. The dynamic range of a detector as well as the analog-digital converter (ADC) determines the differences in signals, ΔS, that can be distinguished as two independent signals, S and $S + \Delta S$. This value is very important in most pump-probe time-resolved laser spectroscopy measurements. The pumping beam excites only a small fraction of a molecular system. The difference between the perturbed system and the system in equilibrium is characterized just by the signal change, $\Delta S = S - S_0$. If the detector does not distinguish the signals S and S_0, it will record no changes caused by the pumping pulse and monitored by the probing pulse. The dynamic range of a digital detection system is determined by an analog-digital converter, which converts the analog signal of a detector into a digital signal. For a 16-bits converter the full scale of the output signals is $2^{16} = 65536$, which denotes 65536 states that can be distinguished. Therefore, the weakest ΔS that can be measured as a separate component is 1/65536 of the full measured signal range.

Detector linearity denotes that a signal in a detector has to be directly proportional to the incident radiation intensity.

Detector response time, τ, is an important parameter that should be considered when fast dynamic processes are monitored in real-time. It also plays a role in pump-probe measurements with modelocked lasers emitting ultrashort pulses. In many cases, particularly in infrared measurements, gated detectors are used to avoid thermal background noises. The gating time depends on a detector's response time, τ. The response time can be determined by studying the response of a detector to modulation frequency changes. If light illuminating a detector is modulated with frequency f, the signal $S(f)$ is a function of the modulation frequency

$$S(f) = \frac{S(0)}{\sqrt{1 + (2\pi f \tau)^2}}. \tag{12.5}$$

Eq. (12.5) indicates that the signal modulated with frequency f is smaller than the signal $S(0)$ without modulation. When the frequency reaches $f = 1/(2\pi\tau)$, the modulated signal value $S(f)$ decreases $\sqrt{2}$ times with respect to $S(0)$. The higher the modulation frequency at which a detector responds, the shorter is the response time τ.

12.2. PHOTOEMISSIVE DETECTORS

Comparison of parameters such as responsivity, R, noise equivalent power, NEP, detectivity, D, or normalized detectivity, D^*, of different detectors shows that photomultipliers (PMT) belong to very sensitive and very versatile detectors, placing them at the top of radiant energy detectors.

The fundamental principle of operation for all types of photomultipliers is presented in Fig. 12.2. Incoming light crosses through the input window and strikes into the photocathode. If the energy of the incoming photon, $E = h\nu$, is larger or equal to the threshold energy, W, needed to pull an electron out of the metal, the electron leaves the photocathode and moves in the vacuum container with kinetic energy, K, defined by

$$E = h\nu = W + K. \tag{12.6}$$

If the voltage V is applied to photocathode, with the positive potential on the anode, the released electrons begin directional traveling towards an anode and a

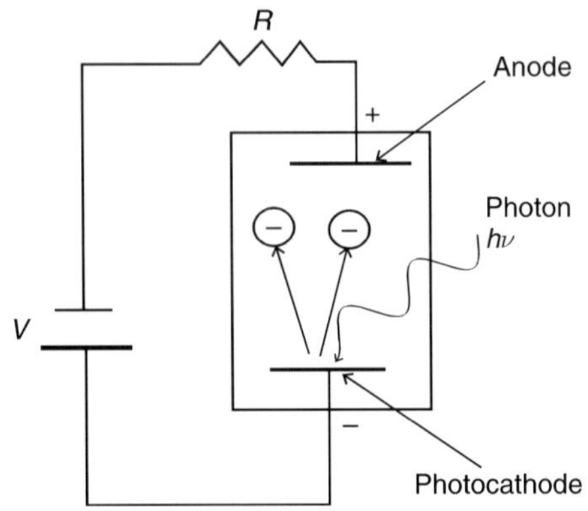

Fig. 12.2 Illustration of photomultiplier operation.

current with an intensity proportional to the number of free electrons will be generated in the circuit.

The threshold energy, W, for typical metals is 4–5 eV. We can calculate from eq. (12.6) the lowest frequency ν of the incoming radiation that can cause an electron to be pulled out of the metal. The lowest frequency ν corresponds to the case when the kinetic energy of the electron, K, is equal to zero. The lowest frequency ν for W equal to 4 eV is $9.7 \cdot 10^{14}$ Hz (310 nm). This number indicates that only photons from the ultraviolet range could eject electrons which would limit considerably the field of photomultiplier applications. In practice, to extend the spectral range of photomultiplier detectivity, the photocathode is covered with different materials characterized by a smaller threshold energy, W (usually a mixture of alkali metals or semiconductors), which permits detection from 115 nm to 1200 nm.

Table 12.1 shows typical photocathode materials and the spectral range they cover.

In practice, the numbers given in Table 12.1 can change over quite a wide range, because the spectral range depends on the photocathode material, but also on the

Table 12.1

Typical materials used as photocathode covering and their spectral ranges

Photocathode material	Spectral range [nm]
CsI	115–200
CsTe	160–320
GaAs	160–930
SbCs	VIS
SbRbCs	VIS
SbKCs	VIS
SbNaKCs	VIS
AgOCs, InGaAs	300–1200

window material as well as on the photomultiplier configuration. The typical windows through which photons enter the photocathode are made of borosilicate, which do not transmit radiation shorter than 300 nm. Windows made of MgF_2, sapphire and glass quartz transmit in the UV range down to 115, 150 and 160 nm, respectively. The detailed specifications of different types of photocathodes can be found in commercial catalogues of companies dealing with the sale of detectors.

In practice, photomultipliers have a much more complicated configuration than that presented in Fig. 12.2. The typical photomultiplier consists of a sealed vacuum column in a glass, ceramic or metal cover. The pressure in the column is on the order of 10^{-4} Pa. Inside the column there is a photocathode and an anode as well as a sequence of dynodes from which the primary electrons eject next electrons, inducing the secondary emission and the avalanche of electrons on the last dynode causing significant growth of anode current. Typical voltages applied between the photocathode and anode are 1–2 kV. This indicates that for a photomultiplier with 10 dynodes the potential difference between the next two dynodes is on the order of 100–200 V.

The box-and-grid type of dynodes configuration presented in Fig. 12.3 is typical for many photomultipliers. Such a configuration is characterized by high quantum efficiency. Other configurations are: a) the *circular-cage* of dynodes resulting in short detector response time τ, b) the *linear focused* configuration characterized by a short response time τ, and signal linearity, c) the *fine-mesh dynodes* employed in a new generation of photomultipliers providing information about the position of primary electrons (*position-sensitive PMT*).

Another way to considerably enhance secondary emission of electrons is a solution applied in the multichannel plate (MCP) presented in Fig. 12.4.

The multichannel plate consists of millions of microscopically small photomultipliers tightly packed into glass capillary tubes (6–20 μm in diameter) (Fig. 12.4a). The capillary matrix surface, coated with a material of low threshold energy W, plays the role of the photocathode. The ejected electrons fall into a capillary where secondary emission takes place since the inside of each capillary is coated with a semiconductive material from which electrons are ejected easily. As a consequence, each individual photoelectron can generate a cloud of secondary electrons. In some MCP configurations the individual photoelectron can generate 10^7 secondary electrons. One of the advantages of MCP detectors, besides high quantum efficiency, is the short response time on the order of 100 ps. The response time of MCP becomes shorter

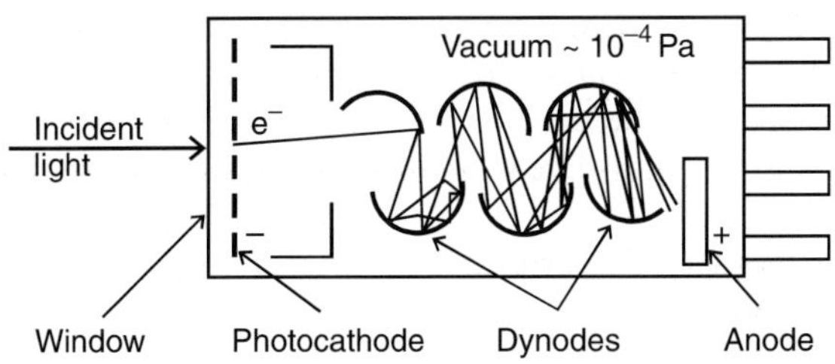

Fig. 12.3 Photomultiplier scheme, box-and-grid type configuration.

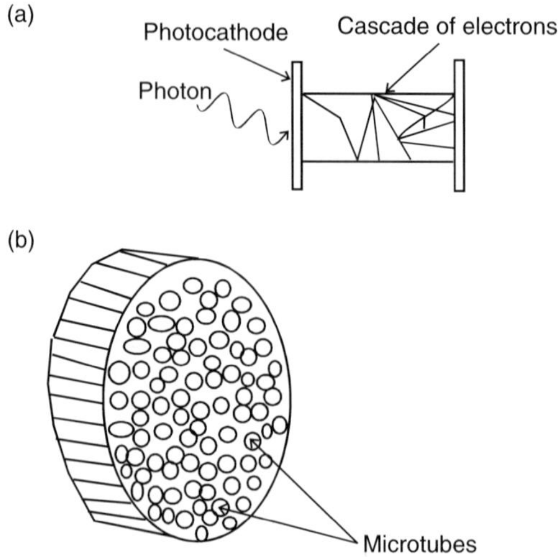

Fig. 12.4 Scheme of multichannel plate (MCP).

in comparison to conventional photomultipliers that is in nanoseconds because the electron path in the capillary is considerably shorter than the time needed to travel between neighboring dynodes in a photomultiplier. Moreover, the multichannel plate MCP is not sensitive to magnetic fields in contrast to the photomultiplier in which a magnetic field can disturb considerably the electron track and reduce the efficiency.

The prices of photoemissive detectors are much higher than the prices of the semiconductor detectors presented in the next section.

12.3. SEMICONDUCTOR DETECTORS

The principle of semiconductor detector operation was presented in Chapter 5, section 5.8.1. The simplest and the cheapest semiconductor detector is a photoconductor, a thin layer (50–100 μm) or bulk form made of semiconductive material (CdS, PbS, PbSe) placed between electrodes. When the material is illuminated by incoming light, electrons are promoted to the conduction band, releasing an electron-hole pair. This causes dramatic decrease of the material resistance and increase of the conductivity resulting in a current flow, which is proportional to the photon irradiance. *Photoconductors*, also called *photoresistors*, are used commercially in simple applications because of their low price, but they are not very precise detectors. They are usually non-linear and have a long response time on the order of 10–100 milliseconds.

To get higher detectivity, linearity and a shorter response time, one should use *p-n* junctions made of semiconductive materials instead of one type of semiconductor as employed in photoconductors. Such a detector employing phenomena occurring on the *p-n* junction is called a *photodiode*. We distinguish two types of photodiodes: the *photoconductive diode* and the *photovoltaic diode*. The principles of their operation were discussed in Chapter 4, section 4.8.1. In photoconductive diodes, a current induced by light begins to flow in the short-circuited case in contrast to the photovoltaic diodes measuring a change in voltage across the open-circuited device. The response time of photoconductive diodes is much shorter than in photoconductors and is about 10 ps. The photoconductive diodes are as cheap as photoconductors. The photovoltaic diodes have a considerably longer response time in comparison with photoconduction diodes as a relatively large charge has to be accumulated on the *p-n* junction to show a change in voltage. Their advantage is the lack of dark current, resulting in a high signal-to-noise (*S/N*) ratio even for small intensities of the incident radiation. Photodiodes permit us to record light from a wide spectral range – from 180 nm (UV) to 3000 nm (IR). Specific materials like HgCdTe can extend this range to 30000 nm. The most commonly used silicon photodiodes are sensitive in the range from 180 nm to 1100 nm. Photodiodes made of gallium indium arsenide, InGaAs, are sensitive in the range of the near infrared, from 800 nm to 1700 nm. Photodiodes made of lead sulfide and lead selenide (PbS and PbSe) cover the range from 800 nm to 3000 nm. Prices of PbS and PbSe detectors are higher than that of silicon detectors. HgCdTe photodiodes are sensitive in the range from 2000 nm to 30000 nm but due to their toxicity and high price they are not commonly used. Employing four-component materials like HgCdZnTe instead of HgCdTe further improves the spectral range and considerably reduces the number of defects in the material, resulting in lower thermal noise. The development in materials technology offers the possibility of far-infrared detectors recording at room temperatures, which so far can work only at low temperatures.

The development of materials engineering has made it possible to produce doped semiconductors that can record radiation from the far-infrared region with wavelengths longer than 20,000 nm. The BIB (*blocked-impurity-band*) type crystals will be probably used in astronomy telescopes, covering the 100,000–200,000 nm range! The discovery of high-temperature *superconductors* (*HTS*) in 1986 stimulated the search for new solutions in the infrared range detection because these materials are characterized by a very small energy gap between the valence and the conduction bands. The superconductor detectors can work as bolometers, recording heat changes due to radiation, or as quantum detectors recording photon-matter interactions. So far commercial bolometers have been made of HTS superconductors ($Co{:}YBa_2Cu_3O_x$) but this situation is changing rapidly due to quantum detectors known as a *Josephson junction* and tunnel junction (*STJ – superconducting tunnel junction*) [3–5].

12.4. MULTICHANNEL DETECTORS PDA AND CCD

Multichannel detectors revolutionized image detection, which for two hundred years was based mainly on photographic techniques. The classical image detectors are the TV camera and vidicon that are still used in some instances. They were recently replaced by devices based entirely on solid-state semiconductive properties.

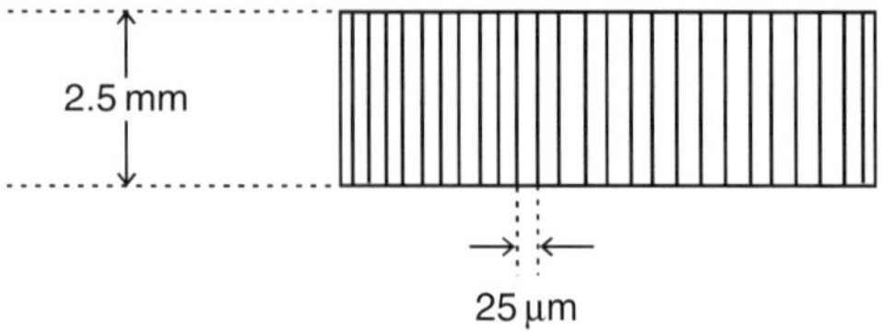

Fig. 12.5 Structure of a diode line.

Diode line PDA and CCD cameras belong to multichannel detectors that can detect multiple wavelengths simultaneously in contrast to single channel detectors (Fig. 12.1) that measure only a single spectrum component.

The diode line is a collection of silicon detectors of typical diode dimension of 25 μm × 2.5 mm placed in a series (Fig. 12.5). The number of the diodes is usually 512, 1024 or 2048. Every diode is a *p-n* junction to which reverse biased voltage is applied (Fig. 4.28) and acts as a charged capacitor. If light illuminates the diodes the electron-hole pairs are created on the junction. The external and internal electrical fields on the junction that have the same direction move electron holes towards the *p* area, and electrons towards *n*, causing charge depletion on the junction like discharging of a capacitor. To refresh the diodes in order to return to the situation before illumination, an additional current signal is needed. The spectral range of a diode line PDA is from 200 to 1100 nm, although below 400 nm the quantum efficiency falls suddenly due to silicon absorption in UV. The detectivity maximum is around 650 nm (about 70% quantum efficiency), and the quantum efficiency at 200 nm is only 14%. The diode line PDA is a collection of individual diodes, and the signal from each of them is read out independently.

The CCD camera is also a collection of microscopic *p-n* junctions that are arranged in an array (Fig. 12.6a) with the common cathode (silicon of type *p*). The anodes made of a metal are isolated from the silicon by a layer of silica dioxide, SiO_2 (Fig. 12.6b).

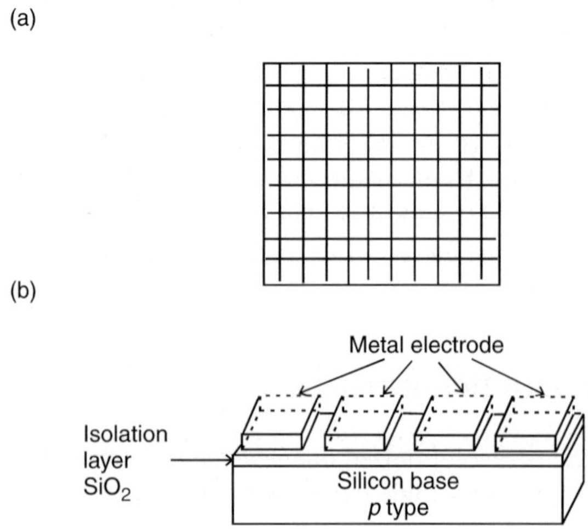

Fig. 12.6 Configuration (a) and architecture of *p-n* junctions (b) in a CCD camera.

The method of signal reading in a CCD is completely different from that in the PDA diode lines. To understand phenomena occurring in a CCD camera, one should recall how a potential well is created and what the effect of the applied external voltage is on the well depth. For this purpose we need to recall some fundamental electron properties in a solid. The probability that an electron occupies a state of energy E_n is determined by the *Fermi distribution*

$$f_0(E_n) = \frac{1}{e^{(E_n - E_F)/kT} + 1}. \tag{12.7}$$

The derivation of this equation can be found in textbooks on statistical physics, e.g., [6]. When spin is introduced, the distribution is called the *Fermi–Dirac distribution*. The expression (12.7) was derived for an ideal gas consisting of particles called fermions to which electrons belong. The model of an ideal gas consisting of free electrons can be applied to describe valence electrons of atoms in crystals. The energy E_F in eq. (12.7) is called the *Fermi energy*. For metals, the Fermi energy is the energy below which all electron states are occupied, and above which all states are empty. The energy of an electron in vacuum is larger than the Fermi energy E_F (Fig. 12.7) and the difference is called the output work, W, or the threshold energy needed to eject the electron from a metal (see eq. (12.6)). At a temperature of absolute zero the Fermi distribution of electrons in a metal is a stepwise function

$$\begin{aligned} f_0(E_n) &= 1, \quad \text{for } E_n < E_F, \\ f_0(E_n) &= 0, \quad \text{for } E_n > E_F. \end{aligned} \tag{12.8}$$

For higher temperatures the Fermi distribution is described by the smoother function presented in Fig. 12.8, because electrons have an additional thermal energy, kT, that can promote them to higher states of energy than the Fermi energy. For a semiconductor at absolute zero, all states in the valence band are occupied ($f_0\,(E_n) = 1$), and all states in the conduction band are empty ($f_0\,(E_n) = 0$). At temperatures higher than absolute zero, a certain fraction of electrons becomes promoted to the conduction band, leaving holes in the valence band. For an intrinsic semiconductor for which the number of electron holes in the valence band is equal to the number of electrons in the conduction band, the Fermi energy has to be at half of the gap between the highest energy of the valence band and the lowest energy of the conduction band, particularly when the density of states near the conduction band edge is the same as near the valence band edge (Fig. 12.9b).

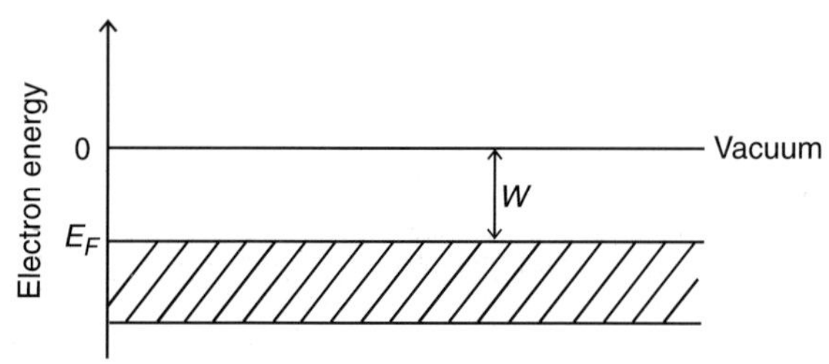

Fig. 12.7 Scheme of energy levels of electrons in a metal.

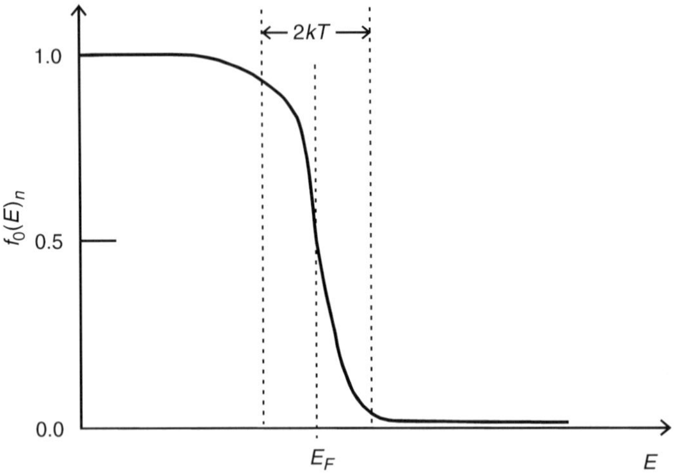

Fig. 12.8 Fermi distribution.

The total energy of electrons in the conduction band is given by

$$E_{el} = \sum_n E_n f_0(E_n) \approx \sum_n E_n e^{-E_n/kT}. \tag{12.9}$$

In n-type semiconductors the bottom edge of the conduction band lies considerably closer to the Fermi energy level, E_F, (Fig. 12.9a) due to a larger number of electron states near the bottom edge of the conduction band caused by introduction of additional electrons as dopants in an intrinsic semiconductor. On the contrary, in p-type semiconductors the top edge of the valence band lies further from the Fermi energy, E_F (Fig. 12.9c).

If we bring semiconductors of type n and p into direct contact, we obtain the schematic diagram of electron energy levels along the n-p junction presented in Fig. 12.10.

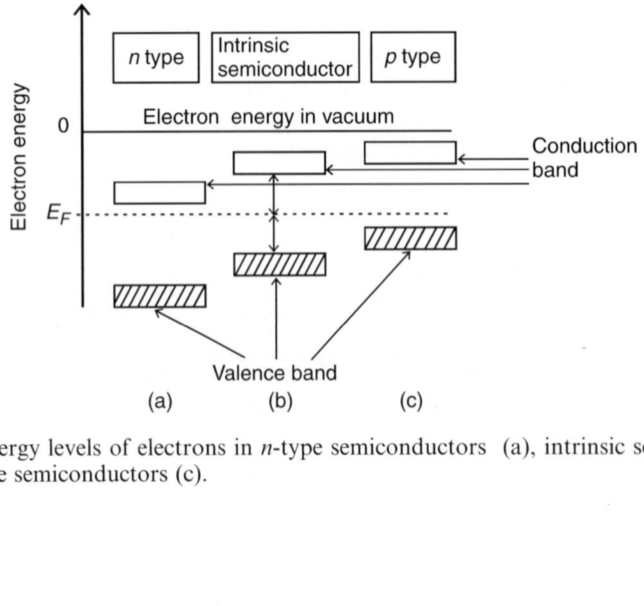

Fig. 12.9 Energy levels of electrons in n-type semiconductors (a), intrinsic semiconductors (b), and p-type semiconductors (c).

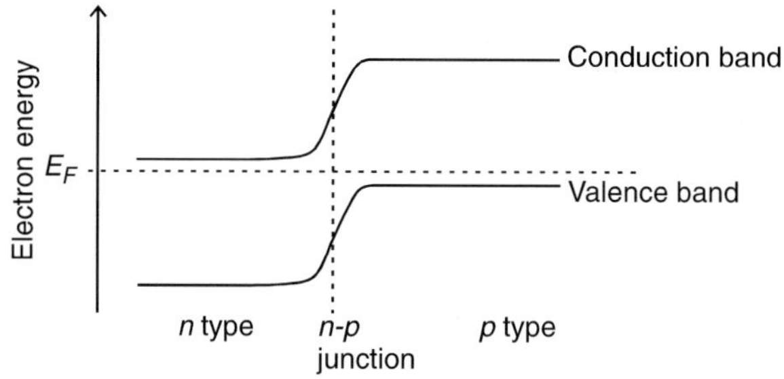

Fig. 12.10 Electron energy along the *n-p* junction.

When one applies the forward-biased voltage V to the *n-p* junction, the energies of the conduction band and the valence band of the *n*-type semiconductor increase, reducing the energy difference by eV (Fig. 12.11). After voltage application the Fermi energy will not be the same in the type *n* and *p* areas. The Fermi energy in the area of type *n* will increase by eV in comparison with the Fermi energy in the area of the type *p* (Fig. 12.11).

When the reverse-biased voltage V is applied to the *n-p* junction, the energies of the conduction and the valence bands in the *n*-type semiconductor decrease, which results in an increase in the energy difference along the *n-p* junction by eV. The Fermi energy E_F in the *n* area will decrease by eV in comparison with the Fermi energy in the area of type *p* (Fig. 12.12).

A similar analysis makes it possible to understand the energy distribution in more complicated semiconductor systems. It is easy to show that the *p-n-p* junction produces an energy well as presented in Fig. 12.13. We can control the depth of the well by applying the external voltage. If we apply the reverse-biased voltage, the depth of the well increases, in contrast to the forward-biased voltage that results in decreasing the well depth.

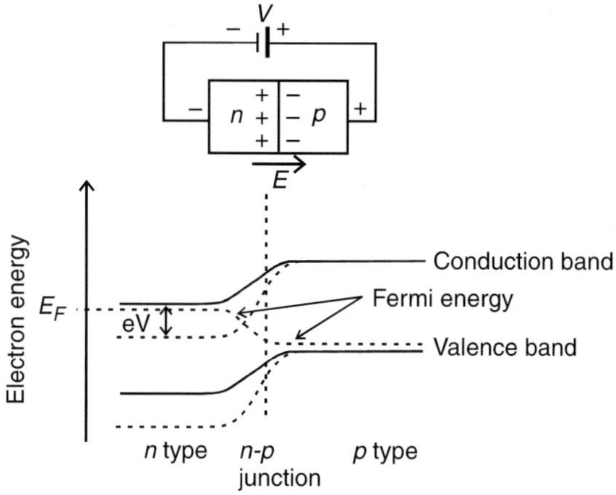

Fig. 12.11 Electron energy along the junction *n-p*, when forward-biased voltage V is applied.

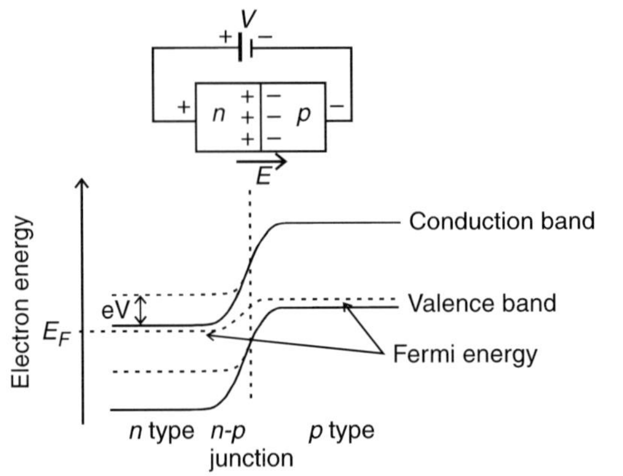

Fig. 12.12 Electron energy along the *n-p* junction when the reverse-biased voltage *V* is applied.

After this brief introduction, let us return now to the principles of operation in the CCD camera presented in Fig. 12.6b. The CCD matrix consists of a set of diodes called pixels. Each pixel can be presented as a MOS capacitor (metal–oxide–silicon) (Fig.12.14). Light incident on the junction of the MOS produces free electron–hole pairs. If one applies the reverse biased voltage, *V*, to the MOS capacitor, the holes from the area *p* under the *n*-type metal layer begin to "run away" towards the negative electrode. Therefore, the number of holes in the area under the metal layer (*A*) becomes smaller than in the area on the left and on the right (*B*) of the silicon base of the *p* type. The distribution of the electrons' energy along *B–A–B* will be similar to the energy distribution along the *p-n-p* junction in Fig. 12.13, since the area (*A*), depleted of holes in comparison with the surrounding areas (*B*), plays the role of an *n*-type semiconductor. This indicates that under the MOS electrode the energy well is created. When incident photons illuminate the MOS, electron–hole pairs are produced with electrons gathering in the energy well. The charge accumulated in the well is proportional to the intensity of the incident radiation. When the forward-biased voltage *V* is applied to the MOS capacitor, the well disappears and the electrons are forced to leave their postion under the metal electrode. The dependence

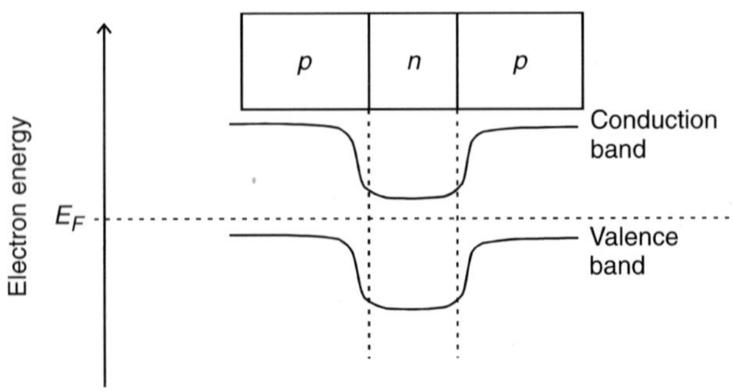

Fig. 12.13 Electron energy along the *p-n-p* junction.

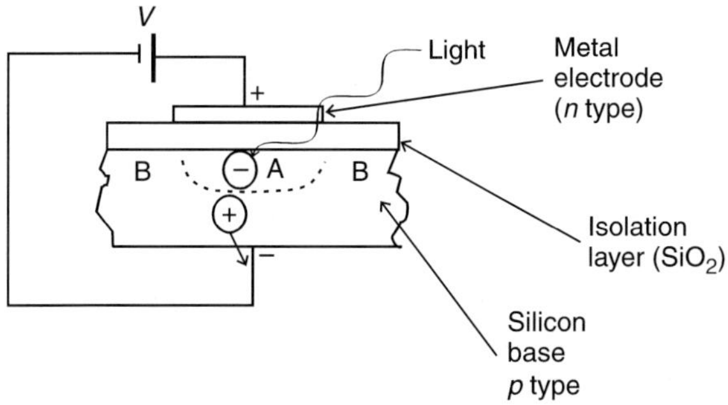

Fig. 12.14 Scheme of an MOS capacitor.

of the well depth on the direction of the applied voltage is employed in the method of reading signals in the CCD matrix during the read-out cycle.

The read-out cycle is described in Fig. 12.15. Let us apply the voltage V of stepwise type changing in a three-phase cycle (Fig. 12.15a) to the next three MOS capacitors (Fig. 12.15b). This indicates that at $t = t_1$ the voltages V_1 (+), V_2 (−), V_3 (−) (Fig. 12.15c) are applied to the capacitors G_1, G_2, G_3 (Fig. 12.15b). For the next three capacitors the situation is repeated, and the voltages V_1 (+), V_2 (−), V_3 (−) are applied to the capacitors G_4, G_5, G_6. The voltages applied in such a way cause the

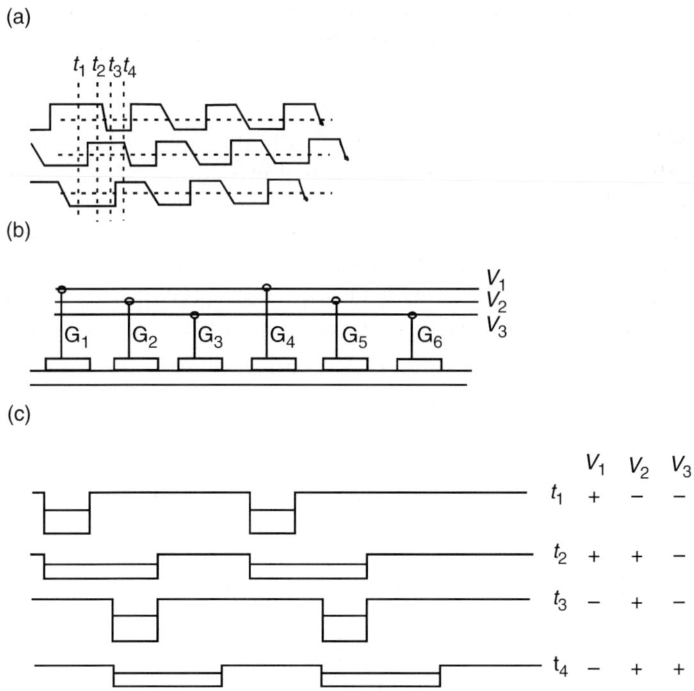

Fig. 12.15 (a) Three-phase voltage applied to the MOS capacitors, (b) voltage applied to the MOS capacitors, (c) change of depth and position of the energy wells as a function of applied voltage.

potential energy well at $t = t_1$ to be produced in the G_1 and G_4 capacitors where the electrons are gathered (Fig. 12.15c), because the reverse-biased voltage V is applied to them, in contrast to the G_2 and G_3 capacitors. At $t = t_2$, the voltages applied to the electrodes change: G_1, G_2, G_3 have voltages $V_1\,(+)$, $V_2\,(+)$, $V_3\,(-)$, respectively. The well under the electrode G_1 becomes shallower because a fraction of the electrons overflows to the neighboring well that was just created under G_2. At $t = t_3$, the voltages applied to the electrodes G_1, G_2, G_3 are $V_1\,(-)$, $V_2\,(+)$, $V_3\,(-)$, respectively. The energy well under the electrode G_1 disappears and all the electrons from G_1 overflow to G_2, and from G_4 to G_5. With the periodic change of the voltage, the cycle repeats and electrons move from one capacitor to the next one until they reach the matrix edge. The registration of the charge is performed along the column. The content of each column is shifted to the serial register, where the total charge is read out. If the light from the spectrograph (Fig. 12.1b) arrives at the CCD detector, each spectral component illuminates a different column and each column is read out individually and simultaneously. Moreover, the distribution of the intensity on the pixels along a single column delivers information about spatial distribution of absorption, emission or scattering centers in the sample.

The method of columnar read out in a CCD detector presented above has a lot of disadvantages. First, if the light intensity is too high and the capacity of the well is too small to gather all produced electrons, the charges "overflow" the well barrier and they are lost. To avoid this undesirable effect various methods of reading are applied. One of them is placing a shutter in every second column to prevent light illumination. The charge from the column, which is illuminated, is shifted to the neighboring column where it waits for reading out to the register. This obviously decreases the active surface of the detector, because the number of pixels is reduced by half. The opposite situation often does take place when the intensity of the incident light is too small in comparison with the noise of the detector to be registered as a signal. One of the most effective procedures of increasing S/N ratio is *binning*. Binning combines charges from several pixels belonging to the same row. The signal increases in proportion to the number of pixels whereas noise increases only with the square root of the number of pixels being binned. Therefore, the binning leads to improvement of the S/N ratio. The procedure of binning makes it possible to receive the signal originating from an individual pixel, a few pixels or the signal from all the pixels of the column that corresponds to the entire height of the monochromator slit.

The standard CCD cameras record light from the 410–1100 nm range. The lower range is limited by strong absorption of silicon below 410 nm. To extend the spectral range towards the UV, special configurations are used in which incoming light arrives directly at the semiconductor (*back-illuminated CCD*) instead of at the electrode (*front side CCD*). In the front side CCD configuration, light arrives at the metal electrode and it has to travel a long optical path before reaching the junction layer. Strong UV absorption in the electrode reduces the intensity of the incident light, which can be avoided when light illuminates the CCD from the side near to the junction as is the case in the back-illuminated CCD configurations. Detection in the UV range creates additional problems. For radiation at 400 nm, the length of the optical path is only 0.3 μm, which means that electrons appear only near the surface of the MOS capacitor and they are not able to penetrate deeper into the well. To avoid problems with penetration, one should reduce the thickness of the capacitor layer – the thickness of the order of 15–20 μm permits electrons to diffuse into the well with comparatively good efficiency. However, such a thin capacitor has a very small efficiency in the near-infrared range because the length of the absorption path is as long as 80 μm at 1000 nm. Covering a detector with a layer of phosphor or a fluorescent dye permits us to extend the range of detection to 200 nm, by employing the effect of conversion of the UV radiation into radiation from the visible range. The typical upper limit of detection is 1100 nm. It can be extended towards the infrared by employing gallium arsenide, GaAs, as an image amplifier. The typical efficiency of CCD cameras is 45–50% at 750 nm, but it can be improved significantly by applying enhanced configurations of matrix architecture.

The capacity of the well in CCD detectors is an important parameter, which determines the dynamic range of a detector. The well capacity controls how many electrons can be accumulated simultaneously in an individual pixel (capacitor MOS). This number depends on methods of silicon doping, size of the capacitor and the matrix architecture. In typical CCD cameras the well capacity is 300,000 electrons. The capacity of the well determines the maximum intensity of a signal that can be measured by a CCD detector. The lower limit of detectivity is determined by the noise of a detector. Both parameters determine the dynamic range of the CCD camera.

The noise of CCD cameras originates from the following sources:

a) noise caused by the incident radiation (*shot noise*). It is proportional to the square root of the incident light intensity,
b) *dark noise* caused by thermal electrons. This noise doubles with temperature, increasing every 10 °C above 25 °C. The significant reduction of dark noise is achieved by cooling with liquid nitrogen or by thermo-electrical cooling,
c) noise produced during reading the charge of individual pixels (*read–out noise*), which is a function of the electronics and the quality of the detector.

We should mention the advantages of multichannel detectors. The primary advantage of modern multichannel detectors is their high detectivity. Diode lines and CCD cameras produced nowadays can compete successfully with PMT and MCP detectors.

The other advantages of multichannel detectors are:

a) the ability to record a broad spectral range simultaneously,
b) all types of accidental fluctuations give the same contribution to all the spectral components, which indicates that the whole spectral range suffers from the same systematic error that can be easily identified and removed,
c) the increase in the signal/noise ratio originating from *Felgette's effect* (*Felgette's S/N advantage*),
d) the elimination of time-consuming spectrum scanning and errors related to step motor operation during the scanning,
e) the possibility to perform kinetic measurements in real-time.

Recording the whole spectral range simultaneously is an unquestionable advantage of CCD detectors. However, one should be aware that faster recording is sometimes at the expense of lower spectral resolution. Spectral resolution in a single channel detector depends mainly on the operator who determines the speed of the scanning. In multichannel detectors the resolution is determined by the number of pixels in a CCD matrix as well as by the dispersion properties of a spectrometer [2]. To illustrate this interplay, let us consider the spectrometer of a focal length of 1 m. The resolution of the spectrometer depends on the number of grooves in the diffraction grating. For a diffraction grating with 1200 grooves per millimeter, the dispersion in a typical spectrometer is of the order of 0.8 nm/mm, for a diffraction grating with 150 grooves per millimeter, the dispersion is around 6.4 nm/mm. Let the light coming out of the spectrometer illuminate a diode line consisting of 1024 elements with the width of an individual element being 25 μm. The total width of the detector is $1024 \cdot 25\,\mu\text{m} = 25\,\text{mm}$. This denotes that the spectrometer characterized by a dispersion of 0.8 nm/mm covers the spectral range of $0.8\,\text{nm/mm} \cdot 25\,\text{mm} = 20\,\text{nm}$, and a $6.4\,\text{nm/mm} \cdot 25\,\text{mm} = 160\,\text{nm}$ range for dispersion of 6.4 nm/mm. The spectral resolution is equal to the spectral range divided by the number of elements, therefore $\frac{20\,\text{nm}}{1024} = 0.02\,\text{nm}$ and $\frac{160\,\text{nm}}{1024} = 0.16\,\text{nm}$, respectively. Usually the average from 4 elements is recorded, so the final resolution is 0.08 nm and 0.64 nm.

The improvement of the signal to noise ratio occurring in multichannel detectors, known as the Felgette effect, results from the fact that the detector "sees" 1024 elements simultaneously. This indicates that a single channel detector needs 1024 times more time to register all spectral components compared to a multichannel detector at the same signal to noise (S/N) ratio. When the integration time increases 1024 times in the multichannel detector, the signal also increases 1024 times. The noise also increases, but only $\sqrt{1024}$ times, as noise is proportional to the square root of the integration time. Finally, recording the signal in both types of detectors over 1024 seconds, the multichannel detector exhibits a higher S/N ratio, $\frac{1024}{\sqrt{1024}} \cdot \frac{S}{N} = \sqrt{1024} \cdot \frac{S}{N}$ which is 32 times larger than that in a single channel detector.

REFERENCES

1. W. Demtröder, *Laser spectroscopy, Basic Concepts and Instrumentation*, 3rd ed., Springer-Verlag, Berlin (2003)
2. Jobin Yvon Spex, *Guide for Spectroscopy* (1994)
3. F.P. Mancini, P. Sodano, A. Trombettoni, *On the Phase Diagram of Josephson Junction Arrays with Offset* in *New Developments in Superconductivity Research*, R.W. Stevens ed., pp. 29–52, Nova Science Publishers, New York (2003)
4. A. Peacock, P. Verhoeve, N. Rando, A. van Dordrecht, B.G. Taylor, C. Erd, M.A.C. Perryman, R. Venn, J. Howlett, D.J. Goldie, J. Lumley, M. Wallis, *Single optical photon detection with a superconducting tunnel junction*, Nature 381 (1996) 135
5. M.A.C. Perryman, A. Peacock, N. Rando, A. van Dordrecht, P. Videler, C.L. Foden, *Optical photon detection using superconducting tunnel junctions* in *Frontiers of Space and Ground-Based Astronomy. The Astrophysics of the 21st Century*, W. Wamsteker, M.S. Longair, Y. Kondo eds., Kluwer (1994) p. 537
6. K. Huang, *Statistical Mechanic*, J. Wiley and Sons, Inc., New York (1963)

Subject Index

absorption
  coefficient 13, 15, 16, 72, 73, 109, 272
  cross section 16, 185, 186, 272
accessible emission limit (AEL) 285
acousto-optic devices 40–3, 52, 53
acousto-optic modulator 43, 44, 49, 50, 54
acousto-optic transducer 41, 49
active medium 5, 7–14, 19, 21, 34, 40, 45, 47–50, 54, 59–62, 64–5, 67, 70, 76, 82, 86, 88, 90, 139, 149, 150, 152, 155, 156, 291
active modelocking 40, 44, 49
alexandrite laser 76
amphibious technique 170
amplifier
  multipass 150, 151, 158
  regenerative 44, 71, 126, 129, 150–3, 155, 156, 158, 187, 188
anti-Stokes Raman scattering 131, 134, 135, 186, 189, 211, 255
argon lasers 43, 282, 288
attosecond pulses 32, 78, 219
autocorrelation
  signal 164–70
  techniques 161–7, 172
autocorrelator 162, 164, 165, 166, 168–71, 188

back-illuminated CCD 188, 308
bacteriorhodopsin 232–5, 240, 274
bandwidth 24, 25, 34, 37, 75–7, 80, 81, 87, 103, 122, 144, 156, 263, 296
beam divergence 25, 89, 289
beam radius 27
biaxial crystal 114
binning 308
biostimulation 274
birefringence 114, 116, 181, 182, 244
blackbody radiation 4, 5
Bloch equation 191, 194, 198
blocked-impurity-band 301
bolometers 294, 301
boxcar integrator 178, 180, 181
Bragg cell 54, 55, 56, 57
Bragg reflection 43, 100, 101
Bragg regime 50, 52–4
Brewster angle 10, 11, 87, 142
broadening 21, 34, 74, 85, 87, 142, 144, 156, 190, 191, 210, 228, 236, 238, 250, 254

carbon dioxide laser (CO_2) 59, 60, 61, 63–7, 82, 272, 273, 279, 281, 282, 291
CARS 135–7, 176, 189–91, 210–11, 249, 250, 253–7
CARS photon echo 191
cataract 287, 289
cavity dumper 43, 50, 53–8
cavity dumping 28, 31, 41, 44, 52, 53–5
CCD 28, 172, 188, 229, 293–6, 301–10
charge coupled device (CCD) 189, 293, 294
chemical lasers 59, 63, 68, 69, 291
chirp 31, 39, 40, 44, 49, 77, 82, 129, 142–5, 150, 152, 156, 157, 159, 169
chirped pulse amplification (CPA) 31, 44, 49, 77, 129, 150, 152, 156–9
circular-cage 299
cladding 79, 80, 158
CO laser 59, 61, 63, 67, 68
coherence 3, 10, 13, 25, 32, 82, 103, 113, 189, 190, 200, 211, 228, 231, 232, 235, 250, 252, 253, 255, 262
coherence length 25, 113
coherent anti-Stokes Raman scattering (CARS) 135–7, 176, 189–91, 210, 211, 249, 250, 253–7
coherent Stokes Raman spectroscopy (CSRS) 136, 137
colliding pulse modelocked (CPM) 77, 85
conduction band 91, 92, 94, 95, 98, 257, 259, 300, 301, 303–6
confocal resonator 25, 26, 28
constant polarization (DC effect) 110, 111

continuous wave lasers (cw) 13, 28, 31, 47, 49, 53, 71, 73–5, 77, 82, 84, 96, 99, 101, 120, 136, 151, 158, 187, 188, 230, 274, 288
CPA 31, 44, 129, 150, 156–9
cross section 16, 115, 118, 135, 138, 149, 185, 186, 272
CSRS 136, 137
cw regime 31, 49

damage
  threshold 53, 112, 120, 128, 152, 156
dark signal 301, 309
Debye–Sears effect 41–3
decay time 179, 190
dephasing time 189, 190, 209–11, 250, 251, 252–7
depopulation 64
detectivity (D) 92, 293, 294, 295–8, 300, 302, 309
detector linearity 297
detector response time 295, 297, 299
detectors 161, 184, 204, 293–302, 309–310
dichroic mirrors 139, 188
difference frequency generation (DFG) 113, 120
difference frequency mixing 110, 129, 130
diode arrays 73, 103, 152, 294
diode lines 294, 295, 303, 309
diode-pumped solid-state lasers (DPSSL) 44, 69, 73, 74
Dirac delta 23, 24, 34, 224
direct band gap 94, 95, 97
direct electric-field reconstruction (SPIDER) 161, 162
dispersion 13, 39, 47, 49, 81, 82, 101, 111–14, 122–3, 125, 139–45, 152, 156, 157, 188, 310
doped semiconductor 90, 301
double heterostructure 96, 97, 98
double refraction 114
double resonance oscillator 127
dye 14, 17, 28, 31, 34, 44–6, 50, 53–5, 57, 59, 60, 71, 73, 76, 77, 84–7, 90, 136, 255, 274, 281, 282, 287, 291, 309
dye lasers 14, 28, 31, 34, 53, 54, 57, 59, 60, 71, 73, 76, 77, 84, 85–7, 90, 136, 255, 281, 287
dynamic range 295, 296, 297, 309

edge-emitting diode lasers 99, 100
"eeo" interaction 116
"eeo" phase matching 116
Einstein coefficients 4, 5
electric susceptibility 108
electromagnetic spectrum 2
electro-optic effect 107
energy relaxation T1 193, 209, 244, 251
energy storage 51, 54
"eoe" phase matching 117
"eoo" phase matching 117
Er:YAG laser 59, 72, 282, 288
erbium laser 73, 74
etalon 76, 87, 165, 168, 169
excess electron 92, 176, 221, 257–64
excimer laser 59, 60, 73, 87–90, 279, 282, 287
excited dimmer 88
excited-state vibrational coherence 231
extraordinary ray 114, 115, 116, 154, 181, 243, 244

Fabry–Perot interferometer 9
far infrared lasers 62, 63
feedback 8, 10, 13, 52, 63, 95, 98, 100, 101, 103, 173
Felgette's effect 309
Felgette's S/N advantage 309
femtochemistry 219, 221, 222, 229
femtosecond lasers 49, 77, 78, 107, 122, 139, 142, 143, 147, 187, 188, 219, 242, 246, 247, 250, 271, 274
Fermi distribution 303, 304
Fermi energy 303–6
Fermi–Dirac distribution 303
fiber cable 78–80
fiber laser 31, 78–82, 145, 158, 159
fine-mesh dynodes 299
fluorescence 2, 34, 54, 70, 75, 84, 86, 90, 136, 176, 177–83, 186, 187, 199, 208, 212–15, 233, 241–3, 247–9, 277, 293
fluorescence decay 176–83, 212–15, 241, 247–9
forbidden transition 12
forward biased 93, 94, 97, 305
Fourier transform 23, 24, 34, 37, 38, 39, 139, 144, 162, 171, 172, 209, 210
Fourier-transform limited pulse 39
four-level lasers 14, 69, 75
four-wave interaction 133, 134, 137

Franck-Condon 223, 234, 261–4, 279
free induction decay (FID) 197, 211, 250
frequency chirp 142–3
frequency doubling 28, 71, 73
frequency-resolved optical gating (FROG) 161, 162, 170–3
front side CCD 308
fundamental mode 28

g (gyromagnetic) parameter 192
GaAlAs 103
gain coefficient
  small signal gain coefficient 13, 150
gain medium 9, 59, 60, 69, 82, 83, 84, 151–3
gain-guided diode 98
gas lasers 34, 59, 60–3, 65, 67, 82, 83, 84, 87–9, 103, 291
Gaussian beam 27, 29, 47
Golay detectors 294
grating-fiber method 144, 145
group delay 122, 124, 125, 142, 143
group delay time 124, 125, 142, 143
group velocity 39, 49, 81, 122, 123, 139, 140–3, 156, 157, 188
group velocity dispersion (GVD) 39, 40, 49, 81, 139, 141–5, 156–9, 188

harmonic generation
  second harmonic generation (SHG) 110, 111, 113, 116, 118, 119, 121, 123, 139, 151, 163, 171
  third harmonic generation (THG) 188, 226
H-bond 235, 236, 238
Heisenberg principle 37, 212, 229
Helium–Neon (He-Ne) lasers 34, 59, 82, 83, 272
Hermite polynominal 27
high reflector 9, 84, 100
high-repetition-rate autocorrelator 90, 164, 165, 169, 170
high-temperature superconductors (HTS) 301
Ho:YAG laser 59, 72, 282, 288
hole-burning 223, 229
holmium 59, 70, 72, 73, 282
homogeneous band broadening 87, 210
homogeneous broadening 34, 85, 190, 210, 250
homogeneous processes 200, 201, 210, 250, 253, 254
hydrogen bond (H-bond) 235, 236, 238, 240
hyper-Raman scattering 131, 132
hyper Raman scattering, anti-Stokes 131–7, 176, 186, 189, 190, 211, 212, 250, 255
hyper-Rayleigh scattering 132

idler wave 127
imaging spectroscopy 295
index matching technique 116
index-guided diode 98
indirect band gap 94, 95
induced dichroism 243, 244
inhomogeneous broadening 34, 190, 210, 250
inhomogeneous processes 200, 210, 250, 253, 254
insulator 89, 90, 91, 98, 103
integer plus 1/2 timing 57, 58
interferometer 9, 87, 142–4, 162–4
intramolecular charge transfer 242
internal conversion 177, 231, 232
intersystem crossing 86, 177, 246, 277
intermediates investigation 86, 177, 245, 277
intrinsic semiconductors 90, 92, 304
inverse Raman scattering (IRS) 137
ion–gas lasers 83
IR photon echo 191
irradiance 25, 107, 147, 271, 274, 280, 282, 285, 288, 300
IRS 137

Jablonski's diagram 277
junction
  Josephson junction 301
  n-p junction 90, 304–6
  p-n junction 93–5, 97, 98, 300–2
  p-n-p junction 305, 306
  p-p-n junction 97, 98

KDP 120, 122, 125, 162, 170, 188
Kerr cell 182, 243
Kerr effect 46, 47, 107, 109, 181, 244
Kerr lens modelocking (KLM) 46–9, 77
Kerr optical effect 243
krypton laser 59, 83, 287
KTP 120, 121, 128

Laguerre polynominals 27
Lambert-Beer formula 15, 109

Larmor frequency 193–5
laser amplifier 151, 271
laser cavity 9, 11, 25, 45, 65, 80
laser diode 71, 73, 79, 82, 83, 95, 96, 99–104, 159
laser hazard 285–91
laser safety 285
laser threshold 10, 95
laser-assisted-in situ-keratomileusis 283
lifetime 3, 12, 14, 16, 17, 45, 46, 50, 52, 61, 71, 73, 84, 86, 88, 89, 101, 147, 177, 178, 180, 182, 183, 185–7, 193, 249, 252, 257
light-gating techniques 178
$LiNbO_3$ 120, 121, 122, 125, 127, 128
linear focused 299
linearly chirped pulse 39
liquid dye lasers 59, 84
lithium triborate (LBO) 120, 121, 188
longitudinal flowing-gas laser 65
longitudinal modes 19, 20, 21, 25, 26, 32, 34, 35, 36, 39, 40, 43, 45, 49, 77, 85, 100, 101
longitudinal relaxation 193
Lorentzian line shape 38, 169, 209, 257
loss 8, 10, 12, 13, 21, 22, 28, 31, 44, 45, 48, 50, 53, 55, 56, 78, 81, 82, 86, 127, 137, 142
luminescence 2

maximum permissible exposure (MPE) 285, 288, 289, 290
metastable level 14, 147
Michelson interferometer 162, 164
mode
  bending 64, 263, 264
  longitudinal 19–21, 25, 26, 32, 34, 35, 36, 39, 40, 43, 45, 49, 77, 85, 100, 101
  transverse 25–9, 97
modelocking
  technique 31
  active 40, 45
  passive 40, 44, 45, 46, 81, 85
  saturable dye 44, 45, 46
multichannel detectors 295, 301, 302, 309, 310
multichannel plate (MCP) 299, 300, 309
multipass amplifier (MPA) 150, 151, 158
multiple-quantum-well (MQW) 96, 98
multiplicity 2, 177

Nd:glass laser 70, 71
Nd:YAG laser 44, 53, 54, 59, 70–2, 76, 272, 279, 282, 287–9
Nd:YLF laser 73
Nd:YVO_4 laser 73, 77
negative crystal 114, 115, 116
negative GVD effect 142
neodymium laser 14, 70–2, 76
nitrobenzene 243
nitrogen laser 59, 88, 90, 287
NMR echo 198
noble gas 88
noise equivalent power (NEP) 295, 296, 297
non-critical phase matching 118, 120, 121
nonlinear optical phenomena 109
nonlinear optics 15, 47, 107, 108, 111, 112, 120, 181
nonlinear refractive index 47
nonresonant Kerr effect 46
non-tunable lasers 69–71, 74, 75
normalized detectivity (D*) 295–7
nuclear magneton 192

"oee" phase matching 117
"oeo" phase matching 117
"ooe" interaction 116
"ooe" phase matching 116
optic disk 287
optical amplifier 79, 81, 103, 138, 158
optical Bloch equation 191
optical cavity 9, 13, 45, 50, 57, 75, 95, 98
optical density OD(λ) 185, 260, 288, 289, 290
optical fiber 39, 68, 72, 79, 81, 82, 101, 137, 145, 281, 293
optical mixing 81, 107
optical parametric amplifier (OPA) 127–30, 184, 250, 255
optical parametric generator (OPG) 127, 128, 184
optical parametric oscillator (OPO) 71, 126, 127–30, 184, 185
optical resonance 7–9, 191, 198, 200, 206
optical resonator 8–13, 19, 20, 22, 25, 32, 33, 40, 41, 43–5, 47, 48, 62, 63, 85–7, 126, 127
optical soliton 81, 145
ordinary ray 114, 115, 116, 120, 154, 155, 181, 243

oscillator 3, 8, 22–4, 47, 49, 64, 71, 81, 108, 125–7, 135, 150–2, 164, 169, 184, 187, 189, 220, 223, 230, 235, 238–9, 250, 252, 255
output coupler 9, 84, 89

parametric conversion 110, 125, 126–8
passive modelocking 40, 44, 45, 46, 81, 85
phase coherence 3, 253
phase matching 111–113, 115–23, 125, 126, 127, 134, 135, 182, 189, 211, 255
phase relaxation T2 209, 250, 251
phase velocity 111, 114, 122, 141
phase-modulation 178, 179
phosphorescence 2, 177, 277
photoablation 77, 273, 274, 279, 281
photochemical interactions 273, 274, 275, 278, 281
photochemistry 7, 107, 246, 277
photoconductive detector 91, 93
photoconductive diode 301
photoconductors 300, 301
photodiode 49, 91, 93, 94, 161, 162, 183, 229, 293, 294, 300, 301
photodiode arrays (PDA) 294, 295, 301–3
photodisruption 77, 273, 274
photodynamic therapy (PDT) 274–6
photoisomerization 219, 232, 240, 241, 242
photokeratitis 287, 289
photomultipliers (PMT) 92, 161, 162, 297, 299, 300, 309
photon echo 108, 176, 191, 198, 199, 200, 201, 207–11, 249, 253, 254
photoreduction 245
photoresistor 91, 300
photovoltaic detector 93
photovoltaic diode 301
photovoltaic semiconductor detectors 294
picosecond lasers 120, 143, 214
planar-cavity surface-emitting diode laser (PCSEL) 96, 99
Planck's law 1, 4, 5
plasma 7, 61, 65, 77, 83, 84, 130, 136, 273, 274, 279, 280
plasma-induced ablation 273, 274, 279, 280
Pockels cell 44, 54, 120, 152–6, 182
polarization 3, 10, 11, 54, 81, 108–12, 114, 116–19, 131, 132, 135, 138, 153–6, 171, 172, 176, 181, 188, 189, 191, 198, 199, 224–7, 243, 244, 251–5, 258, 262
polarizer 10, 11, 54, 127, 152, 153, 155, 156, 181, 243, 244
population inversion 7–17, 45, 50, 53, 54, 60, 61, 64, 69, 71, 80, 82, 83, 88, 90, 95–8, 147, 148, 155
position-sensitive PMT 299
positive crystal 115, 116
positive feedback 8, 13
positive GVD effect 142, 144, 145, 156–9
probe beam 184–9, 253
proton transfer 31, 176, 219, 220, 245–9, 277
pulse π 109, 110, 131
pulsed lasers 13, 17, 31, 77, 108, 120, 136, 139, 175, 274, 282, 288, 289, 291
pump beam 125, 127, 128, 138, 184–9, 255
pump energy 7
pumping 9, 10, 12, 14, 16–17, 21, 28, 45, 46, 50, 51, 61–4, 68, 70, 71, 73, 75, 77, 82, 84, 85, 86–7, 99, 103, 126, 127, 134–5, 147, 148, 151, 153, 155, 158, 185–8, 219, 220, 241, 244, 247, 254, 255, 260, 274, 291, 296, 297
pump-probe method 108, 176, 183–7, 241, 242, 247–51
pure dephasing time 190, 210
pyroelectric detectors 294

Q-switch
  acousto-optic 41, 44, 50, 51, 52, 54, 187
  electrooptical 44, 50, 54, 154
Q-switched 44, 50–4, 61, 71, 73, 81, 120, 153, 178, 187, 280, 282, 291
Q-switched lasers 50, 73, 178, 187
Q-switching technique 31, 50, 52
quality factor 21–4, 50
quantum beats 176, 212–15, 230, 257
quantum efficiency (QE) 295, 296, 299, 302
quantum wells 46, 96, 98, 102, 103
quencher 86, 249

Rabbi frequency 204, 206, 207
radiation hazards 286
Raman photon echo 176, 200, 208
Raman scattering 17, 81, 107, 108, 131–7, 176, 177, 186, 189, 211, 226, 227, 251, 253, 255, 256
Raman–Nath regime 43, 50–2

rare earth ions 70
Rayleigh scattering 131, 132
read out noise 296
regenerative amplifier (RGA) 44, 71, 126, 129, 150–3, 155, 156, 158, 187, 188
regenerative feedback 10
regenerative modelocking 49
relaxation
  dephasing 253
  phase relaxation 189, 190, 200, 201, 209, 250, 251, 253
  pure dephasing 190, 210, 253, 257
  vibrational relaxation 129, 176, 221, 227, 234, 236, 249, 263
  reorientational 176, 227, 244, 253
resonance 2, 3, 7–9, 44, 127, 131, 136, 137, 191, 193, 195, 197, 198, 203, 206, 234, 250, 252, 254, 255, 264
resonant optical waveguide array (ROW array) 99
resonator
  quality factor *Q* 21, 22, 24, 50
  optical 8–13, 19, 20, 22, 25, 32, 33, 40, 41, 43–5, 47, 48, 62, 63, 85, 87, 126, 127
response linearity 295
responsivity 295–7
reverse biased 94, 302, 305, 306, 308
rotating-wave approximation 203, 204
ruby laser 7, 12, 13, 54, 59, 60, 61, 69, 110, 120, 282

safety
  laser safety 285
sapphire
saturable absorbers 40, 45, 46
saturable Bragg reflectors 46
saturable dye absorbers 44, 46
saturation 12, 15, 16, 17, 29, 45, 47, 150, 156, 185, 296
saturation parameter 17
second harmonic generation (SHG) 110, 111, 113, 116, 118, 119, 121, 139, 151, 163, 171
second moment analysis 27
self modelocking 40, 48
self phase modulation (SPM) 81, 107, 139, 144, 184
self focusing 40, 107, 130, 156
semiconductor
  saturable absorber 40, 45, 46, 81, 85, 152
semiconductor detectors 294, 300
semiconductor of *n*-type 92
semiconductor of *p*-type 92
sensitizer 275, 276, 277
SHG 110, 111, 113, 118, 120, 122, 124–8, 163, 170, 171, 172, 188, 226
shot noise 309
single channel detectors 294, 295, 302
single-quantum-well (SQW) 96, 98
single-shot autocorrelator (SSA) 164, 169, 170
small signal gain coefficient 13, 150
solid-state lasers 31, 34, 44, 46, 47, 49, 52, 53, 59, 60, 69, 70, 72–7, 80, 81–3, 85, 87, 99, 101, 103, 145, 147, 151, 153, 158, 187, 255, 287, 301
soliton 81, 144, 145, 158
soliton-effect compressor 144, 145
solvated electron 257–64
spatial coherence 10, 13
spectrally and temporally resolved 162
upconversion technique (STRUT) 162
spin echo 191, 196, 198–201, 207, 208, 250
spin relaxation 191, 194, 246, 247
spin-lattice relaxation 193, 197
spin–spin relaxation 194
spontaneous anti-Stokes scattering 186, 211
spontaneous emission 2–6, 8, 9, 16, 19–21, 148, 158, 193
spot size 25, 27, 29
SRGS 137
stimulated absorption 2–5, 7, 9, 16
stimulated anti-Stokes Raman scattering 134
stimulated emission 2, 3–10, 12, 13, 15, 19–22, 32, 33, 37, 61, 82, 132, 133, 137, 147, 149, 155, 233
stimulated Raman emission 184
stimulated Raman gain spectroscopy (SRGS) 137
stimulated Raman gain Stokes (SRGS) 137
stimulated Raman scattering (SRS) 81, 107, 131, 132, 133, 135–8, 176, 226
stimulated Stokes Raman scattering 135
Stokes Raman scattering 131, 132, 134, 135, 186, 189, 211, 255
strained layers material 96
streak camera 161, 178, 183

stripe-geometry diode 98
sum frequency generation (SFG) 113, 116, 120, 125
sum frequency mixing 125, 129, 130
superconducting tunnel junction (STJ) 301
surface-emitting diode lasers 99

TEM 26–8, 73
temperature-tuned phase-matching 118, 120
temporary analysis by dispersing a pair of light E-field (TADPOLE) 170
temporary coherence 13
theory of the strong field 204
thermal detectors 294, 296
thermal interaction 273, 274, 278, 281
thermal noise 32, 301
thermal radiation 4, 5, 6
thermistor 294
thermocouple 294
THG 188, 226
third harmonic generation (THG) 107, 255, 287
three photon hyper Raman scattering 132
three wave interaction 113
three–level laser 12–14, 61
threshold 10, 13, 20, 21, 31, 34, 45, 50, 53, 74, 95, 97, 98, 107, 112, 120, 121, 125, 128, 132, 133, 135, 152, 156, 259, 297–9, 303
threshold gain 13, 31
Ti:sapphire lasers 31, 44, 47, 49, 71, 72, 74, 76, 77, 81, 101, 129, 141, 142, 144, 147, 152, 156, 164, 187, 188, 279, 280
time-amplitude converter (TAC) 179–81, 183
time-correlated single photon counting 178–80, 182
transient grating 176, 253
transition-state theory 175, 220
transverse electromagnetic mode 26
transverse electromagnetic mode (TEM) 26–8, 73
transverse flowing-gas laser 65
transverse mode 25–9, 97
transversely excited atmospheric lasers (TEA) 65
trigger pulse 179, 181
tunable lasers 69, 71, 74–6, 184, 255
tunable solid-state laser 70, 71, 74, 76
two-points correlation functions 227
two-pulse fluorescence 186, 187
type I^- phase matching 116
type I^+ phase matching 116
type II^- phase matching 117
type II^+ phase matching 117
type I, II phase matching 120
type I, II photoreduction 245

ultrafast coherent spectroscopy 249
ultrashort pulse lasers 32, 240
uniaxial crystal 109, 114, 115, 117, 181
up-conversion method 178
up-conversion signal 182

valence band 91–5, 259, 303–6
vanadate 70
vertical-cavity surface-emitting diode laser (VCSEL) 96, 99, 100, 101
vibrational correlation function 227
vibrational energy relaxation 241, 249
vibrational relaxation 129, 176, 221, 227, 234, 236, 249, 263
vibrational-rotational transitions 63
vibronic band 74, 75
vibronic coupling 85, 262
vibronic lasers 34, 59, 74, 76, 101, 187
visible light 2, 60, 61, 211, 255

waist beam 29, 48
walk-off distance 114, 118
wave packet 32, 120, 122–4, 142, 220–3, 227, 229, 230, 235
wavefront distortion 25
waveguide laser with RF excitation 65
weak field approximation 202, 203, 204

yttrium aluminum garnet (YAG) 14, 59, 70

Zeeman splitting